WITHDRAWN

Accession no.
01001635

WITHDRAWN

Geological Society Special Publications
Series Editor A. J. FLEET

GEOLOGICAL SOCIETY SPECIAL PUBLICATION NO. 96

Island Britain: a Quaternary perspective

EDITED BY

RICHARD C. PREECE
Department of Zoology,
University of Cambridge, UK

1995
Published by
The Geological Society
London

THE GEOLOGICAL SOCIETY

The Society was founded in 1807 as The Geological Society of London and is the oldest geological society in the world. It received its Royal Charter in 1825 for the purpose of 'investigating the mineral structure of the Earth'. The Society is Britain's national society for geology with a membership of 7500. It has countrywide coverage and approximately 1000 members reside overseas. The Society is responsible for all aspects of the geological sciences including professional matters. The Society has its own publishing house, which produces the Society's international journals, books and maps, and which acts as the European distributor for publications of the American Association of Petroleum Geologists, SEPM and the Geological Society of America.

Fellowship is open to those holding a recognized honours degree in geology or cognate subject and who have at least two years' relevant postgraduate experience, or who have not less than six years' relevant experience in geology or a cognate subject. A Fellow who has not less than five years' relevant postgraduate experience in the practice of geology may apply for validation and, subject to approval, may be able to use the designatory letters C. Geol (Chartered Geologist).

Further information about the Society is available from the Membership Manager, The Geological Society, Burlington House, Piccadilly, London W1V 0JU, UK. The Society is a Registered Charity, No. 210161.

Published by The Geological Society from:
The Geological Society Publishing House
Unit 7, Brassmill Enterprise Centre
Brassmill Lane
Bath BA1 3JN
UK
(*Orders:* Tel. 01225 445046
Fax 01225 442836)

First published 1995

The publisher makes no representation, express or implied, with regard to the accuracy of the information contained in this book and cannot accept any legal responsibility for any errors or omissions that may be made.

© The Geological Society 1995. All rights reserved. No reproduction, copy or transmission of this publication may be made without prior written permission. No paragraph of this publication may be reproduced, copied or transmitted save with the provisions of the Copyright Licensing Agency, 90 Tottenham Court Road, London W1P 9HE. Users registered with the Copyright Clearance Center, 27 Congress Street, Salem, MA 01970, USA: the item-fee code for this publication is 0305-8719/95 $07.00.

British Library Cataloguing in Publication Data
A catalogue record for this book is available from the British Library

ISBN 1-897799-32-2

Distributors

USA
AAPG Bookstore
PO Box 979
Tulsa
Oklahoma 74101-0979
USA
(*Orders:* Tel. (918) 584-2555
Fax (918) 584-0469)

Australia
Australian Mineral Foundation
63 Conyngham Street
Glenside
South Australia 5065
Australia
(*Orders:* Tel. (08) 379-0444
Fax (08) 379-4634)

India
Affiliated East-West Press PVT Ltd
G-1/16 Ansari Road
New Delhi 110 002
India
(*Orders:* Tel. (11) 327-9113
Fax (11) 326-0538)

Japan
Kanda Book Trading Co.
Tanikawa Building
3-2 Kanda Surugadai
Chiyoda-Ku
Tokyo 101
Japan
(*Orders:* Tel. (03) 3255-3497
Fax (03) 3255-3495)

Typeset by Bath Typesetting Ltd
Bath, England

Printed by T. J. Press (Padstow) Ltd,
Padstow, Cornwall PL28 8RW, UK

Contents

PREECE, R. C. Introduction. Island Britain: a Quaternary perspective — 1

FUNNELL, B. M. Global sea-level and the (pen-)insularity of late Cenozoic Britain — 3

GIBBARD, P. L. The formation of the Strait of Dover — 15

BRIDGLAND, D. R. & D'OLIER, B. The Pleistocene evolution of the Thames and Rhine drainage systems in the southern North Sea Basin — 27

BELLAMY, A. G. Extension of the British landmass: evidence from shelf sediment bodies in the English Channel — 47

KEEN, D. H. Raised beaches and sea-levels in the English Channel in the Middle and Late Pleistocene: problems of interpretation and implications for the isolation of the British Isles — 63

SCOURSE, J. D. & AUSTIN, R. M. Palaeotidal modelling of continental shelves: marine implications of a land-bridge in the Strait of Dover during the Holocene and Middle Pleistocene — 75

MEIJER, T. & PREECE, R. C. Malacological evidence relating to the insularity of the British Isles during the Quaternary — 89

STUART, A. J. Insularity and Quaternary vertebrate faunas in Britain and Ireland — 111

SUTCLIFFE, A. J. Insularity of the British Isles 250 000–30 000 years ago: the mammalian, including human, evidence — 127

TURNER, A. Evidence for Pleistocene contact between the British Isles and the European continent based on distributions of larger carnivores — 141

LISTER, A. M. Sea-levels and the evolution of island endemics: the dwarf red deer of Jersey — 151

BENNETT, K. D. Insularity and the Quaternary tree and shrub flora of the British Isles — 173

DEVOY, R. J. N. Deglaciation, Earth crustal behaviour and sea-level changes in the determination of insularity: a perspective from Ireland — 181

WINGFIELD, R. T. R. A model of sea-levels in the Irish and Celtic seas during the end-Pleistocene to Holocene transition — 209

COXON, P. & WALDREN, S. The floristic record of Ireland's Pleistocene temperate stages — 243

Index — 269

Introduction
Island Britain: a Quaternary perspective

R. C. PREECE

Department of Zoology, University of Cambridge, Downing Street, Cambridge CB2 3EJ, UK

This volume considers a range of topics under the theme *Island Britain: a Quaternary perspective*. Although it is generally believed that Britain last became separated from mainland Europe about 8500 years ago, earlier stages of the Quaternary are also considered. This review of the subject is particularly timely coinciding, as it does, with the construction and opening of a submarine link between Britain and mainland Europe – compromising our island status!

The evidence for insularity, or otherwise, is based on a wide range of data, reflecting the multidisciplinary nature of Quaternary research. First, there is the physical or geological evidence. **Funnell** sets the scene by outlining the late Cenozoic succession in the North Sea Basin and relating this to global sea-level history. He notes the switch from predominantly biogenic sedimentation in the mid-Pliocene to clastic deposition from the Late Pliocene. This reflects increased erosion by the principal rivers supplying sediment to the southern North Sea following the first northern hemisphere glaciations. The combination of sea-level fall and progressive deltaic progradation resulted in a land-bridge, converting Britain into a European peninsula. By the early Middle Pleistocene, sediment supply to the Great European Delta had dwindled and therefore failed to keep pace with subsidence, allowing interglacial sea-levels to encroach southward across it. **Gibbard** takes up the story and reviews the evidence for the nature and timing of the initial breaching of the Weald–Artois anticline connecting Britain to mainland Europe. He favours the widely held belief that this resulted from overflow from an impounded proglacial lake in the southern North Sea during the Anglian/Elsterian Stage. This theme is further developed by **Bridgland & D'Olier** who discuss the Pleistocene evolution of the Thames and Rhine drainage systems. They attempt to link the onshore Thames record with terraces that continue into the drowned extension of the valley offshore. In contrast, The Netherlands has been an area of marked subsidence throughout the Quaternary, so the lower Rhine sequence takes the form of a sediment stack, rather than terraces. It is assumed that the Thames has been a tributary of the Rhine during every low sea-level episode. Continuing the theme of offshore channels, **Bellamy** describes infilled valley sequences that now lie submerged off the south coast of England. He likewise concludes that such sequences have been subject to predominantly subaerial, rather than submarine, processes for much of the Quaternary. **Keen** discusses the implications of new stratigraphical and dating evidence from certain raised beaches in the critical area of the eastern Channel. Using much of this physical evidence for the location of palaeocoastlines and sea-level history, **Scourse & Austin** have undertaken palaeotidal modelling, from which they assess the palaeoceanographic implications of the existence of a land-bridge across the Strait of Dover during the Middle Pleistocene and Holocene.

The second batch of evidence for insularity or connection comes from biological and palaeontological data. **Meijer & Preece** examine the molluscan faunas which existed during different temperate stages of the Quaternary. The marine faunas during the Middle Tiglian, Eemian and Holocene are diverse (more than 100 taxa) and include many species with southern or 'lusitanian' affinities. It is thought that during these stages the Strait of Dover was open, allowing entry into the southern North Sea. Conversely, temperate stages from the Late Tiglian up to and including the Holsteinian have yielded impoverished faunas (less than 40 taxa) lacking southern elements. A barrier preventing such entry is assumed to have existed during this interval, a conclusion supported by the history of non-marine molluscan faunas.

In general, the faunas and floras of Britain, and especially those of Ireland, are impoverished when compared with those of neighbouring countries of mainland Europe. This is particularly clear from an examination of the vertebrates. **Stuart** provides a review of the Pleistocene vertebrate faunal history of Britain and

Ireland and discusses which of the many faunal absences might plausibly be attributed to the effects of isolation, rather than to banishment by people such as St Patrick! **Sutcliffe** focuses his attention on the record from the Late Pleistocene, pointing out the existence of major changes in faunal composition, in particular the absence of horse and humans and the presence of hippopotamus in the Last Interglacial, which appear to be helpful not only from a biostratigraphical stand-point, but which might also be explicable in terms of the isolation of Britain from Europe. **Turner** discusses the evidence specifically from the larger carnivores, noting, for example, that teeth of spotted hyaena from the Last Interglacial are significantly larger than those from the succeeding Devensian glacial. Such differences, he argues, might well result from the effects of the isolation of Britain. **Lister** describes an even more spectacular example of the ecological and evolutionary consequences of isolation on the red deer populations of Jersey. On this island, red deer from Last Interglacial raised beach deposits are dwarfed when compared with mainland populations. Uniquely, this island has also furnished normal-sized deer in stratified deposits both preceding (in the Saalian deposits at La Cotte) and succeeding the dwarfing episode. From a detailed study of the local sea-level history, he calculates the period that Jersey would have been isolated and estimates that the dwarfing must have occurred in no more than about 6000 years.

Bennett considers whether any tree or shrub species might have been excluded from the British Isles by the effects of insularity. He concludes that, as the overwhelming majority can potentially spread across channels of 10–100 km extent, it is unlikely that insularity would have been a significant factor in controlling the composition of the tree and shrub flora in the present or in previous interglacials. He does note that there have been significant losses of tree taxa from many islands off the British mainland, probably attributable to human activities.

If the post-glacial colonization of Britain was difficult for many plants and animals, then securing a foothold in Ireland must have been infinitely worse. The possible existence of land-bridge connections between Britain and Ireland has been debated since the early days of biogeography. Both **Devoy** and **Wingfield** address this problem by reviewing the latest evidence for the interplay between crustal behaviour and sea-level change, key factors in the determination of insularity. Wingfield's model suggests that land-bridge connections from Britain did exist, first to Ireland and later to the Isle of Man after 11 350 years BP. If so, the faunal absences discussed by Stuart require explanation. In the final chapter, **Coxon & Waldren** review the floristic record of Ireland's temperate stages and highlight its unique biogeographical composition. They discuss how such a distinctive flora may have arisen against a background of glaciations and massive sea-level changes.

Some of the ideas contained in this volume are necessarily speculative. Nevertheless, this volume does contain important facts, many not previously published elsewhere, that bear on the question of when Britain was and when it was not an island. A consensus is emerging that suggests that during the Middle Tiglian, Eemian and Holocene, Britain was isolated. Molluscan evidence suggests that this was not the case from the Late Tiglian up to and including the Holsteinian. If the Strait of Dover was formed by the breaching of the Weald–Artois anticline during the Anglian/Elsterian, then the absence of diverse marine faunas with southern elements in the southern North Sea during the Holsteinian requires explanation. Likewise, events in the late Middle Pleistocene are not entirely clear. If there were one or more additional interglacial stages, then these must surely have been accompanied by high sea-level stands, but these have not yet been recognized in areas of subsidence such as The Netherlands. We not only need to be sure of our correlations but also need to have a much greater understanding of sea-level history and forebulge effects following periods of major glaciation. This applies to the situation in Ireland as well as in Britain. Such problems are identified in this volume; perhaps the answers will be contained in a future one.

[Readers will notice that various authors have placed the Pliocene–Pleistocene boundary at different positions, and I have not tried to enforce any particular party-line on this much debated issue.]

The papers included in this volume arose from the Annual Discussion Meeting of the Quaternary Research Association held in the Department of Zoology, University of Cambridge on 6–7 January 1993. At the meeting invited speakers discussed a range of topics under the theme *Island Britain – a Quaternary perspective.*

I would like to thank all the authors for their forbearance during the preparation of this volume, and the referees who kindly reviewed the papers. Special thanks are due to the Geological Society Publishing House for the production of this volume.

Global sea-level and the (pen-)insularity of late Cenozoic Britain

BRIAN M. FUNNELL

School of Environmental Sciences, University of East Anglia, Norwich NR4 7TJ, UK

Abstract: During mid-Pliocene times, Britain was surrounded by warm temperate seas. In the south, these received relatively little clastic input, accumulated mainly biogenic sediments, and appear to have allowed free circulation of marine waters between the North Sea and the Atlantic Ocean, either around or across southern Britain.

In late Pliocene times, commencing at about 2.5 Ma BP, the first major global sea-level falls associated with northern hemisphere glaciation occurred. At about the same time, climatic changes appear to have accelerated the rate of headwater extension and bedload transport of the principal rivers flowing towards the southern North Sea. The combination of sea-level fall and progressive deltaic progradation into the southern North Sea converted Britain into a European peninsula.

By the mid-Quaternary Cromerian stage, the Great European (Ur-Frisia) Delta top was sufficiently moribund, and subject to subsidence, to allow interglacial sea-levels to encroach southwards across it. During the following Anglian Glaciation, impounded glacio-marine or glacio-lacustrine waters appear to have extended southwards, across both the former delta top and the bounding Weald–Artois chalk ridge, into the English Channel.

Subsequently a marine connection between the North Sea and the Atlantic Ocean, via the English Channel, was alternately established and broken, as global sea-level rose and fell (synchronously with interglacial and glacial periods), cyclically converting Britain from island to peninsula and back again.

Glacio-eustatic variations in global sea-level have strongly influenced the timing and development of insularity in the British Isles during the late Cenozoic. In mid-Pliocene times, relatively high sea-levels ensured well developed insularity, but following the onset of significant northern hemisphere glaciation in late Pliocene times, at about 2.5 Ma BP, marine regression, induced by sea-level fall and deltaic build-out into the southern North Sea, converted the British Isles into a NW European peninsula. In middle and late Pleistocene times major repetitive oscillations in global ice volume and global sea-level led to episodes of insularity during interglacial periods, and peninsularity during glacial periods, culminating in a final return to island conditions in the Post-glacial (Holocene) epoch.

This paper briefly examines the evidence for late Cenozoic global sea-level changes, and the late Cenozoic stratigraphy of the southern North Sea Basin, and puts these together in an attempt to show how the history of British insularity relates to the global record of sea-level change (Fig.1, Table 1). Estimates of variations in global sea-level over the last 2.6 Ma can best be obtained from the stable oxygen-isotope record preserved in deep-sea biogenic carbonates. Estimates of variations in the position of coastlines between Britain and continental Europe, and the depth/altitude of deposition of marine sediments, can be obtained from the stratigraphy of deposits in and around the southern North Sea.

Global sea-level over the last 2.6 Ma

Ultimately, there is no completely dependable method for computing an absolute global sea-level framework for the late Tertiary and Quaternary. Indeed, it is questionable whether there is ever a level at any one time which merits designation as an absolute global sea-level. However, it is well known in general terms that global sea-level during the late Cenozoic has responded overall to the amounts of ice incorporated into continental glaciers, and that global ice-volume is reflected in the ratio of the stable isotopes of oxygen found in ocean waters and recorded in the carbonate skeletons of foraminifera (Shackleton 1987). A detailed calibration of variations in global sea-level (as measured by a 'staircase' of uplifted and dated coral reefs in the Huon Peninsula, Papua New Guinea), with the $\delta^{18}O$ record from deep-sea, benthic foraminifera, was made by Chappell & Shackleton (1986) for the last 140 000 years.

Extrapolation of these calibrations of $\delta^{18}O$ values with sea-level to earlier periods (for which there is no independent reference framework for

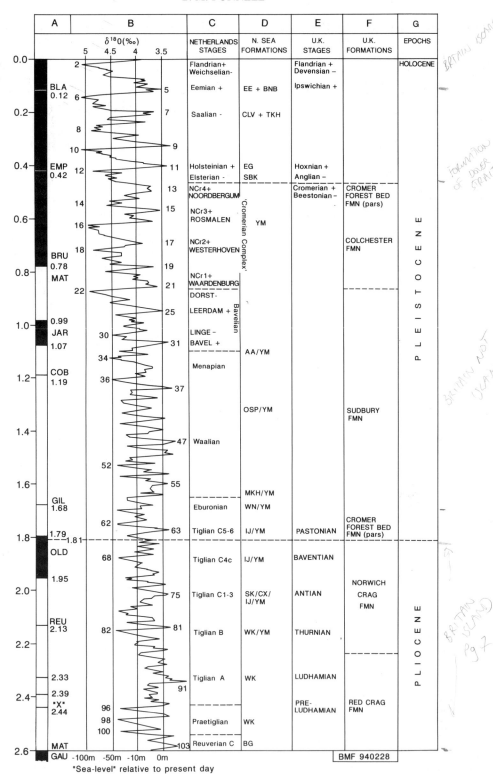

sea-level corresponding to the Huon 'staircase') is fraught with uncertainties and difficulties. Differences in the temporal, geographical and altitudinal development of continental glaciers during earlier glacial periods will have affected both the isotopic composition of their ice, and also, therefore, the precise relationship between global ice-volume and the $\delta^{18}O$ values found in the oceans. Longer-term global and local tectonics will also have affected the volumetric capacity of the ocean basins and their peripheral seas. So the use of the oxygen–isotope curve as a proxy for sea-level in the longer term, cannot be relied upon in any absolute sense. Nevertheless, making the simplifying assumption that the general relationship between oceanic $\delta^{18}O$, global ice-volume and global sea-level has remained essentially the same over the last few million years, it is instructive to compare it with the record of sea-level change preserved in the sediments linking Britain to continental Europe across the southern North Sea Basin.

A number of long Plio-Pleistocene high-resolution records of oceanic $\delta^{18}O$ has become available over the last few years, but the most detailed was obtained from Ocean Drilling Program (ODP) Site 677 in the Panama Basin of the eastern equatorial Pacific. It is this benthic foraminiferal $\delta^{18}O$ record that has been used in constructing Fig. 1. Shackleton et al. (1991) tuned this record directly to a calculated time-scale of astronomical forcing. However, it needs to be emphasized that, although the $\delta^{18}O$ curve from ODP Site 677 is the best available at the present time for providing an independent measure of the sequence and approximate magnitude of late Cenozoic global sea-level changes, its representation of the amplitude of sea-level in any particular $\delta^{18}O$ stage can only be regarded as indicative rather than absolute.

Sea-level changes in the southern North Sea

Sea-level changes in the southern North Sea can be inferred from evidence of changes in the position and elevation of the coastline, supplemented by indications of changes in the depth of deposition of the marine sediments of different ages. A series of palaeogeographical maps summarizing coastline development in the southern North Sea from early Pliocene times (Coralline Crag) to the time of the Anglian Glaciation (Lowestoft Till) was published by Funnell (1991). First presented at the Quaternary Research Association Field Meeting held in Norwich in 1988, and currently in the process of further revision, these maps draw heavily on the work of Zagwijn (1979), and on offshore data published by Cameron et al. (1987) and Zagwijn (1989). For a comprehensive account of the

---

Fig. 1. North Sea Basin Plio-Pleistocene stratigraphy *versus* global sea-level (2.6–0 Ma).
Column A. Astronomically tuned absolute ages have been allocated to the geomagnetic reversal (magnetochron) and event time-scale by reference to the record of geomagnetic field intensity variations and reversals presented by Valet & Meynadier (1993) and Shackleton et al. (1995). Named magnetochrons, subchrons, cryptochrons and events have been selected for Fig. 1 by additional reference to Cande & Kent (1992), Denham (1976) and McDougall (1979). A summary of information on the ages and abbreviations used for the magnetochrons and events in Fig. 1, is given in Table 1.

Column B. The $\delta^{18}O$ stage-scale, based on ODP Site 677, is taken from Shackleton et al. (1991, fig. 7). Selected stages have been numbered by reference to Shackleton et al. (1991, figs. 2 and 4); when inserting numbers of interpolated stages, reference has also been made to the numbering of stages assigned for ODP Site 607 (against a different absolute age-scale) in Ruddiman et al. (1989, fig. 7 and table 6) and Raymo et al. (1989, fig.6 and table 7).

Columns C & E. Interglacial (high sea-level) stages are indicated ' + '; glacial (low sea-level) stages are inidcated '−'.

Column D. Abbreviations and correlations of North Sea formations have been inserted after Cameron et al. (1992, fig. 91). BG = Brielle Ground Formation; WK = Westkapelle Ground Formation; IJ = Ijmuiden Ground Formation; SK = Smith's Knoll Formation; CX = Crane Formation; WN = Winterton Shoal Formation; MKH = Markham's Hole Formation; OSP = Outer Silver Pit Formation; AA = Aurora Formation; YM = Yarmouth Roads Formation; SBK = Swarte Bank Formation; EG = Egmond Ground Formation; CLV = Cleaver Bank Formation: TKH = Tea Kettle Hole Formation; EE = Eem Formation; BNB = Brown Bank Formation.

NB (a) The correlation of the Ijmuiden Ground Formation with the Praetiglian suggested by Zagwijn (1989) is believed to be in error. (b) The Yarmouth Roads Formation is a seismostratigraphic formation corresponding to the delta-top lithofacies of the Great European Delta and is strongly diachronous. In the south these deposits are Tiglian or even Praetiglian in age (Cameron et al. 1992). Only towards the north are they equivalent in age to the Kesgrave Group (Sudbury and Colchester Formations) – mainly post-Tiglian to Cromerian Complex deposits of the onshore successions.

Table 1. *Absolute age attributions of geomagnetic reversals, events and excursions*

Label	Abbreviation	Ma†	ka
‡*Laschamp* excursion			40
Blake event	BLA	0.12	118
Jamaica excursion			195
Biwa II excursion			280
Biwa III excursion			412
Emperor cryptochron/event	EMP	0.42	419
Big Lost excursion			554
Delta excursion			690
Bruhnes chron (base)/	BRU	0.78	780
Matuyama chron (top)	MAT	0.78	780
Kamikatsura excursion			931
Jaramillo subchron (top)	JAR	0.99	990
Jaramillo subchron (base)	JAR	1.07	1070
Cobb Mountain cryptochron/event	COB	1.19	1190
§*Ontong Java 1* excursion			1371
Ontong Java 2 excursion			1444
Gilsa event	GIL	1.68	1680
Olduvai subchron (top)	OLD	1.79	1787
Olduvai subchron (base)	OLD	1.95	1947
Reunion subchron/event	REU	2.13	2133
?event		2.33	2330
?event		2.39	2387
'**X**' cryptochron/event	'X'	2.44	2441
‖ **Matuyama** chron (base)/	MAT	2.60	2600
Gauss chron (top)	GAU	2.60	2600

*Calibrated to an astronomical time-scale established in eastern equatorial Pacific sediment cores, obtained by Ocean Drilling Program Leg 138 (Valet & Meynadier 1993; Shackleton *et al.* 1995)
Only magnetochron labels listed in **bold** are shown in Fig. 1
†Only values in this column, to the nearest 10 000 years, have been entered in Fig. 1. They are least likely to be subject to future change. Values to the nearest 1000 years are clearly more liable to future modification
‡Laschamp to Cobb Mountain dates from table 1 in Valet & Meynadier (1993)
§Ontong Java 1 to 'X' dates read from figure 3 in Valet & Meynadier (1993)
‖ Gauss/Matuyama boundary date from Shackleton *et al.* (1995); the date that can be read from Figure 3 in Valet & Meynadier (1993) is 2.512 Ma

offshore (and onshore) geology of the UK sector of the southern North Sea Basin, and alternative interpretations of the position of shorelines for selected times in the late Pliocene and early to late Pleistocene, see Cameron *et al.* (1992). The stratigraphical correlations between the UK, The Netherlands, Belgium and contiguous parts of the North Sea continental shelf adopted in this paper are based, wherever possible, on our current understanding as recorded in Gibbard *et al.* (1991). In the interests of completeness, other offshore formations recorded in Cameron *et al.* (1992) are included in Fig. 1, according to the correlations adopted in that publication.

North Sea to deep-sea correlations

Correlations between the stratigraphy of the North Sea Basin and that of the deep-sea can best be made on the basis of: (a) 'absolute' (radiometric) age determinations; (b) planktic microfossil datums; and (c) biostratigraphically, radiometrically or astronomically constrained palaeomagnetic field reversals (defining magnetochronozones). In the deep-sea, 'absolute' ages can also be determined from the identification of $\delta^{18}O$ stages tuned to the astronomical time-scale, and independently calibrated in relation to radiometrically dated magnetic reversals (Shackleton *et al.* 1991, 1995; Cande & Kent 1992; Wilson 1993)

Radiometric age dating methods are rarely applicable to both North Sea and deep-sea sequences, except in the case of radiocarbon dating, which has the potential for very high-resolution dating and correlation, but only over the last *c.* 30 000 years. Beyond that time-span, uranium series dating has some potential for selected terrestrial and marine materials, and K/Ar dating is capable of providing reliable dates on volcanic materials older than about 200 000 years.

Planktic datums (first and last appearance events of planktic microfossils) already exhibit some diachroneity and provincialism in the deep-sea (Clement & Weaver 1986). In epicontinental seas, the actual diachroneity and provincialism of such datums may be accentuated, because shallower waters generally support, and their sediments entomb, fewer planktic organisms. However, with care, planktic datums can be used to constrain (and permit identification of), particular magnetostratigraphic events, even in epicontinental seas.

Magnetostratigraphy has been very successfully applied to deep-sea sediments, in which the magnetic field reversal record can be precisely matched (Cande & Kent 1992; Valet & Meynadier 1993; Wilson 1993) with the succession of 'magnetic stripes' recorded in the basaltic ocean crust. Note, however, that sedimentation rates in the deep-sea rarely exceed 100 mm per thousand years, whereas sedimentation rates of shallow marine and fluviatile sediments frequently exceed sedimentation rates of 1000 mm per thousand years. Therefore, short-lived magnetic events or excursions may well be recorded in epicontinental or continental sediments, whilst they appear unreliably or not at all in deep-sea sediments, which are accumulating one or two orders of magnitude more slowly.

Correlations between the North Sea Basin and deep-sea successions (Fig. 1), based on the above principles, are used to evaluate the evidence from North Sea Basin deposits relating to Britain's insularity during the last 2.6 Ma. Since this paper was originally prepared, an independently derived correlation of The Netherlands succession with the $\delta^{18}O$ record of ODP Site 677 has been published by Veldkamp & van den Berg (1993), based on modelling the terraces of the River Maas in relation to their ages as determined by pollen, palaeomagnetics, thermoluminescence and ^{14}C.

The Pliocene island of Britain

All lines of evidence suggest that, until latest Pliocene times, Britain was separated from continental Europe by the North Sea and a broad seaway, which allowed marine floral and faunal exchange between the North Sea and the Atlantic Ocean via a southern route. The marine Coralline Crag Formation of Suffolk contains cysts of the planktic alga *Bolboforma costata*, and coccolith and planktic foraminiferal species, which enable direct correlation with the mid-Pliocene of the deep-sea record (Hodgson & Funnell 1987; Jenkins & Houghton 1987; Jenkins *et al.* 1988). The marine flora and fauna of the Coralline Crag both indicate warm temperate marine conditions. The benthic foraminifera suggest deposition in water depths of around 50 metres or more (Hodgson & Funnell 1987). Large-scale tidal cross-bedding indicates deposition as offshore sandbanks, probably comparable in scale to those found off the East Anglian coast at the present day (Balson *et al.* 1993). No shoreline deposits of the same age as the Coralline Crag have been preserved, but the inferred depth of deposition of the Coralline Crag, together with its present elevation of up to 20 m OD, imply that local sea-level was substantially (at least 70 m) higher than at the present day.

On the opposite side of the Southern Bight of the North Sea, deposits of the Luchtbal Formation in Belgium and the Oosterhout Formation in The Netherlands give place landward to the fluviatile Kieseloölite Formation (Zagwijn 1985; Funnell 1991). The clastic content of the Belgian and Dutch marine formations is higher than that of the Coralline Crag, with evidence of brief tempestites interrupting an otherwise continuous input of fine sand into the margins of the southern North Sea.

In late Pliocene times, the Red Crag Formation of East Anglia commences with sediments which already contain cool-water species of North Pacific origin, such as the gastropod *Neptunea angulata* and the benthic foraminiferan *Elphidiella hannai*. The proportion of Boreal and even 'Arctic' marine molluscan species increases during the accumulation of the formation (Harmer 1900), and both the cross-bedding of the deposits and the benthic foraminifera they contain indicate a change in time from deposition in moderate depths of water (up to 50 m) to shallow-water, near 'beach' deposition (Dixon 1979). This evidence suggests a trend towards overall colder water marine conditions and a corresponding overall fall in sea-level.

In Belgium and The Netherlands, marine deposits of the Kruisschans and Oosterhout Formations contain late Reuverian pollen assemblages and can be palynologically correlated with the Kieseloölite Formation fluviatile deposits (Zagwijn 1985). Pollen obtained from the oldest onshore Red Crag Formation in Britain at Walton-on-the-Naze, Essex has been attributed to the latest Reuverian 'C' Sub-stage by Hunt (1989).

All except the latest Reuverian deposits in the Netherlands exhibit normal palaeomagnetic polarity (van Montfrans 1971*a, b*; Gibbard *et al.* 1991), and a correlation of most of the

Reuverian with the latter part of the Gauss magnetochron has always been assumed. The upper boundary of the Gauss magnetochron has recently been dated at 2.60 Ma BP (Cande & Kent 1992; Shackleton et al. 1995); previous dates for this boundary, used in earlier references to North Sea Basin data, range between 2.40 and 2.48 Ma BP.

Late Pliocene sea-level fall

In The Netherlands, the Reuverian Stage is followed by the Praetiglian Stage, which is characterized in both marine and fluviatile environments by 'glacial' (i.e. high non-arboreal) pollen assemblages, and in the marine facies by major shallowing and marine regression. Because the Gauss/Matuyama magnetochron boundary is observed just below the Reuverian/ Praetiglian boundary, it is highly probable that the Praetiglian can be equated with one or more of the marked $\delta^{18}O$ stages 100, 98 and 96, which correspond with the first major incursions of ice-rafted debris into the North Atlantic. These $\delta^{18}O$ stages immediately post-date the Gauss/ Matuyama magnetochron boundary in the deep-sea record (Shackleton et al. 1991, 1995). No unequivocal equivalents to the Praetiglian deposits have so far been discovered in Britain, and the distinctive benthic foraminiferan *Elphidiella oregonense*, which characterizes Praetiglian marine deposits in The Netherlands, has not yet been found in the UK (Funnell 1987). However, parts of the Red Crag Formation referred to the Pre-Ludhamian Stage, and confined to deep troughs in Suffolk, have been tentatively correlated with the Praetiglian of The Netherlands (Gibbard et al. 1991).

The presence of the planktic foraminiferan *Neogloboquadrina atlantica* in Pre-Ludhamian Stage sediments was taken by Funnell (1987) to indicate that they were likely to correlate with the late Reuverian Stage. At that time, the Last Appearance Datum (LAD) of *N. atlantica* in the North Atlantic was being reported at 2.3 Ma BP (Clement & Weaver 1986). Later, Raymo et al. (1989) reported the LAD of *N. atlantica*, both in terms of $\delta^{18}O$ stages and their revised dating of those stages in the North Atlantic. From their data it is clear that *N. atlantica* actually persisted in the North Atlantic as late as the transition between $\delta^{18}O$ stages 95 and 94. In Fig. 1 these $\delta^{18}O$ stages are positioned close to the boundary between the Praetiglian and the Tiglian Stages. On the basis of this revision of its stratigraphic range, it would have been possible for *N. atlantica* to occur in the Praetiglian, and possibly even the earliest Tiglian Stage, in the North Sea Basin. Furthermore, although the stratotype Pre-Ludhamian of the Stradbroke borehole (in the Stradbroke Trough) is normally magnetized, and all records from the Praetiglian of The Netherlands are reversely magnetized (van Montfrans 1971a, b; Funnell 1987), the possibility exists that the normal polarity recorded from the Pre-Ludhamian and earliest Ludhamian Stages of the Stradbroke sequence could correlate with the 'X' event (dated to between 2.441 and 2.421 Ma BP by Cande & Kent (1992), and to 2.441 Ma BP by Valet & Meynadier (1993)), and has simply not yet been recorded in Praetiglian Stage deposits from The Netherlands. It should be noted that two episodes of normal polarity were recorded from the generally reversed polarity Westkapelle Ground Formation (Cameron et al. 1984), the lower of which may also relate to the 'X' event and the higher to the Reunion (dated to 2.133 Ma BP by Valet & Meynadier (1993)).

Although it seems possible that Britain could have lost its insularity and been temporarily connected to the continent by the magnitude of the Praetiglian sea-level fall, the clearest evidence of a breaking of the southern link between the North Sea and the Atlantic Ocean comes in the following Tiglian Stage.

Latest Pliocene sea-level changes

The uppermost Red Crag Formation member, the Ludham Crag, contains a diverse marine fauna, suggesting the continuing existence of some direct marine linkage between the southern North Sea and the Atlantic Ocean. However, upper Red Crag Formation deposits in general contain significant quantities of material reworked from earlier Red Crag deposits (Beck et al. 1972), and this compromises any unequivocal interpretation of the total fauna as indigenous and necessarily implying a continuing southern connection with the Atlantic Ocean. The $\delta^{18}O$ values shown for $\delta^{18}O$ stage 91 (Fig. 1) suggest the attainment of the highest global sea-levels subsequent to $\delta^{18}O$ stage 103 during the early Tiglian. This would be consistent, not only with the indications of marine transgression (and a possible continuing southern connection with the Atlantic Ocean) exhibited by the Ludham Crag (early type Ludhamian Stage), but also with the evidence of marine transgression shown by the Tiglian A deposits of The Netherlands (Gibbard et al. 1991).

In the Ludham Royal Society borehole, the later part of the type Ludhamian Stage contains marine foraminiferal faunas (Funnell 1987) of Norwich Crag Formation aspect (biofacies). The foraminiferal and molluscan faunas of the Norwich and Chillesford Crag Members of the Norwich Crag Formation, and of the Weybourne Crag Member of the Cromer Forest Bed Formation (Funnell & West 1977), are relatively impoverished compared with those of the Red Crag Formation. These changes suggest the breaking of any residual southern marine connection with the Atlantic Ocean, and a possible reduction of salinities in the Southern Bight of the North Sea. Similar changes in marine foraminiferal and molluscan faunas occur in The Netherlands in the Tiglian Stage (Meijer & Preece 1995).

The Norwich Crag Formation spans several stages defined by pollen assemblage zones: late Ludhamian, Thurnian, Antian, Baventian and Bramertonian (Gibbard *et al.* 1991). The pollen of these stages suggests alternately warm and cool conditions. A similar pattern of alternating warm and cool conditions is indicated by the pollen of the sub-stages of the Tiglian of The Netherlands. Comparisons of the British and Dutch successions, based on pollen, vertebrates and molluscs, suggest (Gibbard *et al.* 1991) correlations between: the Thurnian Stage and the Tiglian B Sub-stage (both cool); the Antian (Bramertonian) Stages and the Tiglian C1–3 Sub-stages (all warm). The Thurnian Stage/Tiglian B Sub-stage provide evidence of shallower marine conditions/marine regression, as well as cooling, and could correspond with the stronger cooling implied by the oxygen–isotope values of $\delta^{18}O$ stage 82. The Antian (Bramertonian) Stages/Tiglian C1–3 Sub-stages provide evidence of deeper water/marine transgression, as well as warming, and could correspond with $\delta^{18}O$ stages 75 and/or 73. Changes in global sea-level and climate that were less marked than those represented by the more extreme $\delta^{18}O$ values of $\delta^{18}O$ stages 82, 75 and 73 might not be detectable or distinguishable in the marine and peri-marine record of the Southern Bight of the North Sea of this period.

The Plio-Pleistocene boundary

The current international definition of the base of the Pleistocene (Aguirre & Pasini 1985) has been identified as occurring just below the top of the Olduvai magnetosubchron (Hilgen 1991) and at 1.81 Ma BP on Hilgen's astronomically tuned time-scale. (Because of very small differences between the astronomically tuned timescales of Hilgen (1991) and Shackleton *et al.* (1995), the age for the Plio-Pleistocene boundary, as identified by Hilgen (1991), is probably equivalent to 1.805 Ma BP on the Shackleton *et al.* (1995) timescale.) The normally magnetized Olduvai magnetochron appears to be represented in Britain in deposits of the Baventian and possibly some part of the Pastonian Stage, and in The Netherlands in the Tiglian C4c Sub-stage (von Montfrans 1971*a,b*; Zagwijn 1985). Recent results from Sheringham, on the Norfolk coast, have found Pastonian Stage sediments bearing a primary reversed polarity, assumed to relate to the immediately post-Olduvai portion of the Matuyama magnetochron (Hallam & Maher 1994). Identification of the Olduvai magnetosubchron as corresponding to the Baventian Stage of Britain and to the Tiglian C4c Sub-stage of The Netherlands would enable the international Plio-Pleistocene boundary to be determined as occurring at the base or within the Pastonian Stage in Britain, and at the base or within the Tiglian C5–6 Sub-stages in The Netherlands. In the deep-sea, the Olduvai magnetochron includes a marked cold episode at $\delta^{18}O$ stage 68, and is followed by a marked warm episode at $\delta^{18}O$ stage 63. These $\delta^{18}O$ stages could well correspond respectively to the cold, marine regression of the Baventian Stage/Tiglian C4c, and the warm, marine transgression of the Pastonian Stage/Tiglian C5–6 Sub-stages. In both Britain (Pastonian Stage) and The Netherlands (Tiglian C5–6 Sub-stages) this is essentially the last occasion that marine deposits are seen in and around the Southern Bight of the North Sea until the Cromerian. (Marine deposits correlated with the Waalian Stage have, however, been recorded from a limited area of the SW Netherlands by Meijer (1988).)

Early Quaternary peninsularity

As we have seen above, the internationally defined (Aguirre & Pasini 1985; Hilgen 1991) Plio-Pleistocene boundary, and the beginning of the Quaternary period, appears to correlate with the onset of the Pastonian Stage/Tiglian C5–6 Sub-stages of the British and Dutch successions. The early Quaternary period extends until the end of the Matuyama magnetochron at 0.78 Ma BP.

From late Pliocene to early Pleistocene times, the major north European rivers appear to have generally extended their headwaters and increased bedload transport (Urban 1982; Zagwijn 1986; Whiteman & Rose 1992; Veldkamp & van den Berg 1993), probably, at least partly, as a response to the widespread climatic changes associated with the inception of northern hemisphere glaciation. Resultant massive growth of the Great European (Ur-Frisia) Delta, across the southern and into the northern North Sea, during this period (Zagwijn 1979, 1989; Bijlsma 1981; Cameron et al. 1987, 1989, 1992; Funnell 1991) had already radically changed geographical conditions in the southern North Sea Basin by the early Quaternary. Delta growth and global sea-level fall finally excluded the sea from the southern areas it had previously occupied (until the Pastonian Stage/Tiglian C5–6 Substage), and Britain became not just a peninsula of Europe, but broadly connected to the continent across the Great European Delta top. For an extended period there follows a major hiatus in the record of marine conditions in and around the southern North Sea Basin.

In The Netherlands, alternating early Quaternary cold and warm periods have been identified, based on pollen assemblages in freshwater sediments, and used to define the Eburonian, Waalian, Menapian, Bavelian and early 'Cromerian Complex' Stages. The $\delta^{18}O$ stages corresponding with this sequence commence with 64, the first cold episode of the middle (post-Olduvai) Matuyama magnetochron, and conclude with 20, the last cold episode of the late Matuyama magnetochron. Altogether, the middle and late Matuyama sequence comprises 22 obliquity-forced $\delta^{18}O$ cycles. It is clear that only the stronger oscillations have registered as the generally *cold* periods of the Eburonian and Menapian Stages, and the Linge and Dorst 'glacials' of the Bavelian Stage, or as the generally *warm* periods of the Waalian Stage, the Bavel and Leerdam 'interglacials' of the Bavelian Stage, and the Waardenberg (NCr1) 'interglacial' of the Cromerian Complex.

The strong, typical middle and late Quaternary 100 000 year cycle of glacials and interglacials actually appears to commence with (a) the Dorst 'glacial', of the Bavelian Stage, which is probably equivalent to $\delta^{18}O$ stage 22, and (b) the following Waardenburg (NCr1) 'interglacial' of the Cromerian Complex of the Dutch succession, which is probably equivalent to $\delta^{18}O$ stage 21 of the deep-sea record.

Modelling of the River Maas terraces in the south of The Netherlands (Veldkamp & van den Berg 1993) has suggested that the Eburonian to early 'Cromerian Complex' Stages encompass nine principal aggradation phases, corresponding only to the more marked $\delta^{18}O$ cold stages indicated by the deep-sea record. In Britain, the six main aggradation events of the River Thames terraces of the Sudbury Formation (Whiteman & Rose 1992) may similarly relate only to the more extreme of the middle and late Matuyama $\delta^{18}O$ cold stages (e.g. 62, 52, 36, 34, 32 and 22), with the warmer intervening $\delta^{18}O$ stages (e.g. 55, 54, 47, 37, 31 and 25) mainly contributing to weathering of the terrace aggradations. Unfortunately, the British terrace deposits are remarkably unfossiliferous (and consequently difficult to date) over this period of time, and nothing can therefore be concluded about the migration of fauna and flora across the Great European Delta top between Britain and Europe during this time.

Middle Quaternary glaciations

The Middle Quaternary is taken to commence at the Matuyama/Bruhnes magnetochron boundary at 0.78 Ma BP, corresponding to $\delta^{18}O$ stage 19.

Veldkamp & van den Berg (1993) attribute four River Maas terrace aggradational episodes to the later Cromerian Complex. Three Bruhnes magnetochron (normal polarity) Cromerian Complex interglacials have so far been recognized in The Netherlands: Westerhoven (NCr2), Rosmalen (NCr3) and Noordbergum (NCr4).

Whiteman & Rose (1992) identify four River Thames terrace aggradational episodes within their pre-Anglian Colchester Formation. Organic remains from the Ardleigh (second earliest) terrace indicate correlation with Westerhoven (NCr2) or Rosmalen (NCr3) of The Netherlands succession (Gibbard et al. 1991). The Cromerian interglacial at the West Runton type site in Britain appears to pre-date the Noordbergum interglacial of the Dutch Cromerian Complex by at least a short interval (Gibbard et al. 1991), but marine conditions appear in both localities (Zagwijn 1985; Gibbard et al. 1991), suggesting that in both cases the sea had now returned southwards over the moribund Great European Delta close to its former Plio-Pleistocene limits.

Nevertheless, this marine transgression can have done no more than reduce Britain to a peninsula. That there was ready access to Britain during the Cromerian interglacial for small and large mammals has recently been highlighted by the discovery of the major part of a skeleton of an early form of mammoth, *Mammuthus trogontherii*, within the type West Runton section of the Cromerian Stage (Stuart 1993, 1995).

Late Quaternary restoration of island status

Ice sheets sourced in Britain and Scandinavia occupied much of the southern North Sea Basin during the Anglian/Elsterian glacial stage. Meltwater from these ice sheets, impounded by grounded ice to the north and rising ground to the south, appears to have overflowed southwards (Gibbard, 1995), perhaps catastrophically (Smith 1989), into the English Channel. It is not clear whether a marine connection was established between the southern North Sea and the Atlantic Ocean along this route during the following Hoxnian–Holsteinian interglacial stage, but by the time of the Ipswichian–Eemian interglacial stage, at the beginning of the late Quaternary, there is clear evidence of a well established marine connection through the Dover Strait, bringing warm-water faunas with Lusitanian associations into the Southern Bight of the North Sea (Meijer & Preece, 1995).

Following the inevitable loss of island status by glacio-eustatic lowering of global sea-level during the Devensian–Weichselian glacial stage, the return of marine waters to the southern North Sea following deglaciation is well documented by radiocarbon dates obtained on cockle shells (Cerastoderma edule) which lived in the initially brackish waters of the Southern Bight, from c. 9560 ^{14}C years BP onwards (Eisma et al. 1981). The marine component of this brackish Southern Bight water was probably initially sourced from the English Channel through the Dover Strait (Eisma et al. 1981). Not until about 7000 ^{14}C years BP did the Southern Bight waters become fully saline, presumably as they linked up with marine waters transgressing from the north. It was at this later date that Britain again became a true island. In eastern England the sea continued to transgress across coastal and estuarine freshwater peats to initiate the present-day coastal belts of intertidal sedimentation, shortly after 6600 to 6300 ^{14}C years BP (Funnell & Pearson 1989; Brew et al. 1992). The preservation of estuarine sediments on the sea bed, up to 6 km offshore, suggests that the flanking soft-sediment cliffs of Suffolk, England, have retreated by approximately that distance during the last 6500 ^{14}C years BP (Brew et al. 1992)

Summary and conclusions

The stratigraphy of the late Cenozoic deposits of eastern Britain, the Netherlands and the southern North Sea has been reviewed and compared with the deep-sea benthic foraminiferal δ^{18}O proxy record of global sea-level change, in order to provide a considered account of the evidence for the timing of the (pen-)insularity of Britain during the late Cenozoic (Fig. 1).

During early to late Pliocene times, relatively high sea-levels connected the southern North Sea directly to the Atlantic Ocean via a southern (English Channel) route, and Britain was isolated as an island. Global sea-level fall associated with δ^{18}O stages 100, 98 and 96, and correlated with the Praetiglian of the Dutch succession, may have produced an initial linkage of Britain to mainland Europe during the period 2.5 to 2.4 Ma BP.

Expansion of the Great European (Ur-Frisia) Delta into the southern North Sea probably closed the southern connection of the North Sea to the Atlantic Ocean, and definitively ended Britain's insularity during the early part of the Tiglian Stage at about 2.3 Ma BP. Continued expansion of the delta consolidated the terrestrial connection of Britain and excluded marine influences (including evidence of the effects of global sea-level changes) from the Southern Bight area of the North Sea Basin from about 1.7 Ma (early Pleistocene) until about 0.5 Ma BP (early Middle Pleistocene).

The re-occupation of the delta-top area by marine waters, which was initially achieved during the early Middle Pleistocene, seems to have been followed, during the Anglian–Elsterian stage (δ^{18}O stage 12 or 10, at around 0.45 or 0.35 Ma BP), by breaching of the land barrier between the southern North Sea Basin and the English Channel by overflowing glacial meltwaters. Whether this route was occupied continuously, or at all, as a seaway by the interglacial sea-levels of the ensuing Hoxnian–Holsteinian stage is not clear. However, the evidence of southern marine species in the southern North Sea Basin during the later Ipswichian–Eemian interglacial stage indicates that island status was clearly well established by the high sea-levels of that time, and that a southern connection was re-established between the North Sea and the Atlantic Ocean, probably for the first time since late Pliocene times. Following the Devensian/Weichselian glacial stage, the progress of marine flooding of the southern North Sea is clearly preserved in sediments at or close to the present sea-floor surface. A connection between northern North Sea waters and English Channel waters coming through the Dover Strait was established at around 7000 ^{14}C years BP.

I am greatly indebted to Dennis Jeffery and Jim Rose for their detailed comments on my initial draft of this paper. These have helped me considerably in reconsidering and improving it during its revision.

References

AGUIRRE, E. & PASINI, G. 1985. The Plio-Pleistocene boundary. *Episodes*, **8**, 116–120.
BALSON, P. S., MATHERS, S. J. & ZALASIEWICZ, J. A. 1993. The lithostratigraphy of the Coralline Crag (Pliocene) of Suffolk. *Proceedings of the Geologists' Association*, **104**, 59–70.
BECK, R. B., FUNNELL, B. M. & LORD, A. R. 1972.

Correlation of Lower Pleistocene Crag at depth in Suffolk. *Geological Magazine*, **109**, 137–139.

BIJLSMA, S. 1981. Fluvial sedimentation from the Fennoscandian area into the North-West European basin during the Late Cenozoic. *Geologie en Mijnbouw*, **60**, 337–345.

BREW, D. S., FUNNELL, B. M. & KREISER, A. 1992. Sedimentary environments and Holocene evolution of the lower Blyth estuary, Suffolk (England), and a comparison with other East Anglian coastal sequences. *Proceedings of the Geologists' Association*, **103**, 57–74

CAMERON, T. D. J., BONNY, A. P., GREGORY, D. M. & HARLAND, R. 1984. Lower Pleistocene dinoflagellate cyst, foraminiferal and pollen assemblages in four boreholes in the southern North Sea. *Geological Magazine*, **121**, 85–97.

——, CROSBY, A., BALSON, P. S., JEFFERY, D. H., LOTT, G. K., BULAT, J. & HARRISON, D. J. 1992. *United Kingdom Offshore Regional Report: The Geology of the Southern North Sea*. HMSO, London.

——, LABAN, C. & SCHÜTTENHELM, R. T. E. 1989. Pliocene and Lower Pleistocene stratigraphy in the Southern Bight of the North Sea. *In:* HENRIET, J. P. & DE MOOR, G. (eds) *The Quaternary and Tertiary Geology of the Southern Bight, North Sea*. Belgian Geological Survey, Brussels, 97–110.

——, STOCKER, M. S. & LONG, D. 1987. The history of Quaternary sedimentation in the U.K. Sector of the North Sea Basin. *Journal of the Geological Society, London*, **144**, 43–58.

CANDE, S. C. & KENT, D. V. 1992. A new Geomagnetic Polarity Time Scale for the Late Cretaceous and Cenozoic. *Journal of Geophysical Research*, **97**, 13917–13951.

CHAPPELL, J. & SHACKLETON, N. J. 1986. Oxygen isotopes and sea level. *Nature*, **324**, 137–140.

CLEMENT, B. M. & WEAVER, P. P. E. 1986. Synchroneity of Pliocene planktonic Foraminiferal datums in the North Atlantic. *Marine Micropalaeontology*, **10**, 295–307.

DENHAM, C. R. 1976. Blake polarity episode in two cores from the Greater Antilles Outer Ridge. *Earth and Planetary Science Letters*, **29**, 422–434.

DIXON, R. G. 1979. Sedimentary facies in the Red Crag (Lower Pleistocene, East Anglia). *Proceedings of the Geologists' Association*, **90**, 117–132.

EISMA, D., MOOK, W. G. & LABAN, C. 1981. An Early Holocene Tidal Flat in the Southern Bight. *International Association of Sedimentologists, Special Publication*, **5**, 229–237.

FUNNELL, B. M. 1987. Late Pliocene and Early Pleistocene stages of East Anglia and the adjacent North Sea. *Quaternary Newsletter*, **52**, 1–11.

—— 1991. Palaeogeographical maps of the southern North Sea Basin: Pliocene (Coralline Crag) to Anglian (Lowestoft Till). *Bulletin of the Geological Society of Norfolk*, **41**, 53–66.

—— & PEARSON, I. 1989. Holocene sedimentation on the north Norfolk barrier coast in relation to relative sea-level change. *Journal of Quaternary Science*, **4**, 25–36.

—— & WEST, R. G. 1977. Preglacial Pleistocene deposits of East Anglia. *In:* SHOTTON, F. W. (ed.) *British Quaternary Studies: Recent Advances*. Oxford University Press, Oxford, 247–265.

GIBBARD, P. L. 1995. The formation of the Strait of Dover. *This volume*.

—— WEST, R. G., ZAGWIJN, W. H. ET AL. 1991. Early and Early Middle Pleistocene correlations in the southern North Sea Basin. *Quaternary Science Reviews*, **10**, 23–52.

HALLAM, D. F. & MAHER, B. A. 1994. A record of reversed polarity carried by the iron sulphide greigite in British early Pleistocene sediments. *Earth and Planetary Science Letters*, **121**, 71–80.

HARMER, F. W. 1900. The Pliocene deposits of the east of England. Part II, The Crag of Essex (Waltonian) and its relation to that of Suffolk and Norfolk. *Quarterly Journal of the Geological Society of London*, **56**, 705–744.

HILGEN, F. J. 1991. Astronomical calibration of Gauss to Matuyama sapropels in the Mediterranean and implication for the Geomagnetic Polarity Time Scale. *Earth and Planetary Science Letters*, **104**, 226–244.

HODGSON, G. E. & FUNNELL, B. M. 1987. Foraminiferal biofacies of the early Pliocene Coralline Crag. *In:* HART, M. B. (ed.) *Micropalaeontology of Carbonate Environments*. Ellis Horwood, Chichester, 44–73.

HUNT, C. O. 1989. The palynology and correlation of the Walton Crag (Red Crag Formation, Pliocene). *Journal of the Geological Society, London*, **146**, 743–745.

JENKINS, D. G. & HOUGHTON, S. D. 1987. Age, correlation and palaeoecology of the St. Erth Beds and the Coralline Crag of England. *Mededelingen van de Werkgroep voor Tertiaire en Kwartaire Geologie*, **24**, 147–156.

——, CURRY, D., FUNNELL, B. M. & WHITTAKER, J. E. 1988. Planktonic foraminifera from the Pliocene Coralline Crag of Suffolk, England. *Journal of Micropalaeontology*, **7**, 1–10.

McDOUGALL, I. 1979. The present status of the geomagnetic polarity time scale. *In:* McELHINNY, M. W. (ed.) *The Earth, its Origin, Structure and Evolution*. Academic Press, London, 543–566.

MEIJER, T. 1988. Mollusca from the borehole Zuurland-2 at Brielle, The Netherlands (an interim report). *Mededelingen Werkgroep Tertiaire en Kwartaire Geologie*, **25**, 49–60.

—— & PREECE, R. C. 1995. Malacological evidence relating to the insularity of the British Isles during the Quaternary. *This volume*.

RAYMO, M. E., RUDDIMAN, W. F., BACKMAN, J., CLEMENT, B. M. & MARTINSON, D. G. 1989. Late Pliocene variations in northern hemisphere ice sheets and North Atlantic deep water circulation. *Paleoceanography*, **4**, 413–446.

RUDDIMAN, W. F., RAYMO, M. E., MARTINSON, D. G., CLEMENT, B. M. & BACKMAN, J. 1989. Pleistocene evolution: northern hemisphere ice sheets and North Atlantic Ocean. *Paleoceanography*, **4**, 353–412.

SHACKLETON, N. J. 1987. Oxygen isotopes, ice volume

and sea level. *Quaternary Science Reviews*, **6**, 183–190.

——, BERGER, A. & PELTIER, W. R. 1991. An alternative astronomical calibration of the lower Pleistocene timescale based on ODP Site 677. *Transactions of the Royal Society of Edinburgh*, **81**, 252–261.

——, CROWHURST, S., HAGELBERG, T., PISIAS, N. G. & SCHNEIDER, D. A. 1995. A new Late Neogene time scale: application to ODP Leg 138 Sites. *Proceedings of the Ocean Drilling Program, Scientific Results*, in press.

SMITH, A. J. 1989. The English Channel – by geological design or catastrophic accident? *Proceedings of the Geologists' Association*, **100**, 325–337.

STUART, A. J. 1993. An elephant skeleton from the West Runton Freshwater Bed (Early Middle Pleistocene; Cromerian Temperate Stage). *Bulletin of the Geological Society of Norfolk*, **41**, 75–90.

—— 1995. Insularity and Quaternary vertebrate faunas in Britain and Ireland. *This volume*.

URBAN, B. 1982. Quaternary paleobotany of the Lower Rhine Basin. *Zeitschrift Geomorphologie, N. F. Suppl.*, **42**, 201–213.

VALET, J.-P. & MEYNADIER, L. 1993. Geomagnetic field intensity and reversals during the past four million years. *Nature*, **366**, 234–238.

VAN MONTFRANS, H. M. 1971a. *Palaeomagnetic Dating in the North Sea Basin*. PhD Thesis, University of Amsterdam, Princo N.V., Rotterdam.

—— 1971b. Palaeomagnetic dating in the North Sea Basin. *Earth and Planetary Science Letters*, **11**, 226–235.

VELDKAMP, A. & VAN DEN BERG, M. W. 1993. Three-dimensional modelling of Quaternary fluvial dynamics in a climo-tectonic dependent system. A case study of the Maas record (Maastricht, The Netherlands). *Global and Planetary Change*, **8**, 203–218.

WHITEMAN, C. A. & ROSE, J. 1992. Thames River sediments of the British Early and Middle Pleistocene. *Quaternary Science Reviews*, **11**, 363–375.

WILSON, D. S. 1993. Confirmation of the astronomical calibration of the magnetic polarity timescale from sea-floor spreading rates. *Nature*, **364**, 788–790.

ZAGWIJN, W. H. 1979. Early and Middle Pleistocene coastlines in the southern North Sea basin. *In:* OELE, E., SCHÜTTENHELM, R. T. E. & WIGGERS, A. J. (eds) *The Quaternary History of the North Sea. Acta Universitatis Upsaliensis, Symposia Universitatis Upsaliensis Annum Quingentesimum Celebrantis*, **2**, 31–42.

—— 1985. An outline of the Quaternary stratigraphy of the Netherlands. *Geologie en Mijnbouw*, **64**, 17–24.

—— 1986. The Pleistocene of the Netherlands with special reference to glaciation and terrace formation. *Quaternary Science Reviews*, **5**, 341–345.

—— 1989. The Netherlands during the Tertiary and the Quaternary: A case history of Coastal Lowland evolution. *Geologie en Mijnbouw*, **68**, 107–120.

The formation of the Strait of Dover

P. L. GIBBARD

Subdepartment of Quaternary Research, Botany School, University of Cambridge, Downing Street, Cambridge CB2 3EA, UK
Present address: Godwin Institute of Quaternary Research, Department of Geography, Downing Place, Cambridge CB2 3EN, UK

Abstract: The Strait of Dover (Pas-de-Calais) is a narrow sea passage that links the North Sea and the English Channel between Britain and France. Much evidence suggests that the Dover Strait did not exist throughout most of the Pleistocene. Instead, a Chalk barrier was present, formed by the Weald–Artois anticline. Advance of the continental ice-sheet across the North Sea in the Middle Pleistocene Elsterian/Anglian Stage apparently dammed the southern part of the basin and water discharging into it was prevented from reaching the Atlantic to the north. The resulting lake apparently drained by spilling over the barrier and initiated the gap. This scenario is supported by sediments at Wissant, France. This theory contrasts with one, proposed by French geologists, that favours a structural origin.

Overflow from the lake has been interpreted as having been catastrophic. This was invoked to explain the origin of the complex anastomosing system of valleys or channels that occurs on the floor of the Channel cut into bedrock, predominantly of Mesozoic age. These valleys are interlinked to form a drowned drainage system and can also be linked to present rivers that enter the Channel on both the British and French sides. However, it is highly probable that the drowned valley system is multigenetic, resulting from repeated fluvial and marine erosion and deposition rather than representing a single, short-lived event. The valley system probably originated in the late Early to early Middle Pleistocene as a consequence of uplift of the Weald–Artois region.

Once the Dover Strait was formed, the rivers Thames and Scheldt were deflected through the gap and into the Channel River system during periods of low relative sea-level. Subsequently these rivers were probably joined by the Maas and Rhine, diverted by glaciation in the Drenthe Sub-stage (Saalian/Wolstonian Stage). The narrows seem to have been progressively enlarged by marine tidal scour and coastal erosion during high sea-level events.

The island of Great Britain is separated from France by the Strait of Dover (Pas-de-Calais) (Fig. 1). This sea passage is 34 km wide at its narrowest point, between Dover and Cap Gris Nez, and reaches a maximum depth here of −55 m. The Strait is cut into Cretaceous bedrock, predominantly Chalk and underlying Gault Clay. These rocks have been gently anticlinally folded along an axis trending WNW–ESE and form part of the Weald–Artois anticlinorium of probable Miocene age. The present Strait is predominantly an erosional feature and includes few Quaternary deposits except for a sequence of silts and clays that fill palaeovalleys of the Fosse Dangeard in the Strait (Destombes *et al.* 1975) and the nearby Sangatte raised beach of Late Pleistocene age (Sommé 1979; Keen 1995). There is therefore very little evidence for the age of this passage, except that it was open in the Eemian/Ipswichian (see below and Keen 1995; Meijer & Preece 1995).

The Strait of Dover occurs at a crucial position between two substantial shallow marine embayments: the Channel and the southern North Sea. On the basis of our knowledge of the sediments and features associated with these two embayments, they have strikingly different geological histories (cf. Gibbard 1988).

The Channel

The Channel floor is generally smooth and the surface is gently inclined from the Dover Strait towards the continental shelf margin. Close to the coasts, steeper, ramp-like slopes occur (Smith 1985*a*). Over most of the Channel, this smooth surface is floored by a thin, mobile sediment cover beneath which bedrock, predominantly of Mesozoic age, is present (Larsonneur *et al.* 1979; Smith & Curry 1975). Excavated into this bedrock in the central and eastern Channel is a complex network of narrow valleys or channels, many of which are infilled

From Preece, R. C. (ed.), 1995, *Island Britain: a Quaternary perspective*
Geological Society Special Publication No. 96, pp. 15–26

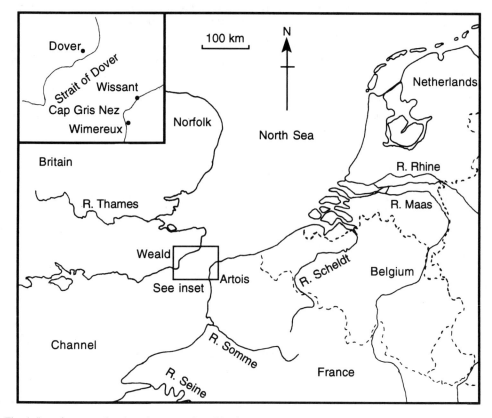

Fig. 1. Location map showing places mentioned in the text.

by unconsolidated gravels, sands and fine sediments (Dingwall 1975). The best known example of these is the Hurd Deep, that was shown to contain a multiple infilling and have a composite origin by combined fluvial deposition and tidal scour (Hamilton & Smith 1972). Dingwall (1975) and Smith (1985a), have shown that the valleys are interlinked to form a drowned drainage system, presumably formed during periods of lowered sea-level. Evidence suggests that sea-level minima of 90–130 m below present occurred locally during the cold stages of the Pleistocene (Hamilton & Smith 1972). Fluvial downcutting and infilling during these periods, followed by tidal scour during repeated periods of marine transgression and regression, appear to have been the major formative processes of these features.

In detail, the valleys form a complex anastomosing system and many are over-deepened, particularly at confluence points, a feature also found beneath fluvial valleys on land (Berry 1979; Hutchinson 1991). The valleys can be shown to be continuations of those of coastal rivers including the Seine, Somme, Béthune, Solent and Arun (Smith 1985a). The central or Lobourg valley can be followed into, and indeed through, the Dover Strait where it merges into the Fosse Dangeard of Destombes et al. (1975), a series of scour hollows on the Gault–Lower Chalk contact (Smith 1985a). These hollows are associated with structural weaknesses arising from NW–SE aligned faults of the Pas-de-Calais Fault Zone (Colbeaux et al. 1980).

The suggestion that the valley system is mainly of fluvial origin and that it functioned during periods of low sea-level is now almost universally accepted. This is contrary to former suggestions that the system originated directly from glaciation (Destombes et al. 1975; Kellaway et al. 1975). Further discussion is offered by Zagwijn (1979) and Oele & Schüttenhelm (1979).

Evidence for the long-term existence of the Channel as a marine embayment is fragmentary. However, it is certain that it existed in the Pliocene to earliest Pleistocene on the basis of marine sediments at St Erth, in Cornwall

(Jenkins et al. 1986), at La Londe in Normandy and at several sites in Brittany (Lautridou et al. 1986) at various elevations above sea-level. These sediments accumulated during both warm and cold periods. The river system developed on the Channel floor post-dates these marine sediments. The earliest fluvial sediments identified in the present palaeovalleys have been found in the offshore extension of the Seine valley. Here sediments attributed to the 'Cromerian Complex' occur (Alduc et al. 1979). However, high sea-levels certainly occurred during temperate (interglacial) events in the later part of this stage, on the basis of the evidence of marine and estuarine sedimentation in the contemporary Solent Estuary at Boxgrove and other sites in Sussex, and Bembridge on the Isle of Wight (Preece et al. 1990). It would seem likely, therefore, that for most of the Early Pleistocene the area was one of net erosion; the lack of preserved sediments is probably a consequence of tectonic activity both in the Western Approaches Basin (Smith & Curry 1975) and northern France (Bourdier & Lautridou 1974; Pomerol 1978; Colbeaux et al. 1980; Smith 1985b) during this time. Preece et al. (1990) have indicated that relationships of late Neogene and early Pleistocene sediments in southeast England-northeastern France suggest a local updoming of over 400 m through the past few million years. This doming, they maintain, covers a far larger area than the Weald–Artois anticline and instead may be a response to tectonic downwarping in the southern North Sea.

River downcutting and deposition presumably took place throughout the late Early and early Middle Pleistocene during periods of low sea-level, since contemporary periglacial fluvial deposits were laid down by coastal rivers such as the Seine (Alduc et al. 1979) and the Solent (Allen 1991; Allen & Gibbard 1993) systems. A proto-Channel River aligned broadly ENE–WSW along the axis of the basin must have existed at this time (Gibbard 1988). It appears that this river was formed by the confluence of the southern British and northern French rivers (Hull 1912; Dingwall 1975; Jones 1981) (Fig. 2).

Southern North Sea

In contrast to the Channel, the southern North Sea Basin has been being infilled by sediment from Neogene and indeed earlier times (Zagwijn 1974, 1979; Zagwijn & Doppert 1978). This major tectonically downwarping basin has been an important element in northwest European geology throughout the Cenozoic. From the Pliocene to the Middle Pleistocene substantial delta sediments derived from the North German rivers, the Rhine, the Maas, the Scheldt and the Thames were progressively deposited into the basin to form the important offshore Winterton Shoal and Yarmouth Roads Formations (Balson & Cameron 1985; Cameron et al. 1987). By the Middle Pleistocene these sediments had infilled most of the southern North Sea Basin; the rivers drained broadly northwards into the sea and the areas between became dry land for much of the time (Fig. 2). The area was apparently not invaded by the sea during periods of high eustatic sea-level throughout most of this interval, since sedimentation outpaced downwarping of the basin (Zagwijn 1979).

As Gibbard (1988) has emphasized, this pattern of river flow and consequent delta progradation continued into the Elsterian/Anglian Stage (see also Funnell (1995)). However, this stage saw the first major extension of the continental ice-sheet from southern Scandinavia into temperate regions of central Europe and Britain. Advance of the ice towards the south and southwest is well established (Ehlers et al. 1984; Ehlers & Gibbard 1991). This direction would have carried the ice-sheet across the very gently graded delta plain in the North Sea Basin and thus would almost certainly have blocked the free northward drainage of the rivers into the Atlantic Ocean. This could only have occurred once the ice had either impinged on higher ground on the British side and/or met the British ice on the eastern basin margin. The latter is generally accepted to have taken place throughout most of the period (cf. Hart 1987; Ehlers & Gibbard 1991; Hart & Boulton 1991). The presence of an ice-dam would therefore have caused an immense glacial lake to have formed in the basin south of the ice-front (Fig. 3).

Evidence for the occurrence of glacio-lacustrine sedimentation in the southern North Sea Basin is well known. In Britain, particularly in the coastal sequences of northern and eastern East Anglia, the North Sea Drift Formation includes both glacial diamictons (tills) and associated meltwater sediments. Banham (1988), Hart (1990) and Hart & Boulton (1991) have concluded that the glacial sediments were deposited on land, although they recognized that the associated thick fine sediment units are of glacio-lacustrine origin (Hart 1992). In contrast, several authors have concluded that some of the tills are waterlain and the meltwater sediments represent delta and lake-bottom deposits (Kazi & Knill 1969; Banham 1970, 1988; Gibbard

 Main river course

Tributary course

Fig. 2. Palaeogeography in the late 'Cromerian Complex' (modified from Gibbard 1988).

1980; Lunkka 1988, 1991, 1994; Hart 1992; Eyles et al. 1989). Indeed Eyles et al. (1989) considered the sequence to be of glacio-marine origin. This latter point is not, however, generally accepted, particularly by Lunkka (1991, 1994), whose detailed sedimentological investigations have clearly demonstrated that the sequence represents a series of ice-lobe oscillations in a continuously present freshwater body.

On the eastern side of the North Sea, there is also extensive evidence for glacio-lacustrine sedimentation during this period. In the northern Dutch provinces of Friesland and Groningen, deep valleys cut into pre-existing Quaternary sediments contain thick freshwater glacial lake sediments of the Peelo Formation (Ter Wee 1962, 1983a). These widespread deposits comprise clays in the north and grade into sand-dominated facies in the south where they include sediment of Rhine origin (Zagwijn 1974; Ter Wee 1983a). The clays occur not only on land but extend beneath large areas of the sea-floor adjacent to The Netherlands (Oele 1969, 1971). In the British offshore sector, the similarly occurring deposits of the Svarte Bank Formation are almost certainly of the same origin.

In general, therefore, there is strong evidence for the presence and evolution of a substantial freshwater body in the southern North Sea during the Elsterian/Anglian glaciation. The present distribution of these sediments at or above present sea-level in some areas implies either isostatic depression of the basin or high lake water-level or an interplay of both factors.

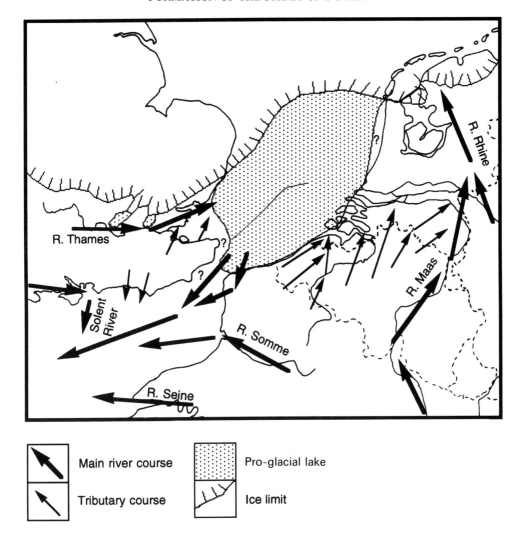

Fig. 3. Palaeogeography during the Elsterian/Anglian glacial maximum (modified from Gibbard 1988).

Formation of the Strait of Dover

During much of the late Early and Middle Pleistocene, the southern North Sea Basin was dry land being drained by major rivers, and the fluvial system in the Channel region was developing. Whether the Dover Strait existed at this time is problematic because almost no pre-Elsterian/Anglian deposits are known in the immediate vicinity. Colbeaux et al. (1980) have concluded that the Strait formed by graben-like movements on the Pas-de-Calais Fault Zone. On the basis of geomorphological arguments, they assumed that the Strait formed at the end of the Early Pleistocene but provided no sedimentary evidence in support of this interpretation. The exception is the Herzeele Formation, a sequence of shallow marine sands present in France, northeast of the Strait (Paepe & Sommé 1975; Sommé et al. 1978; Paepe & Baeteman 1979; Sommé 1979). These deposits include evidence of a marine transgression in the Holsteinian (Hoxnian) Stage (Vanhoorne 1962). The occurrence of a possible earlier transgression beneath assigned to the 'Cromerian' by Paepe & Sommé (1975) and 'Pastonian' (= Tiglian Tc5–6) by Zagwijn (1979) is disputed, the lower sediments probably being of freshwater origin (T. Meijer, pers. comm.). A sea-passage may have existed, therefore, during high sea-level periods up to and including the Tiglian. In the Channel in general, evidence for high sea-level during an early Middle Pleistocene interglacial (?'Cromerian Complex') is found in southern England, but evidence for an open Dover Strait is equivocal. It therefore appears that throughout most of the Pleistocene (post-Middle Tiglian), up to and including the early Anglian/Elsterian Stage, the Strait did not exist.

The precise form of the pre-existing topography in the Strait area is not known, but it seems reasonable to presume that it closely resembled the adjacent areas of northern France and Kent, i.e. chalklands with dry valley topography and short rivers. The ridge would certainly have been an interfluve in common with the areas on each side at present. Indeed, on the basis of the modern drainage patterns, Stamp (1927, 1936) even suggested the courses of two streams on opposing sides of the interfluve that might have been expected to occur in the area before the breaching of the ridge began.

Once the southern North Sea Basin was dammed by the ice-sheets, it is implicit that the water entering from much of western Europe would have caused the level to rise rapidly until an outlet col was reached. There is a general consensus that the outlet was at what is now the Strait of Dover (Belt 1874; Reid 1913; Stamp 1927, 1936; Gullentops 1974; Roep et al. 1975; Jones 1981; Smith 1985a, 1989; Gibbard 1988; Zalasiewicz & Gibbard 1988; Ehlers et al. 1991) (Fig. 3).

The precise mechanism by which the breach formed was not addressed until very recently. Roep et al. (1975) and Smith (1985a, 1989) independently presented the case for a catastrophic overflow from an ice-dammed lake in the southern North Sea. The former authors based their interpretation on the presence of coarse conglomerates at Wissant, France (see below). To explain the form and distribution of the valley system on the Channel floor, the latter author invoked a comparison with the so-called 'scablands' channelled topography of Washington State area in the NW United States (Bretz 1969), a comparison also drawn by Roep et al. (1975) to explain the mode of occurrence of the Wissant conglomerates. Smith considered that the overspill of an hypothetical lake, comparable to the outburst of Glacial Lake Missoula at the end of the last glaciation, would have formed a very similar system of channels and associated scour phenomena. There is indeed a superficial similarity in the form and morphology of the Channel floor valleys with those of eastern Washington, but some details do not match.

Firstly, the Channel valleys contain multiple infillings, indicating that some, at least, were formed during the early Middle Pleistocene – the valleys relating to existing rivers on each side of the Channel. Later infill, such as the Early Devensian sequence of silts and clays within palaeovalleys and scour hollows on the floor of the Strait, e.g. Fosse Dangeard (Destombes et al. 1975), suggest that the hollows are later features unrelated to the initial formation of the valley.

Secondly, the catastrophic drainage of Lake Missoula took place when a confining ice-dam collapsed, allowing the whole lake to drain in a single event (Bretz 1969). In contrast, a lake in the southern North Sea would have been confined by a substantial bedrock ridge in the Strait area. Spilling of the lake over a confining ridge, would potentially have been dramatic during the initial phases, given the volume of water available. The volume of overflowing water must have been initially equivalent to the input, thus when the bedrock col or cols were reached, the resulting erosional capacity of the water would been considerable. Since it is known that chalk is highly susceptible to dissolution by cold water, downcutting during the overflow may have been very rapid, even 'catastrophic' in the sense of the Missoula flood. Nevertheless, Smith's concept that the Channel floor valley system, together with its associated erosional features, formed at a single time is not supported by the evidence. On the contrary, the valley system seems to be the net product of polycyclic fluvial and marine scour events caused by alternating low and high sea-level (Hamilton & Smith 1972), the latter driven by the climate history of the later Pleistocene. Further support for this is the evidence that glacio-lacustrine conditions clearly persisted in the southern North Sea until the end of the Elsterian/Anglian Stage. If the lake had drained in a single event, no evidence of its later persistence should be found.

Detail of the timing of the overspill is very difficult to obtain because evidence has largely been removed. Smith (1985a) originally offered no date, but he has subsequently accepted Gibbard's (1988) explanation that the overspill represented that of the Elsterian/Anglian southern North Sea lake (A. J. Smith, pers. comm.).

Evidence of events during and immediately following the breakthrough are available from the thick channel-fill sequence thoroughly described from near Wissant and Wimereux by Roep et al. (1975). Here they identified a three-fold sequence infilling a deep, irregularly shaped valley cut into the local Mesozoic bedrock. The basal unit comprised a coarse to very coarse boulder gravel of fluvial origin infilling trough- or pot-hole-like depressions in the bedrock surface to at least −18 m NGF (Nivellement General Français). This was over-lain by green laminated clays and sands, that coarsened slightly upwards. They, in turn, were succeeded by medium to fine gravel and sand that showed facies typical of normal, braided-river deposition. A comparable sequence was exposed in the cliffs on the south side of Cap Gris Nez. Pebble counts indicated that the gravel was overwhelmingly of local origin (De Heinzelin 1964) but with very rare, possibly northerly derived (possibly ice-rafted) erratics. Conversely, the clays and sands contained abundant heavy minerals, particularly high frequencies of epidote, that indicate a combined Fennoscandian and Rhine source (Roep et al. (1975) and references cited therein). Palaeocurrent measurements from the gravel and sand units unquestionably indicated a flow towards the south and west. The basal gravel has yielded vertebrate remains including *Mammuthus meridionalis* and '*Hippopotamus major*'. They have been interpreted as of 'Mindel or immediately pre-Mindel age' by Sommé (1969) but are probably of Early to early Middle Pleistocene age (Stuart 1982). The remains were presumably reworked because they were mostly eroded, but were apparently not available for study (D. A. Hooijer, pers. comm. to Th. B. Roep, 1975).

Overall, therefore, the Wissant sequence indicates extremely rapid initial deposition accompanied by severe erosion, under highly turbulent flow from the north. This was followed by a period of sedimentation in a quiet water, lacustrine environment and then a return to fluvial conditions. This probably represents an initial breakthrough of high velocity drainage, a period of stasis, possibly reflecting local uplift or stabilization of lake-level. Then more consistent fluvial drainage returned through this valley, possibly reflecting a second period of relatively higher lake water-level. In the light of this sequence it seems highly likely that the sediment-fill represents an overspill channel that carried water from the southern North Sea lake into the Channel. It cannot, however, be the main channel, since that must have been situated in the area of what is now the Strait gap (Fig. 3).

Initially, the overspill river might be expected to have followed existing local stream valleys into the Channel River system. However, the volume of water would have completely overwhelmed these valleys and probably quickly caused them to be incised and enlarged. This precursor of the present central or Lobourg valley would have evolved rapidly as discharge varied in response to water supply in the lake. Structural weaknesses in the Strait area (Destombes et al. 1975; Colbeaux et al. 1980) would have assisted in the incision process. As the col was lowered, lake water-level would have fallen.

Again, the precise timing is unknown but there is considerable evidence for the continuation of lacustrine conditions until very late in the stage. The distribution of laminated glacio-lacustrine sediments in The Netherlands (Peelo Formation) and the similar Lauenburg Clay Formation (Lauenburger Ton) in NW Germany and Denmark (Zagwijn 1974, 1985; Ehlers 1983), together with the glacio-lacustrine sediments filling the deep glacial valleys in East Anglia (Ventris 1985) and offshore (Balson & Cameron 1985; Balson & Jeffery 1991), strongly support this interpretation. Clearly, it is uncertain whether all these deposits accumulated in a single lake or whether the basin became fragmented into a complex of lakes in individual valleys, particularly during the deglacial phase. Nevertheless, the North Sea lake or lake complex does seem to have continued as a major feature throughout the Late Elsterian/Anglian Stage (Gibbard 1988).

The fall of the lake water-level seems to have caused rivers close to the northern side of the Dover Strait col to have been deflected southwards. This phenomenon is seen in both the Scheldt and Thames systems (Vandenberghe & de Smedt 1979; Vandenberghe et al. 1986; Gibbard 1988; Bridgland et al. 1993). Indeed, no later deposits of either of these rivers are found in the southern North Sea further north than their present valleys. In contrast, it is much less certain that the Rhine and Maas rivers also followed the Strait course at this time. Indeed, Holsteinian-age Rhine sediments in the eastern and northern Netherlands (e.g. at Neede) are aligned north to north-northwestwards.

This deflection of the Thames and Scheldt rivers southwards seems surprising against a background of probable isostatic depression of the North Sea basin, beneath the ice-sheets. This depression would be expected to have drawn river flow northwards. In contrast, the southward flow may reflect the rapid opening of the Strait col to a level below that of the contemporary floor of the southern North Sea between central East Anglia and The Netherlands. This is a zone of potential substantial sedimentation throughout the existence of the glacial lake. Such a lowering might not have arisen solely from erosion of a bedrock valley, but also by rapid isostatic recovery, potentially accompanied by collapse of a peripheral forebulge in the Channel region, following retreat of the Elsterian/Anglian ice-sheets further north. A rapid forebulge collapse is known to occur following ice-sheet retreat (cf. Walcott 1972; Wingfield 1995). If this happened in the Dover Strait area, it would certainly have encouraged the Thames

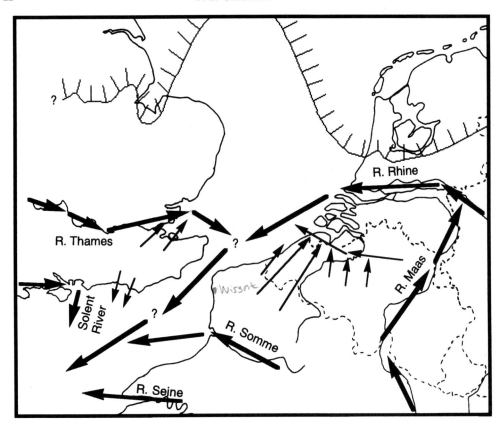

Fig. 4. Palaeogeography in the Saalian/Wolstonian Stage during the glacial maximum (modified from Gibbard, 1988).

and Scheldt to turn southwards and continue to flow through the Strait into the Channel River, but this remains unproven. Additionally, the possible occurrence of ice-marginal sediments in the area between East Anglia and The Netherlands might have further encouraged such a course. Adoption of the course would have ensured the continued incision of both the Strait valley and those of the rivers entering the Strait (Vandenberghe et al. 1986; Gibbard 1994) until arrival of the sea following eustatic transgression in the subsequent Holsteinian/Hoxnian Stage. The Wissant valley was clearly abandoned once the lake-level fell, since quantities of northerly derived pebble material are absent from the sequence.

The evolution of the Strait during the Holsteinian/Hoxnian Stage is unclear, but the occurrence of marine deposits at Herzeele, suggests that a sea-passage might have been open during the period of maximum interglacial sea-level. This is suggested by the occurrence of a 'southern' marine mollusc *Mytilaster minima*, a species unknown from other contemporary localities in the North Sea Basin (Meijer & Preece 1995). However, this does not unequivocally prove the existence of a marine connection. If a passage did occur, it would probably have been relatively narrow, perhaps 1–4 km (Keen 1995).

Following the eustatic fall of sea-level at the beginning of the Saalian/Wolstonian Stage, the rivers seem to have re-established their courses. Events similar to those of the Elsterian seem to have recurred at this time. Smaller-scale glaciolacustrine sedimentation may have developed in the southern North Sea off the Dutch coast (Oele & Schüttenhelm 1979). These deposits are associated with the glaciation during the Drenthe Sub-stage. The advance into the central Netherlands forced the Rhine and Maas to flow westwards (de Jong 1965; Zagwijn 1974; Ter Wee 1983b) and they probably became confluent with the Thames and Scheldt (Fig. 4). Such a drainage diversion would certainly require a barrier or ice-dam across the southern North Sea. Although some writers have suggested that this is unlikely (Cameron et al. 1987; Joon et al. 1990), indirect evidence, particularly ice-lobe movement directions in The Netherlands, implies the presence of British ice (Rappol et al. 1989). Recent investigations in northwest Norfolk (Gibbard et al. 1992) lend support to this theory, since they indicate that an ice-lobe did indeed move down the eastern British coast. Moreover, the presence of British gravel in Drenthe Till, 80 km west of Texel, suggests interaction of ice from each side of the North Sea (A. W. Burger, pers. comm.).

Whatever the mechanism of the diversion, the distribution of the Saalian to Weichselian Kreftenheye Formation sediments indicates that the Rhine and Maas drained southwards after advance of the Scandinavian ice-sheet in the Drenthe Sub-stage (Jelgersma et al. 1979; Oele & Schüttenhelm 1979; Bridgland & D'Olier 1995). Moreover, there is little doubt that the Strait was open in the Eemian/Ipswichian Stage on the basis of the occurrence of Mediterranean molluscs including *Angulus distortus* (Spaink 1958) in the southern North Sea region (Meijer & Preece 1995).

Conclusion

On the basis of the evidence assembled from the Channel, southern North Sea and Strait regions, it therefore seems highly probable that the Dover Strait was initiated during the Elsterian/Anglian Stage by overflow of a substantial ice-dammed lake in the southern North Sea and not as the result of structural phenomena, as previously thought. A seaway may have existed during the late Neogene and possibly sporadically during high sea-level phases in the Tiglian Stage. Nevertheless, for most of the Early and Middle Pleistocene the area was high ground with a morphology comparable to that of the adjacent areas of Kent and the Boulonnais today.

The potential effects of glacio-isostatic depression of the North Sea basin in the Elsterian/Anglian Stage, possibly accompanied by uplift of the eastern Channel, may have combined to produce increased incision of the Strait ridge once water-level in the lake was able to overtop the ridge. The nature of the initial overspill was forceful, if not catastrophic. Later overspill was more moderate and probably persisted for a long period, since the lake or lakes continued to exist until the end of the glaciation.

The form of palaeovalleys on the Channel floor is not solely the consequence of the single catastrophic event that initiated the Strait, but of the polycyclic interaction of fluvial incision and tidal scour during alternating respectively low and high eustatic sea-level events throughout the Middle and Late Pleistocene.

Drainage of the lake(s), possibly accompanied by rapid isostatic recovery and collapse of a peripheral forebulge uplift in the Strait region, encouraged the rivers Thames and Scheldt to become confluent in the latest Elsterian/Anglian and to flow into the Channel River system. Subsequent sea-level rise would have drowned the narrow Strait passage, but later eustatic sea-level fall caused the rivers to readopt their course into the Channel by the beginning of the Saalian/Wolstonian Stage. Following glaciation in The Netherlands in the Drenthe Sub-stage, the Rhine and Maas rivers were diverted to join the Thames and Scheldt through the Strait.

I would particularly like to thank Th. B. Roep (Vrije Universiteit, Amsterdam) and Helen Roe (University of Cambridge) for stimulating discussions on aspects of this topic. I am also indebted to D. R. Bridgland (University of Durham), T. Meijer, A. Berger and P. Cleveringa (Rijks Geologische Dienst) and R. C. Preece for critically reading the manuscript. The diagrams were drawn by S. Boreham.

References

ALDUC, D., AUFFRET, J.-P., CARPENTIER, G., LAUTRIDOU, J.-P., LEFEBVRE, D. & PORCHER, M. 1979. Nouvelles données sur le Pléistocène de la basse vallée de la Seine et son prolonguement sous marin en Manche Orientale. *Bulletin d'information des géologues du Bassin de Paris*, **16**, 27–34.

ALLEN, L. G. 1991. *The Evolution of the Solent River System during the Pleistocene*. PhD thesis, University of Cambridge.

—— & GIBBARD, P. L. 1993. Pleistocene evolution of the Solent River of southern England. *Quaternary Science Reviews*, **12**, 503–528.

BALSON, P. S. & CAMERON, T. D. J. 1985. Quaternary mapping offshore East Anglia. *Modern Geology*, **9**, 221–239.

—— & JEFFERY, D. H. 1991. The glacial sequence of the southern North Sea. *In:* EHLERS, J., GIBBARD, P. L. & ROSE, J. (eds) *Glacial Deposits in Great*

Britain and Ireland. Balkema, Rotterdam, 245–253.
BANHAM, P. H. 1970. North Norfolk. *In:* BOULTON, G. S. (ed.) *East Anglia Field Guide*. Quaternary Research Association, 11–17.
—— 1988. Polyphase glaciotectonic deformation in the Contorted Drift of Norfolk. *In:* CROOT, D. G. (ed.) *Glaciotectonics: Forms and Processes*. Balkema, Rotterdam, 27–32.
BELT, T. 1834. The Glacial Period. *Nature*, **10**, 25–26
BERRY, F. G. 1979. Late Quaternary scour hollows and related features in central London. *Quarterly Journal of Engineering Geology*, **12**, 9–29.
BOURDIER, F. & LAUTRIDOU, J.-P. 1974. Les grands traits morphologique et structuraux des régions de la Somme et de la Basse-Seine. *Bullétin de l'Association Française pour l'Étude Quaternaire*, **40–41**, 129–135.
BRETZ, J. H. 1969. The Lake Missoula floods of the channelled scablands. *Journal of Geology*, **77**, 505–543.
BRIDGLAND, D. R. & D'OLIER, B. 1995. The Pleistocene evolution of the Thames and Rhine drainage systems in the southern North Sea. *This volume*.
——, ——, GIBBARD, P. L. & ROE, H. M. 1993. Correlation of Thames terrace deposits between the Lower Thames, eastern Essex and the submerged offshore continuation of the Thames–Medway valley. *Proceedings of the Geologists' Association*, **104**, 51–57.
CAMERON, T. D. J., STOKER, M. S. & LONG, D. 1987. The history of Quaternary sedimentation in the UK sector of the North Sea. *Journal of the Geological Society, London*, **144**, 43–58.
COLBEAUX, J.-P., DUPOIS, C., ROBASZYNSKI, F., AUFFRET, J.-P., HAESAERTS, P. & SOMME, J. 1980. Le Détroit du Pas-du-Calais: un élément dans le tectonique de blocs de l'Europe Nord-Occidentale. *Bullétin d'information des géologues du Bassin de Paris*, **17**, 41–54.
DE HEINZELIN, J. 1964. Cailloutis de Wissant, capture de Marquise et percée de Warcove. *Bullétin de la Société Belge de Géologie, Palynologie et Hydrologie*, **73**, 146–161.
DE JONG, J. 1965. Quaternary sedimentation in the Netherlands. *Geological Society of America, Special Papers*, **84**, 95–123.
DESTOMBES, J.-P., SHEPHARD-THORN, E. R. & REDDING, J. H. 1975. A buried valley system in the Strait of Dover. *Philosophical Transactions of the Royal Society of London*, **A285**, 243–256.
DINGWALL, R. G. 1975. Sub-bottom infilled channels in an area of the eastern English Channel. *Philosophical Transactions of the Royal Society of London*, **A279**, 233–241.
EHLERS, J. 1983. The glacial history of north-west Germany. *In:* EHLERS, J. (ed.) *Glacial Deposits in North-west Europe*, Balkema, Rotterdam, 229–238.
—— GIBBARD, P. 1991. Anglian glacial deposits in Britain and adjoining offshore regions. *In:* EHLERS, J., GIBBARD, P. L. & ROSE, J. (eds) *Glacial Deposits in Great Britain and Ireland*, Balkema, Rotterdam, 17–24.

——, —— & ROSE J. 1991. Glacial deposits of Britain and Europe: general overview. *In:* EHLERS, J., GIBBARD, P. L. & ROSE, J. (eds) *Glacial Deposits in Great Britain and Ireland*, Balkema, Rotterdam, 493–501.
——, MEYER, K. D. & STEPHAN, H. J. 1984. The pre-Weichselian glaciations of north-west Europe. *Quaternary Science Reviews*, **3**, 1–40.
EYLES, N., EYLES, C. H. & McCABE, A. M. 1989. Sedimentation in an ice-contact subaqueous setting: the Mid-Pleistocene 'North Sea Drifts' of Norfolk, U.K. *Quaternary Science Reviews*, **8**, 57–74.
FUNNELL, B. M. 1995. Global sea-level and the (pen)-insularity of late Cenozoic Britain. *This volume*.
GIBBARD, P. L. 1980. The origin of stratified Catfish Creek Till by basal melting. *Boreas*, **9**, 71–85.
—— 1988. The history of the great northwest European rivers during the past three million years. *Philosophical Transactions of the Royal Society of London*, **B318**, 559–602.
—— 1994. *Pleistocene History of the Lower Thames Valley*, Cambridge University Press.
——, WEST, R. G., ANDREW, R. & PETTIT, M. E. 1992. The margin of a Middle Pleistocene ice advance at Tottenhill, Norfolk, England. *Geological Magazine*, **129**, 59–76.
GULLENTOPS, F. 1974. The southern North Sea during the Quaternary. *In: L'évolution Quaternaire des Bassins Fluviaux de la Mer du Nord Méridionale*. Centenaire de la Société Géologique de Belgique, Liège.
HAMILTON, D. & SMITH, A. J. 1972. The Origin and Sedimentary History of the Hurd Deep, English Channel, with Additional Notes on other Deeps in the Western English Channel. *Mémoire de la Bureau recherche Géologique et Minéralogique*, **79**, 59–78.
HART, J. K. 1987. *The Genesis of the North East Norfolk Drift*. PhD thesis, University of East Anglia, UK.
—— 1990. Proglacial glaciotectonic deformation and the origin of the Cromer Ridge push moraine complex, north Norfolk, UK. *Boreas*, **19**, 165–180.
—— 1992. Sedimentary environments associated with glacial Lake Trimingham, Norfolk, UK. *Boreas*, **21**, 119–136.
—— BOULTON, G. S. 1991. The glacial drifts of Norfolk. *In:* EHLERS, J., GIBBARD, P. L. & ROSE, J. (eds) *Glacial Deposits in Great Britain and Ireland*, Balkema, Rotterdam, 233–243.
HULL, E. 1912. On the interglacial gravel beds of the Isle of Wight and south of England and the conditions for their formation. *Geological Magazine*, **9**, 100–105.
HUTCHINSON, J. N. 1991. Theme lecture: periglacial and slope processes. *In:* FORSTER, A., CULSHAW, M. G., CRIPPS, J. C., LITTLE, J. A. & MOON, C. F. (eds) *Quaternary Engineering Geology*. Geological Society, London, Engineering Special Publication, **7**, 283–331.
JELGERSMA, S., OELE, E. & WIGGERS, A. J. 1979. Depositional history and coastal development in

the Netherlands and adjacent North Sea since the Eemian. *In:* OELE, E., SCHÜTTENHELM, R. T. E. & WIGGERS, A. J. (eds) *The Quaternary History of the North Sea. Acta Universitatis Upsaliensis Symposia Universitatis Upsaliensis Annum Quingentesimum Celebrantis*, **2**, 115–142.

JENKINS, D. G., WHITTAKER, J. E. & CARLTON, R. 1986. On the age and correlation of the St Erth Beds, SW England, based on planktonic foraminifera. *Journal of Micropalaeontology*, **5**, 93–105.

JONES, D. K. C. 1981. *Southeast and Southern England. The Geomorphology of the British Isles.* Methuen, London.

JOON, B., LABAN, C. & VAN DER MEER, J. J. M. 1990. The Saalian glaciation in the Dutch part of the North Sea. *Geologie en Mijnbouw*, **69**, 151–158.

KAZI, A. & KNILL, J. L. 1969. The sedimentation and geotechnical properties of the Cromer Till between Happisburgh and Cromer, Norfolk. *Quarterly Journal of Engineering Geology*, **2**, 63–86.

KEEN, D. H. 1995. Raised beaches and sea-levels in the English Channel in the Middle and Late Pleistocene: problems of interpretation and implications for the isolation of the British Isles. *This volume*.

KELLAWAY, G. A., REDDING, J. H., SHEPHARD-THORN, E. R. & DESTOMBES, J.-P. 1975. The Quaternary history of the English Channel. *Philosophical Transactions of the Royal Society of London*, **A279**, 189–218.

LARSONNEUR, C., VASLET, D. & AUFFRET, J. P. 1979. *Carte des Sediments Superficiel de la Manche.* Bureau recherche Géologique et Minéralogique. 1:50 000 map.

LAUTRIDOU, J-P., MONNIER, J. L., MORZADEC, M. T., SOMMÉ, J. & TUFFREAU, A. 1986. The Pleistocene of northern France. *Quaternary Science Reviews*, **5**, 387–393.

LUNKKA, J. P. 1988. Sedimentation and deformation of the North Sea Drift Formation in the Happisburgh area, North Norfolk. *In:* CROOT, D. G. (ed.) *Glaciotectonics*, Balkema, Rotterdam, 109–122.

—— 1991. *Sedimentology of the Anglian Glacial Deposits in Northeast Norfolk, England.* PhD thesis, University of Cambridge.

—— 1994. Sedimentation and lithostratigraphy of the North Sea Drift and Lowestoft Till Formations in the coastal cliffs of NE Norfolk, England. *Journal of Quaternary Science*, **9**, 209–234.

MEIJER, T. & PREECE, R. C. 1995 Malacological evidence relating to the insularity of the British Isles during the Quaternary. *This volume*.

OELE, E. 1969. The Quaternary geology of the Dutch part of the North Sea, north of the Frisian Islands. *Geologie en Mijnbouw*, **48**, 467–480.

—— 1971. The Quaternary geology of the southern area of the Dutch part of the North Sea. *Geologie en Mijnbouw*, **50**, 461–474.

—— SCHÜTTENHELM, R. T. E. 1979. Development of the North Sea after the Saalian glaciation. *In:* OELE, E., SCHÜTTENHELM, R. T. E. & WIGGERS, A. J. (eds) *The Quaternary History of the North Sea. Acta Universitatis Upsaliensis Symposia Universitatis Upsaliensis Annum Quingentesimum Celebrantis*, **2**, 191–215.

PAEPE, R. & BAETEMAN, C. 1979. The Belgian coastal plain during the Quaternary. *In:* OELE, E., SCHÜTTENHELM, R. T. E. & WIGGERS, A. J. (eds) *The Quaternary History of the North Sea. Acta Universitatis Upsaliensis Symposia Universitatis Upsaliensis Annum Quingentesimum Celebrantis*, **2**, 143–146.

—— & SOMMÉ, J. 1975. Marine Pleistocene transgressions along the Flemish coast (Belgium and France). *In: Quaternary Glaciations in the Northern Hemisphere.* International Geological Correlation Project, **24**, no.2, 108–116.

POMEROL, C. 1978. Evolution paléogéographique et structurale du Bassin de Paris, du Précambrien à l'Actuel, en relation avec les régions avoisantes. *In:* VAN LOON, A. J. (ed.) *Keynotes of the MEGS-II (Amsterdam). Geologie en Mijnbouw*, **57**, 533–543.

PREECE, R. C., SCOURSE, J. D., HOUGHTON, S. D., KNUDSEN, K. L. & PENNEY, D. N. 1990. The Pleistocene sea-level and neotectonic history of the eastern Solent, southern England. *Philosophical Transactions of the Royal Society of London*, **B328**, 425–477.

RAPPOL, M., HALDORSEN, S., JØRGENSEN, P., VAN DER MEER, J. J. M. & STOLTENBERG, H. M. P. 1989. Composition and origin of petrographically-stratified thick till in the northern Netherlands and a Saalian glaciation model for the North Sea basin. *Mededelingen van de Werkgroep voor Tertiare en Kwartaire Geologie*, **26**, 31–64.

REID, C. 1913. *Submerged Forests.* Cambridge University Press.

ROEP, T. B., HOLST, H., VISSERS, R. L. M., PAGNIER, H. & POSTMA, D. 1975. Deposits of southwards flowing Pleistocene rivers in the Channel region, near Wissant, NW France. *Palaeogeography, Palaeoclimatology, Palaeoecology*, **17**, 289–308.

SMITH, A. J. 1985*a*. A catastrophic origin for the palaeovalley system of the eastern English Channel. *Marine Geology*, **64**, 65–75.

—— 1985*b*. The English Channel: a response to geological events after the Variscan orogeny. *Annales de la Société Géologique Polonaise*, **55**, 3–22.

—— 1989. The English Channel – by geological design or catastrophic accident? *Proceedings of the Geologists' Association*, **100**, 325–337.

—— CURRY, D. 1975. The structure and geological evolution of the English Channel. *Philosophical Transactions of the Royal Society of London*, **A279**, 3–20.

SOMMÉ, J. 1969. Le Calisis: pleine maritime et arrière-pays. Congrès INQUA 8e. Paris, 1969. *Livret-guide de l'excursion A10, Littoral Atlantique*, 4–8.

—— 1979. Quaternary coastlines in northern France. *In:* OELE, E., SCHÜTTENHELM, R. T. E. & WIGGERS, A. J. (eds) *The Quaternary History of the North Sea. Acta Universitatis Upsaliensis Symposia Universitatis Upsaliensis Annum Quingentesimum Celebrantis*, **2**, 147–158.

——, PAEPE, R., BAETEMAN, C. *ET AL.* 1978. La

Formation d'Herzeele: un nouveau stratotype du Pleistocène moyen marin de la Mer du Nord. *Bulletin de l'Association Française pour l'étude du Quaternaire*, **1,2,3**, 81–149.

SPAINK, G. 1958. *De Nederlandse Eemlagen deel I. Koninklijk Natuurhistorische Vereniging* **29**.

STAMP, L. D. 1927. The Thames drainage system and the age of the Strait of Dover. *Geographical Journal*, **70**, 386–390.

—— 1936. The geographical evolution of the North Sea basin. *Journal de la Conseil Internationale de l'Exploration de la Mer*, **11**, 135–163.

STUART, A. J. 1982. *Pleistocene Vertebrates in the British Isles*. Longman, London.

TER WEE, W. M. 1962. The Saalian glaciation in the Netherlands. *Mededelingen van de Geologische Stichting* (nieuwe serie), **15**, 57–77.

—— 1983a. The Elsterian glaciation in the Netherlands *In:* EHLERS, J. (ed.) *Glacial Deposits in North-west Europe*, Balkema, Rotterdam, 413–415.

—— 1983b. The Saalian glaciation in the northern Netherlands. *In:* EHLERS, J. (ed.) *Glacial Deposits in North-west Europe*, Balkema, Rotterdam, 405–412.

VANDENBERGHE, J. & DE SMEDT, P. 1979. Palaeomorphology of the eastern Scheldt Basin (central Belgium) – the Dijle-Demer-Grote Nete confluence area. *Catena*, **6**, 73–105.

——, KROOK, L. & VAN DER VALK, L. 1986. On the provenance of the Early Pleistocene fluvial system in the southern Netherlands. *Geologie en Mijnbouw*, **65**, 3–12.

VANHOORNE, R. 1962. Het interglaciale veen te Loo (België). *Natuurwetenschappelijk Tijdschrift*, **44**, 58–64.

VENTRIS, P. A. 1985. *Pleistocene Environmental History of the Nar Valley, Norfolk*. PhD thesis, University of Cambridge.

WALCOTT, R. I. 1972. Late Quaternary vertical movements in eastern North America: quantitative evidence of glacio-isostatic rebound. *Reviews of Geophysics and Space Physics*, **10**, 849–884.

WINGFIELD, R. 1995. A model of sea-levels in the Irish and Celtic seas during the end Pleistocene to Holocene transition. *This volume*.

ZAGWIJN, W. H. 1974. Palaeogeographic evolution of the Netherlands during the Quaternary. *Geologie en Mijnbouw*, **53**, 369–385.

—— 1979. Early and Middle Pleistocene coastlines in the southern North Sea Basin. *In:* OELE, E., SCHÜTTENHELM, R. T. E. & WIGGERS, A. J. (eds) *The Quaternary History of the North Sea. Acta Universitatis Upsaliensis Symposia Universitatis Upsaliensis Annum Quingentesimum Celebrantis*, **2**, 31–42.

—— 1985. An outline of the Quaternary stratigraphy of the Netherlands. *Geologie en Mijnbouw*, **64**, 17–24.

—— & DOPPERT, J. W. C. 1978. Upper Cenozoic of the southern North Sea Basin: palaeoclimatic and palaeogeographic evolution. *In:* VAN LOON, A. J. (ed.) *Keynotes of the MEGS-II (Amsterdam 1978). Geologie en Mijnbouw*, **57**, 577–588.

ZALASIEWICZ, J. A. & GIBBARD, P. L. 1988. The Pliocene to early Middle Pleistocene of East Anglia: an overview. *In:* GIBBARD, P. L. & ZALASIEWICZ, J. A. (eds) *Pliocene–Middle Pleistocene of East Anglia. Field Guide*. Quaternary Research Association, Cambridge, 1–31.

The Pleistocene evolution of the Thames and Rhine drainage systems in the southern North Sea Basin

D. R. BRIDGLAND[1] & B. D'OLIER[2]

[1] *Department of Geography, University of Durham, South Road, Durham DH1 3LE, UK*
[2] *D'Olier Associates, 64 Hill Rise, Rickmansworth, Hertfordshire WD3 2NX, UK*

Abstract: During most of the Middle and Late Pleistocene, global climate was colder than at present. At such times, sea-level was very much lower and the southern North Sea was land, drained by the same rivers that today issue into the sea from Britain and north-west Europe. Evidence from onshore in Britain suggests that the Thames did not drain into the southernmost part of the North Sea Basin until the early Middle Pleistocene; prior to that it flowed across East Anglia and northwards across the area of the present north Norfolk coast. It is thought that the lowland between Kent, East Anglia and the Continent was at this time drained by the 'Lobourg River', which arose on the northern flanks of the Wealden ridge and drained northwards into what is now the southern North Sea. The early Medway is thought to have joined this Lobourg River in the area that now lies offshore from south-east Suffolk. It seems that by the early Middle Pleistocene the Thames had adopted this course of the Medway and with it, therefore, the confluence with the Lobourg.

It is widely believed that Anglian/Elsterian ice-sheets caused proglacial ponding in the southern North Sea basin and that a large lake formed, fed by the Thames–Medway and Rhine river systems. It has also been suggested that this lake overflowed to the southwest into the English Channel Basin, breaching the Chalk ridge and initiating a through route via the Strait of Dover, a route that was used by the Rhine–Thames drainage system during all subsequent low-sea-level episodes. Middle and Upper Pleistocene deposits of both the Thames and Rhine systems are preserved in the areas immediately offshore from Britain, The Netherlands and Belgium and can be traced from both sides towards the axial channel of the North Sea Basin, the formation of which was perhaps initiated by the Lobourg River. Although the Rhine's offshore deposits, in particular, appear to follow a route that heads southwards, towards the Strait of Dover, no sedimentary evidence has yet been found that demonstrates a late Pleistocene fluvial course either through this strait or, conversely, into the North Sea.

This paper, in discussing the Pleistocene evolution of the most important rivers that drain into the southern North Sea Basin, also explores the role of the drowned valley through the Strait of Dover as a fluvial route (see also Gibbard 1995). This provides a linkage with the theme of this volume, that of the insularity of Britain. It must be emphasized at the outset that the separation of Britain from the Continent is far from the norm, even during the latter part of the Quaternary. For most of this time cold or glacial conditions have prevailed, with sea-levels at least 100 m lower than at present. During these lengthy phases of low sea-level the rivers flowing into the North Sea Basin had extended lower valleys, since coastlines were many kilometres downstream from their positions at the present day.

It is widely believed that falls in sea-level during cold episodes caused rivers to cut down, thus giving rise to the systems of terraces that exist in the valleys of rivers such as the Thames. This may not be the case, however, as has been discussed elsewhere (Bridgland 1994; Bridgland & Allen 1995). It can be demonstrated that the major rivers of North-west Europe are flowing, except in their lowest reaches, on floodplains that were established during the last glacial (Devensian/Weichselian) as a result of deposition, primarily of gravel, in a braided-river environment. During the Holocene, variable thicknesses of fine-grained fluviatile alluvium have been laid down above the last-glacial gravels, so that the modern rivers flow in channels cut into the upper surface of the alluvium. In the vicinity of the modern coasts, the gravel bodies are overlain by estuarine and/or marine alluvium. Onshore, these latter deposits have a near-horizontal upper surface formed in relation to modern sea-level. As the underlying gravels have continuous downstream gradients, the estuarine deposits form wedge-shaped bodies of sediment that thicken downstream. A return to a low sea-level would clearly result in the dissection of these estuarine sediments, but there is no reason why it should cause rivers to cut below their previous cold-climate floodplain levels. Their valleys would

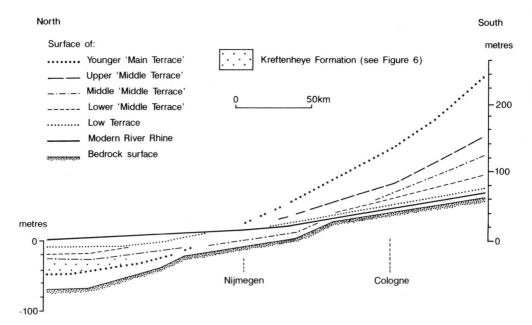

Fig. 1. Long profiles of selected Pleistocene deposits of the River Rhine, showing preservation of terraces upstream and superimposed sediments in the Lower Rhine. Modified from Brunnacker et al. (1982).

merely be extended beyond the present coastline as the land area expanded and the sea retreated.

A mechanism that can explain the formation of terrace systems has recently been proposed (Bridgland 1994; Bridgland & Allen 1995) in which the downcutting that leads to terrace formation is attributed to a combination of two factors. The first is a climatic trigger, by which climatic fluctuation and its direct and indirect influences on fluvial activity translate into the various phases of aggradation and incision that lead to terrace formation. Secondly, this climatically driven cycle of changing river activity is set against a background of isostatic adjustment to the effects of sediment transfer from areas of net erosion to areas of net accumulation. This adjustment, effecting a progressive uplift of source areas and a corresponding subsidence of sedimentary basins (most of which lie offshore at the present day), provides the progressive lowering of base-level that is needed to form flights of terraces. The existence of well preserved terrace systems in the lower reaches of the Thames system, including the area immediately offshore (Bridgland et al. 1993), indicates that this southeastern corner of Britain has been an area of net erosion, and therefore uplift, during the Pleistocene, although it is thought that the lower reaches of the Thames valley are subsiding at the present time (e.g. Devoy 1979; Shennan 1989). This contrasts with the situation in the Lower Rhine valley, on the opposite side of the North Sea, where terraces have not been formed downstream from the Dutch–German border. Instead, the Rhine has deposited a series of deposits one on top of the other, thus giving rise to a considerable thickness of superimposed fluvial sediments beneath The Netherlands and revealing this to be an area of net subsidence (Brunnacker et al. 1982; Ruegg 1994; Fig. 1). The climatically driven alternation between erosion and sedimentation may well have occurred in the Lower Rhine in a similar way to that in the Thames, but each successive aggradation has been superimposed upon the previous set of deposits. The axis of the sedimentary basin is thus not coincident with the centre of the modern southern North Sea, but instead extends beneath The Netherlands (Caston 1977; Balson et al. 1992, section 1).

The history of the Thames and Rhine drainage systems

The Thames and Rhine are both rivers of considerable antiquity (Gibbard 1988). The Thames, however, did not always flow into the extreme southern end of the North Sea. Hey

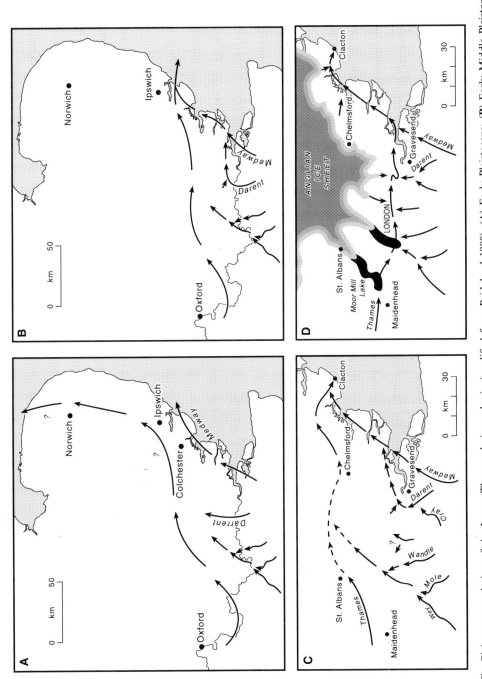

Fig. 2. The Pleistocene evolution of the Lower Thames drainage basin (modified from Bridgland 1988). (**A**) Early Pleistocene; (**B**) Early Middle Pleistocene (early Cromerian Complex); (**C**) Early Anglian (immediately prior to advance of ice sheets); (**D**) Anglian glacial maximum, showing initiation of post-Thames diversion drainage pattern.

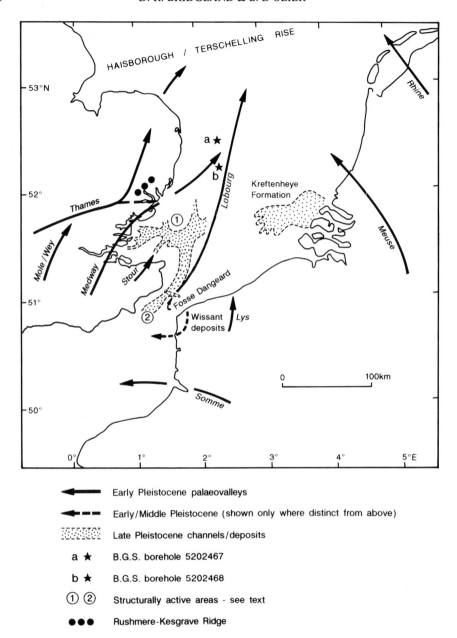

Fig. 3. Map showing the various early courses of the rivers draining into the southern North Sea. Other geological and geomorphological features discussed in the text are also shown.

(1980) showed that Thames deposits, now believed to be of early Pleistocene age, can be traced across East Anglia, following a course that would have taken the river into the central North Sea Basin by crossing the present coastline of Norfolk in the region west of Cromer (Fig. 2A). A substantial valley feature was probably in existence in the southern North Sea Basin by Pliocene times, however. Pliocene (Coralline Crag and Red Crag) sediments have been recovered by the British Geological Survey (BGS) in boreholes within over-deepened hollows beneath the sea-floor in this area (Balson 1989; Fig. 3). This valley is believed to have been occupied in times of low sea-level by a river that arose on the northern flanks of the Wealden anticlinorium and ran northwards to the east of the Kent coast, a neighbour of the Kentish Stour and the French/Belgian Lys (Fig. 3). This was named the Lobourg River by Stamp (1927, after Briquet 1921). Its upper catchment has been obliterated by the breaching of the Wealden ridge and the formation of the Strait of Dover. Its lower reaches are represented amongst the network of offshore drainage channels that has been recognized beneath the southern North Sea (D'Olier 1975), where it presumably coincides with the 'Axial Channel' (Mostaert et al. 1989), into which the Thames and Rhine have both flowed during phases of low sea-level and which has been much modified by marine erosion during the high-sea-level phases of the late Pleistocene and the Holocene.

During the Early Pleistocene, the Medway is thought to have maintained a course entirely separate from that of the Thames, flowing from Kent across eastern Essex and joining with the Lobourg in the area now offshore from southeast Suffolk (Bridgland 1988; Fig. 2A). A buried valley, cut into London Clay bedrock, trends WSW–ENE from a point some 6 km northeast of Orford Ness. This joins the Axial Channel of the southern North Sea close to the position of BGS borehole No. 5202468 (Fig. 3) with its Pliocene sediments, suggesting that the tributary valley could also be of this antiquity. The Pliocene/Early Pleistocene Medway is estimated, from evidence provided by the gravels of eastern Essex (Bridgland 1988), to have flowed in a course across the northeast of the county that would have taken it to the approximate position of this offshore valley (Fig. 2A).

The Rhine, at this time, flowed northwards into the southern North Sea (Fig. 3), carrying abundant clastic material from the European Continent. Evidence of this river's huge influence as a redistributor of sediment is apparent in the considerable accumulations of fluvial, deltaic and prodeltaic sediments of the IJmuiden, Winterton Shoal and Yarmouth Roads Formations, which now fill the southern North Sea Basin, with the greatest thicknesses occurring on the continental side (Cameron et al. 1984). This latter fact may help to explain why the lower reaches of the Rhine valley have been subsiding throughout the Pleistocene, unlike the valleys on the British side of the North Sea. Other continental rivers, particularly the Weser and the Meuse, were important contributors to the huge accumulation of sediments on the eastern side of the southern North Sea that took place during the Early and early Middle Pleistocene (Gibbard 1988).

By the early Middle Pleistocene, the Thames had abandoned its course across Norfolk and had adopted the route into the southern North Sea formerly occupied by the Medway (Bridgland 1988; Fig. 2B). This may have arisen through the capture of the Thames by a tributary of the Medway the result being that the Thames was confluent with the Lobourg for the first time. Allen (1983, 1984) demonstrated that an interfluve, his Rushmere–Kesgrave Ridge, existed between the Early Pleistocene and early Middle Pleistocene courses of the Thames (Fig. 3), confirming that separate valleys were occupied at these times, rather than different terrace levels within a single broad valley system. Interglacial deposits interbedded with the Thames and Thames–Medway gravels laid down in the area of northeast Essex indicate that the later route was in use during the 'Cromerian Complex' (Bridgland et al. 1988, 1990; Bridgland 1994). This situation, in which the Thames and Medway joined in the area east of Colchester and flowed northeastwards towards the southern North Sea Basin, persisted until the Anglian Stage (Bridgland 1988, 1994).

Glaciation during the Anglian Stage had a considerable effect on the Thames. Its upper reaches may have been drastically influenced by earlier glaciation (Whiteman & Rose 1992), but it was during the Anglian that the river was diverted by ice into its modern valley through London (Fig. 2C & D). The eastern part of this valley was, by the Anglian Stage, probably already occupied by a tributary of the Medway, so that the diversion had the immediate effect of transferring Thames drainage into the old Medway route across eastern Essex to the Colchester–Clacton area, probably the second time that the Thames had adopted a route first established by the Medway. Thus, in terms of the catchment as a whole, this diversion was of fairly minor significance; the river was merely re-routed between the Maidenhead and Clacton areas. The western and eastern ends of its old route were subsequently occupied by the Middlesex Colne and the Essex Colne, respectively, something of a coincidence in the naming of rivers.

The Medway valley was entirely untouched by the glaciation in the present onshore area, as is demonstrated by the existence in the Clacton area of a deposit that comprises almost pure Medway gravel, with the addition of a small component of glacial outwash. This deposit, the Upper Holland Gravel (Bridgland 1988, 1994; Bridgland et al. 1988), is thought to represent a period during the course of the Anglian glacia-

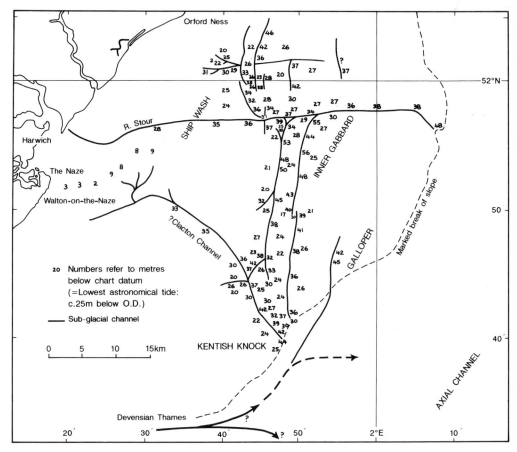

Fig. 4. Map of (submerged) channels offshore from the Essex and Suffolk coasts, with spot depths in metres below chart datum (in this area chart datum is c. 2.5 m below OD).

tion when the Thames was ponded to the north of London and was replaced in the unglaciated lower reaches of its valley by an outwash stream (as illustrated in Fig. 2D). This situation can only have persisted for a brief interval before the overflow of proglacial lakes in the London area brought about the diversion of the Thames (Fig. 2D). It is unlikely that the offshore continuation of the pre-glaciation Thames–Medway valley towards the central North Sea was also unglaciated, however. On the contrary, it is thought that the drainage of the southern North Sea lowland was ponded up by Elsterian ice, which presumably formed an unbroken barrier between Britain and Scandinavia (Charlesworth 1957; Gibbard 1988). There is supporting evidence for this theory from glacio-lacustrine sediments within the Peelo Formation beneath the northern Netherlands and the area immediately offshore. The evidence for the existence of a huge lake in the southern North Sea at the time of the Elsterian glacial maximum has been discussed at length by Gibbard (1988, 1995). If this North Sea lake was contemporaneous with the smaller-scale ponding of the Thames north of London, it would seem that the main source of water supplying the former from Britain was the Medway catchment.

In his reconstruction of an Anglian/Elsterian ice-dammed lake in the southern North Sea, Gibbard (1988) showed a maximum ice-front trending SW–NE, crossing the English coast north of Orford Ness and passing to the north of The Netherlands, before sweeping southwards over northern Germany. In East Anglia, this ice-margin was based on the southeastward limit of preserved till. Channels of apparent subglacial origin, partly filled with Hoxnian sediments, occur to the south of this limit, however, at Shottisham, Hollesley, Aldeburgh and Snape (Mathers & Zalasiewicz 1986; Mathers et al. 1991, 1993), indicating that Anglian ice did in

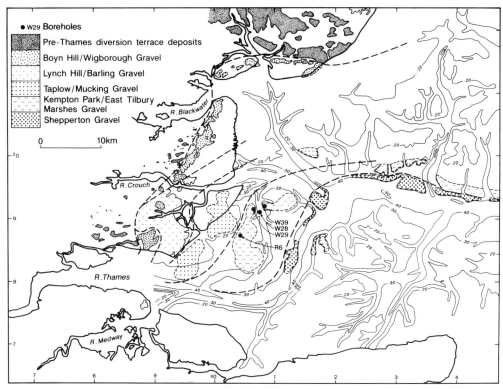

Fig. 5. Early courses and sediments of the Thames–Medway river system in eastern Essex and beneath the southern North Sea. The locations of borehole samples from Foulness Sands are shown. Contours are in metres below O.D.

fact extend slightly further south than is indicated by the till limit. A possible indication that this was also the case offshore is provided by a complex system of channels in the area southeast from Orford Ness (D'Olier & Bridgland 1994; Fig. 4).

Overflow of the Elsterian proglacial lake has been suggested as a mechanism for the formation of the Strait of Dover (Gibbard 1988, 1995). Smith (1985) thought that the deeply excavated Fosse Dangeard feature, which occurs beneath the strait (Fig. 3), might have been formed catastrophically by water exiting from a lake in the southern North Sea, this water also cutting a system of anastomosing channels beneath the eastern English Channel (see Gibbard 1995). Evidence for correlating the opening of the Strait of Dover with the Elsterian glaciation is largely circumstantial. It hinges on the acceptance of the lake overflow theory. Contradictory evidence from palaeontology suggests that Britain remained a peninsula of the northwest European land mass during most, if not all, of the first interglacial to follow the Elsterian. This apparently contrasting interpretation is based on assemblages of marine Mollusca from Holsteinian sediments in The Netherlands and elsewhere (Meijer & Preece 1995) and from similarities between British Hoxnian mammalian assemblages and those from the continental Holsteinian (Stuart 1995; Sutcliffe 1995). It should be noted, however, that the cutting of a channel or valley through the former Chalk ridge between Dover and Calais does not mean that Britain was an island during all subsequent high-sea-level episodes, nor even that Thames–Rhine drainage was routed southwards during all subsequent low-sea-level episodes, so the palaeontological evidence described above is not necessarily incompatible with an Elsterian age for the formation of the new channel. It is possible, for instance, that differential isostatic response to Elsterian and post-Elsterian glaciations and deglaciations may have repeatedly altered drainage directions in this area between the Elsterian and the last glacial.

The onshore terrace sequence in eastern Essex shows that the Thames, in the late Middle Pleistocene, migrated progressively southwards and eastwards from its immediately post-An-

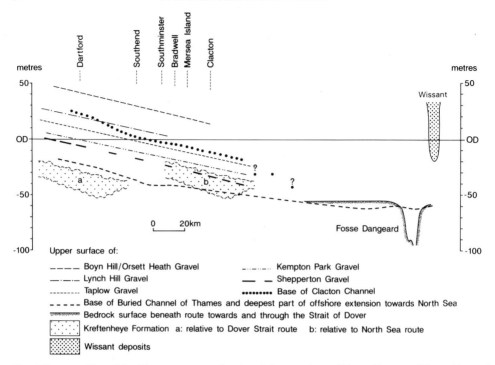

Fig. 6. Long profiles of the Thames terraces, showing their extensions offshore. Two possible positions of the Kreftenheye Formation of the Rhine are plotted, one for a North Sea and the other for an English Channel course. The Wissant deposits are also plotted.

glian route (Fig. 5). It is uncertain whether this migration was accelerated by minor diversions, the result of river capture (Bridgland 1988), or was merely part of the progressive shift in this direction that has continued throughout the Pleistocene (Bridgland 1985). What can be said is that the onshore gravel terraces between Southend and Clacton are part of the submerged valley system originally recognized by D'Olier (1975; Fig. 5). The Southchurch/Asheldham/Mersea Island/Wigborough Gravel (Fig. 5), which traces the immediately post-Anglian course, remains at least partly above sea-level all the way to the northeast Essex coast at Clacton. This is the downstream continuation of both the late Anglian Black Park Gravel and the early Saalian Boyn Hill Gravel of the Middle Thames. It incorporates interglacial deposits attributed to the Hoxnian (*sensu* Swanscombe – believed to be equivalent to Oxygen Isotope Stage 11) at various localities from Swanscombe, in north Kent, to Clacton in Essex, beyond which its continuation immediately offshore has been removed by marine erosion. These interglacial sediments occupy a substantial channel feature excavated to well below the general base-level of the overlying terrace gravels, falling below OD in northeast Essex (Fig. 6) where it is represented by the well known Clacton Channel. An attempt to trace an offshore extension of this channel, based on seismic profiling, was made by Bridgland & D'Olier (1989), who suggested a continuation northeastwards to the area of the Shipwash Bank, and then an eastward extension by way of the southernmost west–east trending (supposed) subglacial channel described above. Further work has led to a revised interpretation, however, allowing a more likely extension of the Clacton Channel to be proposed. This follows the route suggested in 1989 until turning southeastwards off Walton-on-the-Naze towards the Kentish Knock sandbank, following a largely sediment-filled channel (Fig. 4).

Later Thames–Medway gravels onshore are confined to the area south of the Blackwater. A recent re-evaluation of the stratigraphy between the Thames and Blackwater estuaries has led to the conclusion that only one further terrace aggradation can be recognized there, the Barling/Dammer Wick Gravel (Bridgland *et al.* 1993; Fig. 5). Deposits formerly attributed to a terrace formation intermediate between the Southchurch/Asheldham and Barling/Dammer Wick Gravels (see Bridgland 1988) are now thought to be either erosional remnants of the former or products of reworking by tributary streams. The Barling Gravel of the Southend area is now thought to represent the downstream continuation of the Lynch Hill Gravel of the Middle Thames (Bridgland *et al.* 1993). All subsequent terrace deposits fall below sea-level before reaching the coast of southeast Essex. They are represented within the submerged terrace system that lies on the northwestern side of the drowned Thames–Medway valley, which continues northeastwards beyond the modern estuary (Bridgland *et al.* 1993; Fig. 5). This valley system curves eastwards at a point about 10 km east of the modern Crouch estuary, mirroring the earliest post-diversion course that was partly inherited from the Medway (Fig. 5). At 51° 35' N 01° 45' E, the Thames–Medway valley joins the Axial Channel of the southern North Sea, the channel that may mark the route of the erstwhile Lobourg River. Unfortunately this channel is entirely devoid of sediments, either on its floor or as terraces, possibly as a result of marine scouring during the Holocene and/or earlier high-sea-level episodes. It is therefore impossible to determine whether the Middle and Late Pleistocene Thames turned to the north or to the south on reaching this channel. The floor of this empty valley feature has, at the present day, very little slope (Fig. 6). What slope there is, is to the north, but this could result from modification by marine scouring or differential subsidence/uplift.

On the European side of the southern North Sea, Upper Pleistocene cold-climate fluviatile sediments of the Rhine and Meuse systems can be traced to *c.* 50 km offshore from the Netherlands (Balson *et al.* 1992). These form part of the Kreftenheye Formation, which also extends inland beneath the Rhine–Meuse and Rhine–IJssel courses, beneath the northern and central Netherlands (Jelgersma *et al.* 1979). According to Jelgersma *et al.*, the top of the formation lies about 20 m below sea-level near to the coast, but falls to 40 m below sea-level 50 km offshore.

Recognizing gravels of the Thames system offshore

The sediments infilling or on the flanks of the offshore channels of the Thames system are relatively easily distinguished from bedrock by seismic reflection profiling. The bedrock largely comprises over-consolidated London Clay and has very different acoustic properties from the overlying unconsolidated sands and gravels.

Thus channel-fills and terrace remnants can be mapped and former river courses plotted. There have been few opportunities to examine samples of Pleistocene deposits from beneath the southern North Sea, but material from offshore boreholes has occasionally been available for analysis. An important source of information has been an archive of borehole samples from the area of Foulness Sands (Fig. 5), put down in the 1970s as part of the site exploration for the ill-fated Maplin Airport project. These samples included examples of gravels from both channel-fill and terrace situations. They have been subjected to clast-lithological analysis, by which the proportions of different rocks making up the gravel fraction of the deposits have been ascertained, providing information about the provenance of the sediments and, by comparison with other gravels of known origin, an indication of the likely agent of their deposition (Table 1). In addition, the degree of angularity of the flint component of these gravels has also been determined, using an analytical method in which each clast is assigned to one of six angularity/roundness classes (Table 2). This method is based on that (for sand grains) proposed by Powers (1953) and modified by Schneiderhöhn (1954; in Pryor 1971), but adapted for flint clasts in the 16–32 mm size range. The six categories recognized by Schneiderhöhn were redefined for the application to flint clasts (Fisher & Bridgland 1986) and a type collection was assembled to provide a standard for comparison.

A database of angularity/roundness characteristics of flint-dominated gravels of various types has been assembled over the past 15 years (Bridgland 1995). It has been found that for comparisons between samples, the frequencies of certain categories are more meaningful than others. The very angular category, comprising freshly fractured flint, is of extremely variable frequency. Post-depositional frost-shattering has sometimes produced considerable amounts of very angular flint, particularly in the upper levels of deposits, within what would have been the permafrost active layer during periods of periglaciation. Breakage of clasts can also occur during sampling, particularly if from percussion boreholes. For these reasons it is often advantageous to disregard the very angular class and recalculate the remainder of the data with this excluded (Table 3). In some analyses two populations of flint clasts can be recognized, indicated by a bimodal frequency distribution. This is typically the case in fluvial gravels in southeast England, since these contain, in addition to fresh flint that has been shaped by river transport, abundant reworked rounded

Table 1. Clast lithological data from offshore gravel samples.

Locality and sample no.	Flint				Chert			Quartz/ qtzite	Igneous	Other	Non-durables‡	Total
	Nodular	Beach†	Broken	Total	Gsd	Carb	Rhaxella					
Offshore samples from Maplin Sands												
Maplin R6	12.9	35.3	12.9	94.0	2.6			3.5				116
Maplin W28	*	40.6	*	94.3	5.1			0.6			7.4	175
Maplin W29	14.8	49.0	4.5	91.0	5.8			1.9			3.9	155
Maplin W39	17.7	35.4	6.3	92.4	3.8			1.3		0.74		79
Maplin W41												
Post-diversion Thames–Medway gravels from sites near the Essex coast												
Barling 1	*	33.7	*	80.4	18.6			1.0				306
Barling 2A	*	53.9	*	89.0	8.4		0.3	1.0		1.3	0.7	308
Barling 2B	12.9	43.0	*	89.2	9.1		0.1	0.9		0.4		751
Barling 3	*	50.4	*	91.8	6.6			1.2		0.4		512
Dammer Wick 1	14.5	52.0	*	88.3	10.6	0.3		0.8		0.3	1.2	256
Pre-diversion Thames–Medway gravels from sites near the Essex coast												
Holland 1A	*	24.9	*	84.0	2.4	1.6	0.3	11.0	0.3	0.5	1.8	382
Haven 1B	14.6	34.6	*	83.1	2.3	0.3		12.7		1.5		260
2	*	25.3	*	82.0	3.0	1.7	0.2	12.4	0.2	0.6		534
Holland-on-Sea 2D	13.9	26.7	6.7	81.5	2.0	1.1		14.8	0.3	0.3	0.2	655
Great Holland 1	19.1	25.5	6.4	84.0	1.7	0.7		14.3		1.0		419
Little Oakley KA	15.5	30.3	5.2	80.2	2.0	0.6	0.2	15.8	0.3	0.9		653
Dovercourt DA	12.1	30.3	7.4	79.2	2.1	1.6		14.5	1.1	1.6		379
Tributary gravels (Essex rivers - **B** = Blackwater;§ **Co** = Colne; **Cr** = Crouch; **H** = Holland Brook; **S** = Stour)												
Little Hayes 1 **Cr**	*	57.3	*	91.8	1.8	0.3	0.3	5.7		0.1	4.2	707
Little Hayes 2 **Cr**	3.4	61.2	*	93.2	0.8	0.3		5.1	0.2	0.5		613
G.Totham 1 **B**	10.0	31.7	5.9	80.5	0.3	1.8	0.7	15.9	0.2	0.7		609
E.Mersea 1 **B**	12.7	35.4	3.1	85.2	0.3	0.3	0.3	6.6			0.2	393
Restaurant 2 **B**	12.9	42.2	3.3	83.7	7.6	1.1	0.2	8.0		0.8	0.5	630
Tollesbury 1A **B**	12.9	37.8	*	83.6	6.2	1.6	0.1	12.9	0.2	1.1	0.6	805
Brightlingsea 1 **Co**	12.9	26.4	4.7	80.5	0.3	1.9		19.2	0.6	0.3		364
U. Dovercourt 1 **S**	14.5	29.5	4.6	80.2	0.7	1.5	0.2	16.9	0.5	0.2		414
Dakins Pit 1 **H**‖	17.1	25.8	4.0	82.5	1.1	1.8		13.1	0.7	0.7		275
Dakins Pit 2 **H**	16.3	21.1	7.3	81.9	0.9	2.1	0.6	12.1	0.6	0.6		331
Anglian glacial gravels												
Denham Castle 1	34.3	0.4	26.7	93.2		0.9	0.4	3.8	0.9	0.9	103.0	236
Ingham 1	37.5	2.3	18.9	90.9		0.8	0.4	5.7	0.4	1.9	33.7	264
Fingringhoe 1A¶	13.0	15.4	8.4	80.8	2.4	4.1	1.4	8.4	0.8	1.6		369
Fingringhoe 1B	13.7	15.9	6.6	81.7	0.7	1.3	1.1	12.6	0.9	1.8		453

Sources: (Data from onshore fluvial gravels are included, for comparison). Bridgland (1983, 1988, 1994), Bridgland *et al.* (1990), Bridgland & Lewis (1991)
Abbreviations: Gsd = Greensand (chert + other lithologies) Carb. = Carboniferous chert; *Rhaxella* = *Rhaxella* chert (Portlandian/Oxfordian); qtzite = quartzites (sedimentary and metamorphic)
*, Not separately recorded
†, In most samples beach flint is derived entirely from the Palaeogene of the London Basin
‡, Non-durables are excluded from totals and expressed as % total durable material
§, Blackwater, East Mersea Restaurant Gravel samples: No.1 is from the Restaurant Site, No.2 is from the Hippopotamus Site (see Bridgland 1994)
‖, For location of site, see Wymer (1985)
¶, The gravel at Fingringhoe is interpreted as distal outwash, with material introduced from non-glacial sources by tributary rivers (see Bridgland 1994)

pebbles from the marine Palaeogene of the London Basin. The presence of such reworked material very considerably distorts the angularity/roundness data. To minimize this effect, clasts are always categorized according to the most recent damage they have sustained; thus a small angular frost-pit in a well-rounded reworked pebble results in its classification as angular. This means that the results of the analysis do not always characterize the general shapes of the gravel clasts. They are not designed to do this, but rather to give information about the environment of deposition of the gravel deposit. Reworked rounded pebbles that are unbroken and have therefore not been modified during Pleistocene reworking can be excluded from the analysis; so too can whole, unbroken nodules of fresh flint. The shape of these clasts owes nothing to the Pleistocene depositional environment and would thus distort the analysis if included (this point can be illustrated by a comparison of the data in Tables 2 and 3).

The results of clast-lithological analyses of samples from Foulness Sands have allowed comparison of these offshore gravels with the deposits of the Thames–Medway river system, including tributary deposits, previously examined in samples from onshore (Table 1). The samples from beneath Foulness Sands were found to bear a close resemblance to those from the lowest (post-diversion) terraces of the Southend–Burnham-on-Crouch area (Table 1). Angularity/roundness analyses revealed a significant difference, however, between gravels forming terraces on the northeastern side of the submerged Thames–Medway valley (samples R6, W29 and W39) and those filling channels within the late Devensian/Holocene submerged valley system (sample W28). The former show the overwhelming domination of angular and sub-angular flint that characterizes all fluviatile gravels (Table 2), whereas the latter contain significantly higher proportions of rounded and sub-rounded flint. Analyses of both pre- and post-diversion Thames and Thames–Medway gravels from the onshore area have indicated that fresh flint has been abraded to sub-rounded or rounded forms only very rarely, even by what must have been the largest river in the area of southern Britain. The total of these classes is less than 5% in all analyses of fluviatile deposits in which reworked beach pebbles are excluded (Table 3). The gravels of smaller tributary rivers, such as the Medway and the Crouch, are generally even more angular in nature than those of the Thames; in these the angular category is often the modal class, whereas the modal class in mainstream Thames gravels is usually the sub-angular category. In known beach gravels, on the other hand, the proportion of clasts falling in the sub-rounded and rounded categories is invariably larger than 5%, even in relatively angular beach gravels such as those forming the Aldingbourne Raised Beach (samples from Brooks Farm and Aldingbourne Park, Tables 2 and 3) and the Goodwood (Boxgrove) Raised Beach of West Sussex, and the gravel at Priory Bay on the Isle of Wight, the beach origin of which is less certain (Bridgland 1995). In more rounded Pleistocene beach gravels, such as the Bembridge Raised Beach (Bridgland, in Preece et al. 1990), significant amounts of well-rounded flint are also present and the total from the three rounded categories may exceed that from the angular categories (Tables 2 and 3).

These data strongly suggest that the channel-fill gravel from Foulness Sands, represented by borehole sample W29, is of marine origin. Some 8% of its flints are in the rounded and sub-rounded classes, even when likely reworked Palaeogene pebbles are excluded (Table 3). This interpretation is borne out by the occurrence in the sample of marine shells and calcareous worm-encrustations on clasts. The clast-lithological make-up of this deposit indicates that it was probably formed by the reworking of Thames–Medway terrace gravel by marine currents, during or subsequent to the Holocene transgression. The submerged terrace gravels from beneath Foulness Sands, on the other hand, are confirmed as *in situ* fluviatile deposits, laid down by the Thames–Medway river. They belong to the downstream continuation of the Kempton Park/East Tilbury Marshes Formation (Fig. 5).

Discussion

Much has been learned in recent decades about the geology of the North Sea floor, particularly useful in this respect being the publication of offshore solid and drift maps by the combined British, Dutch and Belgian geological surveys. These maps have been compiled largely from seismic reflection evidence, supported by vibrocore and grab sampling. Only a few boreholes have been drilled because of the very high costs involved. It must be stressed, therefore, that in the interpretation of the offshore sequence, much has to be surmised from shapes of sediment bodies and open erosional features, backed up by relatively few samples.

In the area immediately offshore from the Essex coast, seismic reflection profiling has revealed the submerged valley of the Thames–

Medway, complete with a well developed terrace system, preserved beneath Holocene sediments. No comparable submerged valley system has been recognized on the continental side of the North Sea, where the upper Middle and Upper Pleistocene sediments of the Rhine and Meuse rivers, forming the Kreftenheye Formation, are disposed in a delta-like spread extending westwards from the Dutch coast (Balson et al. 1992). The offshore edges of this sediment body appear to be largely erosional, although the shape may in part reflect channelling into the subjacent and surrounding Eem Formation (Balson et al. 1992). Part of the Kreftenheye Formation is older than the Eem Formation, however, and it also overlies Lower Pleistocene deposits. The shape of the Kreftenheye Formation outcrop, and in particular its apparent extension towards the WSW (Fig. 3), has been an important factor in the development of the view that the late Pleistocene Rhine flowed through the English Channel.

Although there is no direct connection between Thames and Rhine sediments beneath the southern North Sea, Balson et al. (1992) have shown (their inset map showing the distribution of Quaternary channel deposits) that parallel sediment-filled channels extend northeastwards from the Thames–Lobourg confluence area, within the Axial Channel of the southern North Sea, broadly towards the Kreftenheye outcrop (Fig. 3). The critical part of the Axial Channel is also documented by Liu et al. (1993). The report on a Dutch Geological Survey borehole through these channel-fill deposits, cited by Liu et al., shows them to comprise Pleistocene beneath Holocene shelly sediments, although the shells in the former are mostly, if not entirely, reworked; the mineral constituents have yet to be examined (T. Meijer, pers. comm., March 1995). In contrast to these sediment-filled channels, the southward channel by which the Rhine–Thames River is thought to have flowed towards the Dover strait is empty of sediments (excepting a veneer of recent sand waves). Clearly, if Late Pleistocene Rhine–Thames drainage was routed through the English Channel, any fluvial sediments of appropriate age in these channels should be of Rhine origin. Until this has been demonstrated, it remains possible to reconstruct a Late Pleistocene drainage pattern in which the Rhine and Thames joined approximately halfway between northern Essex and northern Holland and turned northwards towards the Viking Graben.

A factor that has lent support to the view that Late Pleistocene Rhine–Thames drainage was routed through the English Channel is the apparent absence of Upper Pleistocene fluvial sediments extending northwards into the North Sea Basin. Such sediments have, however, yet to be identified from the southern route and are certainly absent from the submerged valley through the Strait of Dover, although this may be more easily explained than their absence from the subsiding sedimentary basin to the north.

Rhine- or Thames-derived material has yet to be recognized in sediments beneath the English Channel, although there have been few opportunities for sampling such sediments. However, seismic reflection profiling of the eastern part of the English Channel has revealed a deeply incised channel (Hamblin 1989). The floor of this channel generally declines westwards and passes below -70 m OD some 30 km before crossing the Greenwich Meridian, with many deeper hollows between this point and the Dover strait. The fluvial origin of this substantial channel has yet to be demonstrated unequivocally (see, however, Gibbard 1995). Its longitudinal profile is highly suggestive of a fluviatile origin, nevertheless, and it is joined by the submerged and buried extensions of rivers from northern France and southern England. None of these rivers, nor all of them in combination, seem sufficiently large to have cut this channel, which is already 10 km wide to the west of Cap Gris Nez, upstream from the confluence with its most significant 'tributary', the Seine. The existence of this channel beneath the English Channel is perhaps the most compelling evidence for a Rhine–Thames river routed through the Strait of Dover during the last glacial.

As has already been stated, there are no surviving sediments to provide evidence linking the Thames and Rhine sediments offshore from the mouths of the two river systems with low-sea-level fluvial drainage through the Dover strait or northwards towards the Viking Graben. One method of examining the relative merits of the two potential routes is to project the downstream profiles of the two rivers along both possible courses. An important constraint on possible Weichselian Rhine–Thames courses is the shape of the sub-Pleistocene surface offshore, which is coincident over a considerable part of the area with modern-day bathymetry. Figure 6 shows projected downstream extensions of Thames and Rhine Devensian/Weichselian sediments, constructed assuming no differential subsidence since deposition. These can be compared with the bedrock surfaces beneath the 'Axial Channel' and the Strait of Dover, which are also shown. This diagram indicates that, given no post-Weichselian tectonic movements, courses during the last glacial northwards towards the Viking Graben and southwards through the Strait of Dover are both possible. However, for Devensian Thames deposits (the offshore continuation of the Shepperton Gravel) to have formerly extended through the Strait of Dover would require there to have been a marked decrease in the downstream gradient of this sediment body or substantial differential subsidence/uplift between the Strait of Dover and the southern North Sea during the Holocene. A decrease in downstream gradient here might be expected, as a result of joining the larger Rhine system, which would be likely to have a lower gradient.

The surviving Kreftenheye Formation deposits beneath the Dutch sector of the southern North Sea are plotted in two different positions on Fig. 6. One of these, the furthest left, is plotted relative to a course that continues towards the Strait of Dover, in which the Kreftenheye Formation lies about 200 km upstream from the Fosse Dangeard (Fig. 6). The Rhine sediments in this version are seen to be considerably lower (over 30 m) than equivalent Thames deposits at the same upstream distance from the Dover strait. The alternative position of the Kreftenheye Formation, plotted relative to a course turning northwards after meeting the Thames in the middle of the southern North Sea, shows it at a depth below modern sea-level that is closely comparable with that of equivalent Thames sediments.

Taken in isolation, Fig. 6 would seem to be pointing clearly to a northward route for the Rhine during the Weichselian. At its deepest point, the base of the Kreftenheye Formation is already at the level of the bedrock surface beneath the Dover strait, nearly 200 km downstream by the southerly route. At its present gradient in The Netherlands, the Rhine would be expected to have lost a further 30 m in height over the distance from the Dutch coast to the Strait of Dover. If the Weichselian drainage was indeed to the south, this is perhaps a measure of the amount of differential subsidence between The Netherlands and south-east Kent. This is approximately the height difference between the 'a' plot of the Kreftenheye Formation on Fig. 6 and its expected position, relative to the southerly route, without the effects of differential subsidence/uplift. The rate of subsidence of The Netherlands during the late Holocene has been estimated at 1.5 cm/century (Jelgersma 1966, 1992). At that rate, the lowering of the area of Kreftenheye outcrop by 30 m would have taken 20 000 years. Thus subsidence of the right order of magnitude could indeed have occurred since the last glacial, although differential movements of other relevant areas, such as that offshore from Essex, as well as the Strait of Dover, must also be taken into account. It is also likely that subsidence of The Netherlands has accelerated in response to the rise in sea-level during the late Holocene, as a result of hydro-isostasy.

Gibbard (1988, 1995) has indicated that the overflow from the large Elsterian lake in the southern North Sea established the Strait of Dover as a drainage route to the southwest and that all subsequent Rhine–Thames drainage was by this route. This conflicts with evidence from molluscan and mammalian faunas, already mentioned (Meijer & Preece 1995; Stuart 1995; Sutcliffe 1995), which indicates that Britain remained joined to the Continent during at least part of the first interglacial after the Elsterian. The most important question is whether the marine transgression that inundated the Thames–Medway valley during this interglacial, to at least as far upstream as Clacton (Bridgland 1988, 1994), was from the North Sea, in which case Britain would not necessarily have been separated from the continent. If the transgression was from the Southwest Approaches, however, as it must have been if the late Elsterian (post-glacial) Rhine–Thames valley was routed via the Dover strait, separation from the Continent must surely have occurred by the time the estuarine influence reached Clacton. Even if island status was eventually reached, pollen evidence from Clacton implies that the transgression did not reach there until quite late in the interglacial (Turner & Kerney 1971), so it is possible that Britain remained a peninsula for sufficient time after the Elsterian for temperate-climate faunas to become fully established, hence the close comparison with Holsteinian assemblages from the neighbouring part of Europe.

An Elsterian/Anglian breach of the Wealden ridge and Holsteinian/Hoxnian Rhine–Thames drainage to the North Sea are not mutually exclusive interpretations. It is possible to envisage that a through-valley between the southern North Sea and the English Channel was first created by overflow from an Elsterian lake, but that subsequent drainage patterns continued to include a watershed between the Weald and Artois ridges. As Meijer & Preece (1995) have noted, the post-Elsterian/Anglian sea-level rise would have been greatly influenced by other implications of deglaciation, particularly glacio-isostatic effects. It is possible that differential isostatic rebound had a profound effect on the configuration of post-Elsterian drainage in the southern North Sea. Even without later reoccupation of the through-valley to the English

Table 2. *Angularity/roundness of flint from offshore samples. A, all flints as % total flint (excluding unmodified nodules) Data from onshore beach and fluviatile gravels are included for comparison.*

Locality	wr	r	sr	sa	a	va	Total	
Maplin R6	10.1	3.7	1.8	31.2	28.4	24.8	109	(Fluvial terrace)
Maplin W28	7.9	6.1	4.9	42.4	20.6	18.2	165	(Marine channel-fill)
Maplin W29	11.4	8.5	0.7	41.1	27.0	11.4	141	(Fluvial terrace)
Maplin W39	8.2	5.5	1.4	31.5	16.4	37.0	73	(Fluvial terrace)
Samples from Pleistocene beach gravels								
Bembridge 1	9.6	21.0	30.5	24.6	11.4	2.9	509	(Raised beach)
Bembridge 2	4.6	11.7	30.0	35.9	13.6	4.3	582	(Raised beach)
Brooks Farm 1	2.6	2.5	8.1	37.6	23.6	25.6	774	(Raised beach)
Brooks Farm 2	1.7	5.9	18.7	45.3	21.9	6.6	547	(Raised beach)
Aldingbourne Park 1	2.4	6.0	14.0	45.4	19.1	13.1	335	(Raised beach)
Boxgrove 1	1.9	5.8	23.0	29.9	21.0	18.0	618	(Raised beach)
Boxgrove 2	1.4	7.4	38.5	28.8	19.1	4.8	351	(Raised beach)
Priory Bay 1	2.3	3.2	4.6	51.0	28.4	10.6	349	(?Raised beach)
Priory Bay 2	3.7	3.0	5.1	36.8	32.9	18.6	435	(?Raised beach)
Southwold 1	37.7	27.1	16.9	10.7	3.2	4.4	591	(Westleton Beds)
Samples from Pleistocene river gravels								
Little Hayes 1	9.6	4.9	1.7	22.5	29.3	31.9	648	(Crouch)
Little Hayes 2	10.7	5.1	2.5	24.9	33.5	23.3	570	(Crouch)
Barling 1	6.1	2.5	1.2	38.4	26.1	25.7	245	(Thames–Medway)
Barling 2A	11.4	5.1	1.8	17.2	13.6	50.9	273	(Thames–Medway)
Barling 2B	10.7	6.5	2.1	31.2	22.5	27.0	666	(Thames–Medway)
Barling 3	13.4	5.7	1.7	27.0	21.7	30.4	470	(Thames–Medway)
Dammer Wick 1	12.7	5.2	0.4	24.0	15.3	42.4	229	(Thames–Medway)
Barvills Farm 1	24.8	7.2	3.1	24.1	21.2	19.6	638	(Thames)
Tollesbury 1A	13.0	3.3	2.2	39.4	26.3	15.8	670	(Blackwater)
Shakespeare Pit 2A	24.1	6.9	1.3	18.5	22.3	26.8	622	(Medway)
Holland Haven 1A	4.1	2.5	0.9	25.6	26.2	40.8	321	(Thames–Medway)
Holland Haven 1B	9.7	1.4	1.4	33.8	16.7	37.0	216	(Thames–Medway)
Holland Haven 2	3.9	1.8	0.7	22.4	26.5	44.7	434	(Thames–Medway)
Cooks Green 1A	3.1	1.5	1.9	44.7	29.6	19.2	521	(Thames–Medway)
Cooks Green 1B	4.8	3.2	1.7	41.9	29.8	18.6	413	(Thames–Medway)
Cooks Green 2	11.6	3.1	1.2	41.3	28.4	14.4	327	(Thames–Medway)
Rampart Field 4				18.3	54.5	27.2	226	(Ingham River)

Sources: Bridgland (1983, 1988, 1994, in press), Bridgland *et al.* (1995)
*Categories (modified from Fisher & Bridgland 1986):
wr, well rounded: no flat faces, corners or re-entrants discernible; a uniform convex clast outline
r, rounded: Few remnants of flat faces, with corners all gently rounded
sr, subrounded: poorly developed flat faces with corners well rounded
sa, subangular: strongly developed flat faces with incipient rounding of corners
a, angular: strongly developed faces with sharp corners
va, very angular: as angular, but corners and edges very sharp, with no discernible blunting

Channel by southward drainage, marine erosion during subsequent high-sea-level events could have brought about the flooding and subsequent widening of the strait. The number of such high-sea-level events, a matter of some controversy, is addressed elsewhere (Keen 1995). If Weichselian (and, perhaps, late Saalian) Rhine–Thames drainage was routed through this valley, it may be necessary to invoke a further post-Holsteinian/Hoxnian diversionary event. The possibility of glacial ponding of the southern North Sea drainage basin during the Saalian and/or Weichselian is a subject that remains to be fully explored; the absence of evidence for British east coast ice during these stages further south than northern East Anglia is one potential problem

Table 3. *Angularity/roundness of flint: same samples as in Table 2, but shown as percentage of total broken flint, including broken reworked pebbles (i.e. without whole beach pebbles and nodules), with va category excluded. Reworked unbroken beach pebbles, where separable, are shown together with whole nodules in right-hand part of table, as percentage of total broken*

Locality	Fresh and broken reworked*					Unbroken reworked*			
	wr	r	sr	sa	a	wr	r	sr	nods
Maplin R6			1.5	51.5	47.0	16.7	6.1	1.5	
Maplin W28		1.8	6.2	62.0	30.1	2.4	1.5	0.6	
Maplin W29				60.4	39.6	16.7	12.5	1.0	
Maplin W39			2.8	63.9	33.3	16.7	11.1		
Samples from Pleistocene beach gravels									
Bembridge 1	9.6	21.0	30.4	24.5	11.4				0.2
Bembridge 2	4.6	11.7	30.0	35.9	13.6				
Brooks Farm 1	3.9	3.7	12.3	56.8	35.7				0.2
Brooks Farm 2	2.1	7.5	23.9	57.6	28.1				0.2
Aldingbourne Park 1	3.3	8.1	19.2	62.0	26.1				0.4
Boxgrove 1	1.9	5.8	23.1	30.0	21.1				0.3
Boxgrove 2	1.4	7.4	38.5	28.8	19.1				1.7
Priory Bay 1		2.0	4.7	59.9	33.3	2.7	1.7	0.7	
Priory Bay 2	1.0	2.0	3.7	39.2	35.0	2.9	0.7	1.7	0.7
Southwold 1	39.5	28.3	17.7	11.2	3.4				
Samples from Pleistocene river gravels									
Little Hayes 1			0.9	43.1	56.1	18.3	9.4	2.4	0.3
Little Hayes 2			0.9	42.6	57.4	18.3	8.7	3.3	
Barling 1			1.3	58.8	40.0	9.4	3.8	0.6	
Barling 2A			2.3	54.7	43.0	36.1	16.3	3.5	1.2
Barling 2B		0.3	2.7	56.4	40.7	19.2	11.4	1.1	1.1
Barling 3			2.1	54.3	43.6	26.9	11.5	1.3	
Dammer Wick 1			1.1	60.4	38.5	14.3	13.2		
Barvills Farm 1			4.3	51.0	44.7	52.3	15.2	2.3	1.3
Tollesbury 1A			2.0	58.8	39.2	19.4	4.9	1.3	0.7
Shakespeare 2A			1.2	44.8	54.1	58.4	16.7	2.0	2.0
Holland Haven 1A		0.6	49.1	50.3	7.8	4.8	1.2		
Holland Haven 1B		1.8	65.8	32.4	18.9	2.7	0.9		
Holland Haven 2				45.8	54.3	8.0	3.8	1.4	1.9
Cooks Green 1A			1.5	59.3	39.2	4.1	2.0	1.0	0.3
Cooks Green 1B		0.3	2.0	57.1	40.6	6.6	4.0	0.3	0.3
Cooks Green 2			0.9	58.5	40.6	16.6	4.4	0.9	.4
Rampart Field 4				22.4	66.7				1.1

* See Table 2 for abbreviations of categories; nods, unmodified nodules

facing such a theory. However, it has already been established that sediments of the appropriate age of any type are poorly represented in the southern North Sea, a possible result of Late Pleistocene and/or Holocene marine erosion.

It is interesting to speculate about what would happen if, in some future, 'post-Holocene' glacial episode, sea-level was to fall to the sort of level indicated for the Weichselian. To answer the question of which way the Rhine–Thames system would drain is by no means a straightforward task. Evidence from the southern North Sea and the Strait of Dover, where channel-floor gradients are inclined to the north, would seem to indicate a northward route. Taking a wider view, however, including the eastern English Channel and the central North Sea, would seem to imply that the only viable exit to the Atlantic is via the southerly route, adopting the large valley beneath the English Channel. The bathymetric map of the North Sea between Lincolnshire and northern Holland shows a subdued but continuous ridge, the Haisboro–Terschelling Rise of Gregory (1927), where the depths are shallower than the Dover strait (Fig. 3). Thus a fall in sea-level would seemingly lead to this becoming the first connection between Britain and the continent. The question arises as to whether this would indeed be the case, or whether modification of this area during the regression would allow drainage northwards before the Dover strait emerged. The effects of any isostatic movements, either depression of the northern area resulting from glacial isostasy or uplift of the North Sea Basin resulting from its draining (hydro-isostasy), might be crucial. The Haisboro–Terschelling Rise might well prove sufficient to deflect post-Holocene drainage southwards through the English Channel, but this feature is believed to have been constructed during the Devensian/Weichselian glaciation, since it is associated with the late Weichselian ice margin (Praeg 1994), which would mean that it could not have influenced river courses during earlier low-sea-level episodes. Indeed, it is possible that northward drainage of the Rhine–Thames persisted until this route was blocked by the late Weichselian glaciation, bringing about diversion into the English Channel (Praeg 1994).

Since fluvial valleys require a downstream gradient, it is apparent that the Dover strait–Axial Channel of the southern North Sea cannot represent a submerged Rhine–Thames valley in unmodified form. If the feature was used by the Weichselian river system, its floor must have been flattened as a result either of neotectonic tilting or of tidal scouring during the Holocene transgression. If the Weichselian drainage was from north to south, tilting in response to subsidence of the North Sea sedimentary basin could indeed have effectively cancelled out the original downstream gradient. If differential subsidence of this magnitude is characteristic of the Pleistocene, as well as the Holocene, it is difficult to envisage a fluvial course through the Strait of Dover persisting since the Elsterian. It seems likely that subsidence in the north would eventually have brought about a return to a North Sea course, towards the Viking Graben. Such a reversion is most likely to have occurred following a high-sea-level episode, during which the critical area would have been flooded and the river would have been unable to modify its valley in response to the tectonic movement. Thus even if the southerly route persisted until the Holsteinian/Hoxnian transgression, as represented at Clacton, it might have been abandoned when sea-level fell at the start of the next glacial.

Structure and sedimentation

Within the region under discussion in this paper, there are two principal areas in which there is the possibility of structural control having influenced sedimentation (Fig. 3). The first extends approximately 40 km from the Essex and Suffolk coasts in a southeasterly direction. It comprises a 30 km wide monoclinal feature, extending from Clacton towards Orford Ness, in which the steep southwest-facing limb is in part fault-controlled. Seismic activity has been detected in this area as recently as 1985 and 1994 (15 September) and the feature appears to be associated with differential subsidence; the rate of subsidence within the area of the feature, at Felixstowe, is less than half that beyond its southwestern edge, at Southend (Rossiter 1972). It is possible, though as yet unproved, that the later, post-Anglian courses of the Thames and Medway have been influenced by this structure. It is believed that the Colchester earthquake of 1884 had its epicentre on a north–south branch fault of the structure, discovered by one of the authors in 1981 (D'Olier, unpublished), in the region of Mersea Island.

The second area of possible structural influence is on the northern flank of the Weald–Artois anticlinorium, where a WNW trending series of major basement faults of the Hercynian Front, such as the 'Faille de Landrethun', are intersected by a series of SSW trending wrench faults (Shephard-Thorn et al. 1972). This area, and particularly that lying offshore in the Dover strait, has been seismically active within historical time, with events recorded in 1133, 1247, 1342, 1449, 1580, 1692, 1776, 1938 and 1950 (Neilson et al. 1984). There is some inconclusive evidence that the Fosse Dangeard depression lies close and parallel to WNW trending faults and might owe its over-deepening, in part, to fault activity.

It has been suggested by Roep et al. (1975) that a thick sequence of sediments filling a channel feature south of Wissant, on the north French coast, represents a southward-flowing overflow outlet of an ice-dammed lake in the southern North Sea. The principal source of this information is a gravel pit, the Carrière du Fart. Gibbard (1995) also suggests that the Wissant deposits represent an outflow from an Anglian ice-dammed lake. However, the 'erosion' scarps underlying the sediments of the 'Carrière du Fart' river lie parallel to the SSW wrench fault trend. The presence of large clay slabs adjacent to these scarps and of overfolding, plastic deformation and microfaulting of the sediments also suggests a relationship between faulting and sedimentation. The height of these sediments, extending as they do from -20 m to $+33$ m NGF (Fig. 6), makes correlation with southward-flowing late Elsterian drainage in the

Dover strait extremely unlikely. The chronostratigraphically equivalent late Anglian Black Park Gravel of the Thames is considered to be represented within the envelope of sediment between the basal Clacton Channel Deposits and the terrace surface formed by the Boyn Hill Gravel and downstream equivalents. As can be seen from Fig. 6, projection of this sediment body downstream to the Wissant area would take it several tens of metres below the height range of the Carrière du Fart deposits. A very considerable amount of post-Elsterian differential subsidence/uplift between the Strait of Dover and the southern North Sea would be required for the Carrière du Fart sediments to conform to the projected level of Elsterian drainage through the strait, even if they were deposited in a south-bank tributary valley with a steep downstream gradient. Although Roep et al. (1975) dismissed the idea that the Wissant sediments were deposited by a small river running off higher ground immediately to the north, the possible interpretation of the deposits as products of a fault-controlled localized drainage system should perhaps be given further consideration. Both the height of the deposits and the occurrence in them of *Mammuthus meridionalis* molar teeth would seem to argue for a pre-Elsterian age.

Conclusions

It is suggested in this paper that the Thames and Rhine first came to flow together in a single valley beneath what is now the southern North Sea in the early Middle Pleistocene, although an axial southern North Sea (or Lobourg) valley had possibly originated very much earlier as the route of a northward-flowing stream draining the Weald–Artois ridge. The diversion of this river system via the Strait of Dover to the English Channel, ascribed elsewhere in this volume to the overflow of an Elsterian ice-dammed lake, remains to be proved; if it is accepted that such an event did indeed lead to the cutting of a new channel through the Weald–Artois ridge, initiating the breach that later became the Strait of Dover, important questions remaining to be answered are those of whether, after deglaciation, the rivers maintained the southward route prior to the Holsteinian marine transgression and, if so, whether the route persisted during all subsequent low-sea-level episodes until the last glacial.

An important difference between the Quaternary fluvial records on either side of the southern North Sea has been noted. Offshore from the Thames estuary and the Essex coast, the Middle and Late Pleistocene evolution of the Thames–Medway system is recorded in a system of submerged terraces flanking the drowned pre-Holocene valley. In contrast, the sedimentary record of the lower Rhine–Meuse system is one of progressive accumulation in an area of continuous subsidence, reflecting the continuation of the north–south axis of the North Sea basin beneath The Netherlands. This underlines the fact that the eastern side of the southern North Sea is an area of progressive isostatic downwarping, mainly in response to the weight of sediment accumulating there, whereas the British side of the southern North Sea and the Dover strait are both areas of net uplift during the Pleistocene. Whether this difference will have influenced the fluvial systems since the Elsterian, when the connection with the English Channel is thought to have been made, is another question that remains unanswered at present.

The authors wish to thank Hilary Foxwell of Greenwich University (Department of Geology) and Steven Allan/David Hulme of Durham University (Department of Geography) for their assistance in the completion of the figures. Piet Cleveringa, Philip Gibbard, Cees Laban, Tom Meijer, Dan Praeg, Richard Preece and Wim Westerhoff kindly read and commented on earlier versions of the text.

References

ALLEN, P. 1983. *Middle Pleistocene Stratigraphy and Landform Development in South-east Suffolk.* Unpublished PhD thesis, University of London.
—— 1984. *Field Guide to the Gipping and Waveney Valleys.* Field Guide, Quaternary Research Association, Cambridge.
BALSON, P. S. 1989. Neogene deposits of the UK sector of the southern North Sea. *In:* HENRIET, J. P. & DE MOOR, G. (eds) *Quaternary and Tertiary Geology of the Southern Bight, North Sea.* Belgian Ministry of Economic Affairs Geological Survey, 89–95.
——, LEBAN, C., SHÜTTENHELM, R. T. E., PAEPE, R. & BAETEMAN, C. 1992. *Quaternary Geology, Ostend Sheet 51° N–02° E*, British Geological Survey/Rijks Geologische Dienst/Belgische Geologische Dienst, 1:250,000 offshore map series.
BRIDGLAND, D. R. 1983. *The Quaternary fluvial deposits of north Kent and eastern Essex.* PhD thesis, City of London Polytechnic.
—— 1985. Uniclinal shifting: a speculative reappraisal based on terrace distribution in the London Basin. *Quaternary Newsletter*, **47**, 26–33.
—— 1988. The Pleistocene fluvial stratigraphy and palaeogeography of Essex. *Proceedings of the Geologists' Association*, **99**, 291–314.
—— 1994. *Quaternary of the Thames.* Geological Conservation Review Series, Chapman and Hall,

London.

—— 1995. Composition of the gravel deposits. *In:* ROBERTS, M. B. (ed.) *The Middle Pleistocene site at ARC Eartham Quarry, Boxgrove, West Sussex, UK*. English Heritage Monograph Series, London, in press.

—— & ALLEN, P. 1995. A revised model for terrace formation and its significance for the lower Middle Pleistocene Thames terrace aggradations of north-east Essex, U.K. *In:* TURNER, C. (ed.) *The Early Middle Pleistocene in Europe*. Balkema, Rotterdam, in press.

—— & D'OLIER, B. 1989. A preliminary correlation of onshore and offshore courses of the Rivers Thames and Medway during the Middle and Upper Pleistocene. *In:* HENRIET, J. P. & DE MOOR, G. (eds) *Quaternary and Tertiary Geology of the Southern Bight, North Sea*. Belgian Ministry of Economic Affairs Geological Survey, 161–172.

—— & LEWIS, S. G. 1991. Introduction to the Pleistocene geology and drainage of the Lark valley. *In:* LEWIS, S. G., WHITEMAN, C. A. & BRIDGLAND, D. R. (eds) *Central East Anglia and the Fen Basin*. Field Guide, Quaternary Research Association, London, 37–44.

——, ALLEN, P., CURRANT, A. P. ET AL. 1988. Report of Geologists' Association field meeting in north-east Essex, May 22nd–24th, 1987. *Proceedings of the Geologists' Association*, **99**, 315–333.

——, D'OLIER, B., GIBBARD, P. L. & ROE, H. M. 1993. Correlation of Thames terrace deposits between the Lower Thames, eastern Essex and the submerged offshore continuation of the Thames–Medway valley. *Proceedings of the Geologists' Association*, **104**, 51–75.

——, GIBBARD, P. L. & PREECE, R. C. 1990. The geology and significance of the interglacial sediments at Little Oakley, Essex. *Philosophical Transactions of the Royal Society of London B*, **328**, 307–339.

——, LEWIS, S. G. & WYMER, J. J. 1995. Middle Pleistocene stratigraphy and archaeology around Mildenhall and Icklingham, Suffolk: a report on a Geologists' Association field meeting, 27th June, 1992. *Proceedings of the Geologists' Association*, **106**, 57–69.

BRIQUET, A. 1921. Sur l'origine du Pas-de-Calais. *Annales de la Société Géologique du Nord*, **46**, 141–157.

BRUNNACKER, K., LÖSCHER, M., TILLMANS, W. & URBAN, B. 1982. Correlation of the Quaternary terrace sequence in the lower Rhine valley and northern Alpine foothills of central Europe. *Quaternary Research*, **18**, 152–173.

CAMERON, T. D. J., LABAN, C. & SCHÜTTENHELM, R. T. E. 1984. *Quaternary Geology, Flemish Bight Sheet 52° N–02° E*. British Geological Survey/Rijks Geologische Dienst/Belgische Geologische Dienst, 1:250,000 offshore map series.

CASTON, V. N. O. 1977. *Quaternary Deposits of the Central North Sea: 1. A New Isopachyte Map of the Quaternary of the North Sea*. Report of the Institute of Geological Sciences, **77/11**, 1–8.

CHARLESWORTH, J. K. 1957. *The Quaternary Era, with Special Reference to its Glaciation* Vol 2. Edward Arnold, London.

DEVOY, R. J. N. 1979. Flandrian sea level changes and vegetational history of the Lower Thames estuary. *Philosophical Transactions of the Royal Society of London B*, **285**, 355–407.

D'OLIER, B. 1975. Some aspects of late Pleistocene–Holocene drainage of the River Thames in the eastern part of the London Basin. *Philosophical Transactions of the Royal Society of London A*, **279**, 269–277.

—— & BRIDGLAND, D. R. 1994. An ice-marginal drainage system in the Southern Bight of the North Sea. *In: The Geology of Siliciclastic Shelf Seas*. 2nd International Conference on the Geology of Siliciclastic Shelf Seas, University of Ghent, Abstract volume, 39–40.

FISHER, P. F. & BRIDGLAND, D. R. 1986. Analysis of pebble morphology. *In:* BRIDGLAND, D. R. (ed.) *Clast Lithological Analysis*. Technical Guide **3**, Quaternary Research Association, Cambridge, 43–58.

GIBBARD, P. L. 1988. The history of the great northwest European rivers during the past three million years. *Philosophical Transactions of the Royal Society of London B*, **318**, 559–602.

—— 1995. The formation of the Strait of Dover. *This volume*.

GREGORY, J. W. 1927. The relations of the Thames and Rhine, and the age of the Straits of Dover. *Geographical Journal*, **70**, 52–59.

HAMBLIN, R. J. O. 1989. *Dungeness–Boulogne Sheet 50° N–00°*, British Geological Survey, 1: 250,000 offshore map series.

HEY, R. W. 1980. Equivalents of the Westland Green Gravels in Essex and East Anglia. *Proceedings of the Geologists' Association*, **91**, 279–290.

JELGERSMSA, S. 1966. Sea-level changes during the last 10,000 years. *In:* SAWYER, J. S. (ed.) *World Climate 8000 to 0 BC*. Royal Meterological Society, London, 54–69.

—— 1992. Vulnerability of the coastal lowlands of the Netherlands to a future sea-level rise. *In:* TOOLEY, M. J. & JELGERSMA, S. (eds) *Impacts of Sea-level Rise on European Coastal Lowlands*. Blackwell, Oxford, 94–123.

——, OELE, E. & WIGGERS, A. J. 1979. Depositional history and coastal development in the Netherlands and the adjacent North Sea since the Eemian. *Acta Universitatis Upsaliensis: Symposia Universitatis Upsaliensis Annum Quingentesimum Celebrantis*, **2**, 233–248.

KEEN, D. H. 1995. Raised beaches and sea-levels in the English Channel in the Middle and Late Pleistocene: problems of interpretation and implications for the isolation of the British Isles. *This volume*.

LIU, A. C., BATIST, M. DE, HENRIET, J. P. & MISSIAEN, T. 1993. Plio-Pleistocene scour hollows in the Southern Bight of the North sea. *Geologie en Mijnbouw*, **71**, 195–204.

MATHERS, S. J. & ZALASIEWICZ, J. A. 1986. A sedimentation pattern in Anglian marginal meltwater channels from Suffolk, England. *Sedimentology*, **33**, 559–573.

——, ——, GIBBARD, P. L. & PEGLAR, S. M. 1993. The Anglian–Hoxnian evolution of an ice-marginal drainage system in Suffolk, England. *Proceedings of the Geologists' Association*, **104**, 109–122.

——, —— & WEALTHALL, G. P. 1991. Styles of ice-marginal channel sedimentation: as revealed by a conductivity meter and extendable augers. *In:* EHLERS, J., GIBBARD, P. L. & ROSE, J. (eds) *Glacial Deposits in Great Britain and Ireland.* Balkema, Rotterdam, 405–414.

MEIJER, T. & PREECE, R. C. 1995. Malacological evidence relating to the insularity of the British Isles during the Quaternary. *This volume.*

MOSTAERT, F., AUFFRET, J. P., DE BATIST, M. *ET AL,* 1989. Quaternary shelf deposits and drainage patterns off the French and Belgian coasts. *In:* HENRIET, J. P. & DE MOOR, G. (eds) *Quaternary and Tertiary Geology of the Southern Bight, North Sea.* Belgian Ministry of Economic Affairs Geological Survey, 111–118.

NEILSON, G., MUSSON, R. M. W. & BURTON, P. W. 1984. Macroseismic reports on historical British earthquakes. *ix. Dover Straits.* British Geological Survey/Natural Environmental Research Council Report, **234**.

POWERS, M. C. 1953. A new roundness scale for sedimentary particles. *Journal of Sedimentary Petrology*, **23**, 117–119.

PRAEG, D. 1994. Late Pleistocene glacial-stage drainage through the Dover strait – an appraisal of stratigraphical and geomorphological evidence from the Southern Bight of the North Sea. *In: The Geology of Siliciclastic Shelf Seas.* 2nd International Conference on the Geology of Siliciclastic Shelf Seas, University of Ghent, Abstract volume, 93–94.

PREECE, R. C., SCOURSE, J. D., HOUGHTON, S. D., KNUDSEN, K. L. & PENNEY, D. N. 1990. The Pleistocene sea-level and neotectonic history of the eastern Solent, southern England. *Philosophical Transactions of the Royal Society of London B*, **328**, 425–477.

PRYOR, W. A. 1971. Grain shape. *In:* CARVER, R. E. (ed.) *Procedures in Sedimentary Petrology.* John Wiley, New York, 131–150.

ROEP, Th. B., HOLST, H., VISSERS, R. L. M., PAGNIER, H. & POSTMA, D. 1975. Deposits of southward-flowing, Pleistocene rivers in the Channel region, near Wissant, N.W. France. *Palaeogeography, Palaeoclimatology, Palaeoecology* **17**, 289–308.

ROSSITER, J. R. 1972. Sea-level observations and their secular variation. *Philosophical Transactions of the Royal Society of London A*, **272**, 131–139.

RUEGG, G. H. J. 1994. Alluvial architecture of the Quaternary Rhine–Meuse river system in the Netherlands. *Geologie en Mijnbouw* **72**, 321–330.

SCHNEIDERHÖHN, P. 1954. Eine vergleichende Studie uber Methoden zur quantitativen Bestimmung von Abrundung und Form an Sandkornern. *Heidelberger Beitrage. zur Mineralogie und Petrographie*, **4**, 172–191.

SHENNAN, I. 1989. Holocene crustal movements and sea-level changes in Great Britain. *Journal of Quaternary Science*, **4**, 77–89.

SHEPHARD-THORN, E. R., LAKE, R. D. & ATITULLAH, E. A. 1972. Basement control of structures in the Mesozoic rocks of the Strait of Dover region and its reflexion in certain features of the present land and submarine topography. *Philosophical Transactions of the Royal Society of London A*, **272**, 99–113.

SMITH, A. J. 1985. A catastrophic origin for the palaeovalley system of the eastern English Channel. *Marine Geology*, **64**, 65–75.

STAMP, L. D. 1927. The Thames drainage system and the age of the Strait of Dover. *Geographical Journal*, **70**, 386–390.

STUART, A. J. 1995. Insularity and Quaternary vertebrate faunas in Britain and Ireland. *This volume.*

SUTCLIFFE, A. J. 1995. Insularity of the British Isles 250 000–30 000 years ago: the mammalian, including human, evidence. *This volume.*

TURNER, C. & KERNEY, M. P. 1971. The age of the freshwater beds of the Clacton channel. *Journal of the Geological Society, London*, **127**, 87–93.

WHITEMAN, C. A. & ROSE, J. 1992. Thames river sediments of the British Early and Middle Pleistocene. *Quaternary Science Reviews*, **11**, 363–375.

WYMER, J. J. 1985. *Palaeolithic Sites of East Anglia.* Geobooks, Norwich.

Extension of the British landmass: evidence from shelf sediment bodies in the English Channel

ANDREW G. BELLAMY

Postgraduate Research Institute for Sedimentology, The University, PO Box 227, Whiteknights, Reading, Berkshire, RG6 2AB, UK
Present address United Marine Dredging Limited, Francis House, Shopwhyke Road, Chichester, West Sussex, PO20 6AD, UK

Abstract: Mapping and interpretation of high-resolution shallow seismic and vibrocore sample data from a submerged, infilled valley immediately east of the Owers Bank, at -20 m to -40 m chart datum (CD), eastern English Channel, indicate sediment body formation during multiple cut-and-fill events. Interpretation suggests that these occurred in a combination of diachronous gravel-bed river, peat-land, estuarine and shallow sub-littoral environments with subsequent marine planation (forming ravinement surfaces) and accumulation of a sea-bed sediment veneer. These environments are associated with large-scale relative sea-level and (inferred) climatic changes during the Quaternary which forced subaerial inner continental shelf conditions during cold stages, repeatedly interrupted by interglacial shelf submergence. A model of sediment body formation off the south coast of England, based on the case study findings, shows that subaerial, and not submarine, processes have dominated sedimentary history within present-day, submerged, infilled valleys. Relative to the Quaternary as a whole, global ice-volume is presently unusually low, implying that present eustatic sea-level is unusually high. Subaerial linkage of present-day southern England with the adjacent inner shelf has therefore persisted for most of Quaternary time, with the notion of British insularity applicable to only a relatively small part of the period.

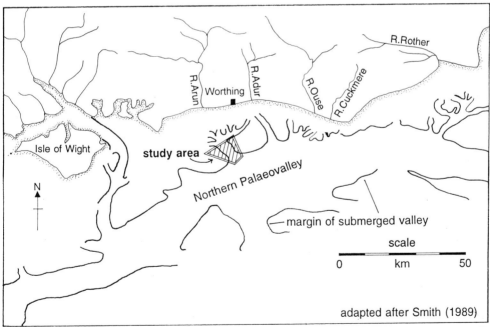

Fig. 1. Submerged valleys in the eastern English Channel off the south coast of England. The submerged, infilled valley of the former Solent River exists immediately to the east of the Isle of Wight. The Northern Palaeovalley is reportedly continuous with the Lobourg palaeovalley in the Strait of Dover and also into the central English Channel where it joins the Hurd Deep, lying 70 km off the Cherbourg Peninsula (Smith 1985). The Northern Palaeovalley lies 30 km off the West Sussex coast and forms the main bathymetric feature of the eastern English Channel, being joined by smaller submerged valleys to the north and south.

From Preece, R. C. (ed.), 1995, *Island Britain: a Quaternary perspective*
Geological Society Special Publication No. 96, pp. 47–62

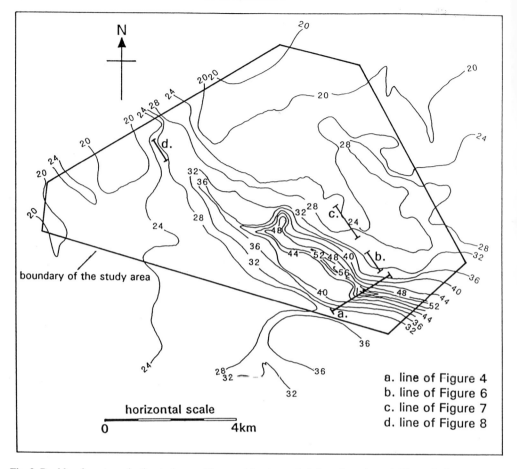

Fig. 2. Rockhead contours in the study area (drawn at 4 m intervals below chart datum). The main feature on the bedrock surface is a linear sub-sea-bed depression which is interpreted as the base of the infilled valley.

The submerged valleys in the eastern English Channel off the south coast of England (Fig. 1) are thought to have formed during the Pleistocene by extension of fluvial systems at times of lower-than-present sea-level when the inner continental shelf was subaerially exposed (e.g. Dingwall 1975; Jones, 1981; Smith 1985, 1989). Owing to a lack of detailed seismic and sample data for specific sites, the thickness, origins and formation of deposits infilling these valleys are poorly understood.

This paper describes a study of one such submerged, infilled valley in the eastern English Channel, using high-resolution shallow seismic and vibrocore sample data obtained during recent surveys by United Marine Dredging Limited. The submerged valley forms the main bathymetric feature immediately east of the Owers Bank, situated 40 km east of the Isle of Wight and 20 km south of Worthing, West Sussex (Fig. 1). The study area occupies about 75 km² of sea-bed, with water depths increasing from around −20 m CD in the north of the area to over −40 m CD at the bottom of a partly infilled valley in the south (all water depths quoted in metres below chart datum, CD).

Seismic and sample data

Seismic data have been used to produce a contour map of the rockhead beneath the infilled valley (Fig. 2). The main feature on the bedrock surface is a continuous linear depression trending NW–SE across and beyond the area, at a depth of up to −56 m. This depression is interpreted as the lower part of the submerged valley, eroded prior to infilling. Seismic data suggest that the valley is completely infilled with up to 20 m of sediments in the north, but only partly infilled in the south, such that the upper

EXTENSION OF THE BRITISH LANDMASS

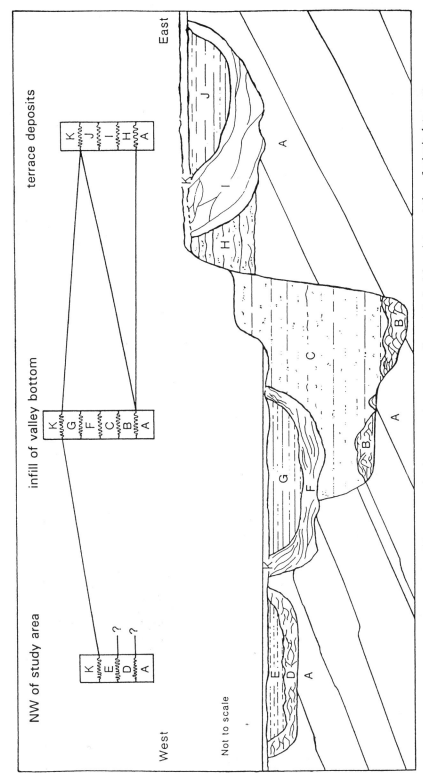

Fig. 3. Seismostratigraphy of the infilled valley. The cross-section is a composite summary diagram derived from interpretation of seismic data.

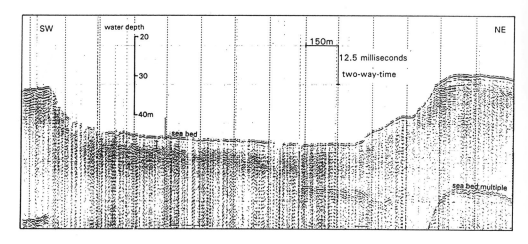

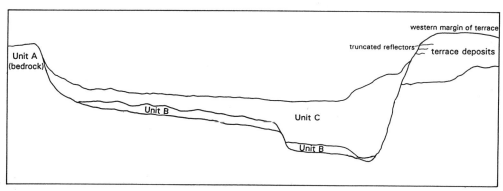

Fig. 4. Interpreted seismic profile across the partly infilled valley in the southeast of the study area. The submerged valley is over 2 km wide and 20 m deep at the sea-bed in the profile. Reflector truncation at the valley sides is interpreted as indicating erosion of sediments during phases of incision in the cut-and-fill history of the valley. Note the considerable vertical exaggeration in this and in the other seismic profiles in the paper.

valley sides are exposed at the sea-bed (omitted from Fig. 2 for clarity). In addition, a terrace exists immediately to the east, underlain by sediments up to 14 m thick.

The seismostratigraphy for the infilled valley bottom and adjacent terrace is shown in Fig. 3. Above the bedrock (Unit A), ten seismic units (B–K) are identified, defined by their external bounding reflectors and internal seismic signatures (seismic facies) (Mitchum et al. 1977; Mitchum & Vail 1977). Bounding reflectors are interpreted as representing sedimentary unconformities in the valley stratigraphy, and different seismic facies are considered to imply discrete and distinct lithological characteristics (including variations in grain-size). The unconformable superimposition of seismic units of contrasting geometry and seismic facies suggests valley-infill evolution during multiple cut-and-fill events, with Unit H, beneath the terrace, the oldest (truncated at the valley side) and Unit K, immediately beneath the sea-bed, the youngest.

The partly infilled valley bottom

A seismic profile across the partly infilled valley in the south of the area (Fig. 4) reveals that four main seismic units occur.

Unit A. Unit A underlies all other seismic units and has no lower bounding reflector on seismic profiles. The undulating upper bounding surface is formed by truncated reflectors and represents a major unconformity. Internally, Unit A is characterized by a series of low- to high-amplitude, continuous, parallel, even and gently dipping reflectors, commonly separated by a homogeneous backscatter. Unit A is continuous across the area and is interpreted as representing

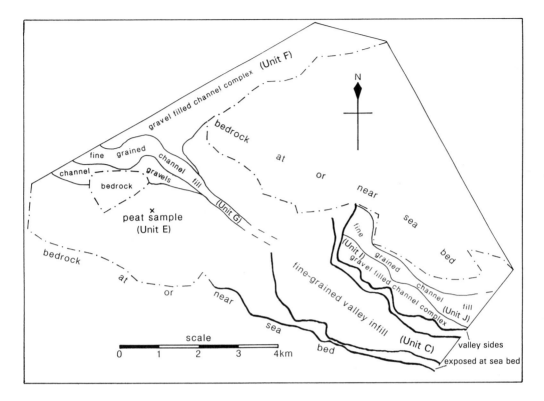

Fig. 5. Simplified surficial geological map of the valley infill.

bedrock, consisting of Palaeogene sands and clays (Smith 1989; Fig. 5).

Unit B. The base of the valley infill is defined by a high-amplitude, discontinuous and irregular seismic unit, <4 m thick, commonly displaying a chaotic internal reflector pattern. This seismic facies is interpreted as representing high-energy (coarse-grained) sedimentation. Unit B is inferred to represent a basal coarse-grained 'lag' deposit, forming an undulating surface upon which the majority of the valley infill has been deposited.

Unit C. Unit C represents the majority of the valley infill (Fig. 5) and is commonly characterized by homogeneous backscatter, although occasional reflectors onlap bounding surfaces. These reflectors are low- to moderate-amplitude, continuous, flat and parallel, separated by reflector-free zones. Vibrocore samples demonstrate that Unit C consists of fine-grained sands and organic-rich muds containing estuarine, intertidal, shallow sub-littoral and open shelf mollusc, ostracod and foraminiferal assemblages, including shells of *Pholas dactylus* and *Mytilus edulis*. Unit C is interpreted from seismic and sample data as having been deposited in low-energy conditions, including former estuarine, intertidal and near-shore environments.

Terrace deposits

The terrace deposits to the east of the valley are also revealed in Fig. 4. The main seismic units beneath the terrace are Units I and J.

Unit I. Unit I is up to 14 m thick and is traceable for over 4 km beneath the terrace as a completely infilled channel, truncated at the valley side (Fig. 5). Internally, irregular and discontinuous sub-units occur (Fig. 6). Within these are steeply dipping point-source reflectors suggestive of gravel-rich sediments – an interpretation verified by core samples of flint-rich sandy gravel. The bases of the infilled sub-units are interpreted as representing surfaces scoured into previously deposited sediments, with subsequent scour-hollow infilling and migration of the zone of

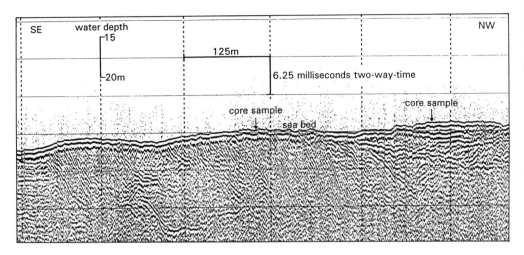

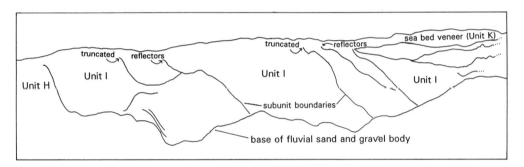

Fig. 6. Interpreted seismic profile of deposits beneath the terrace.

cut-and-fill within the channel as a whole.

These observations are consistent with characteristics described for unstable gravel bed rivers by Miall (1977) and Dawson (1987). It is therefore suggested that Unit I represents the preserved remnants of the cut, fill and lateral extension of dominantly coarse-grained sediments deposited in a fluvial environment.

Unit J. Unit J occurs adjacent to, or overlying, Unit I and is also traceable for over 4 km beneath the terrace surface as a completely infilled channel, truncated at the valley side (Figs 3 and 5). However, the internal seismic signature contrasts with Unit I, being characterized by a relatively uniform series of onlapping, low- to moderate-amplitude, continuous, flat and parallel reflectors (Fig. 7). This seismic response suggests lower-energy environments of deposition relative to Unit I. Core samples demonstrate organic-rich, fine-grained sediments within the channel, perhaps implying deposition in estuarine or near-shore environments during the early stages of a marine transgression.

Units F and G

Units F and G occur in the north of the study area as a completely infilled channel complex, eroded into either Unit C or bedrock (Figs 5 and 8).

Unit F. Unit F displays high-amplitude, continuous, prograding reflectors, interpreted as representing surfaces of lateral accretion or migration within the channel. By contrast, high-amplitude, continuous, low-angle or horizontal and stacked reflectors are interpreted as representing surfaces of aggradation leading to local infilling of the channel complex. Core samples from Unit F comprise flint-rich sandy gravels with a low mud content. On the basis of seismic and sample data, Unit F is interpreted as

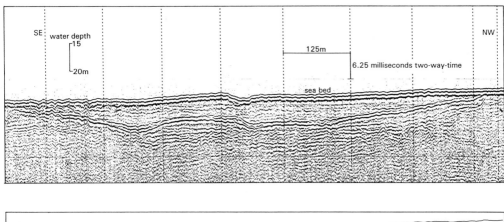
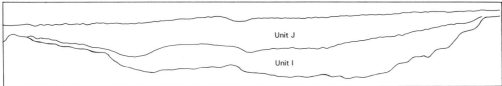

Fig. 7. Interpreted seismic profile of deposits beneath the terrace, immediately to the northwest of those in Fig. 6.

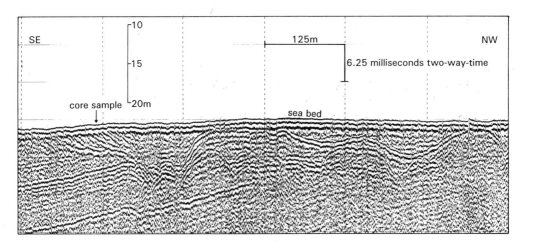
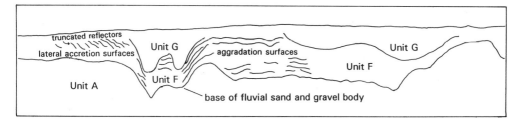

Fig. 8. Interpreted seismic profile of deposits infilling the valley in the north of the study area.

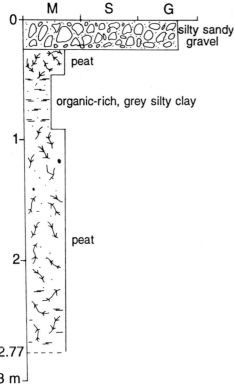

Fig. 9. Lithological log of vibrocore sample from Unit E.

the product of sedimentation in a gravel-bed river system.

Unit G. Unit G infills and onlaps the undulating upper surface of Unit F and displays a lower-amplitude seismic response, typified by a homogeneous backscatter. Samples consist of fine-grained sands and organic-rich muds containing estuarine, intertidal and shallow sub-littoral mollusc, ostracod and foraminiferal assemblages, including shells of *Hydrobia ulvae* and *Pholas dactylus*.

The characteristics of, and relations between, Units F and G and Units I and J are similar, though Units I and J, beneath the terrace, are older (Fig. 3). Both sequences are interpreted as representing a transition from gravel-bed river sedimentation to estuarine, intertidal and nearshore deposition, induced by rising relative sea-level across the continental shelf.

Peat accumulation

Unit E (Fig. 3) infills and onlaps a shallow basin in the northwest of the study area, adjacent to the infilled valley, and displays a low-amplitude seismic response. Core samples indicate that Unit E consists of fine-grained sand, mud and peat. A sample containing peat was extracted at a water depth of −20 m and recovered 2.77 m of sediment (Fig. 9).

The peat sample is overlain by a thin, coarse-grained, marine sea-bed sediment veneer (Unit K, Fig. 3). A sub-sample of peat was extracted from immediately beneath this veneer for pollen and macrofossil analysis and radiocarbon dating. The sub-sample contained well-preserved remains of *Corylus*, common wood fragments and the pollen spectrum shown in Table 1.

Table 1. *Pollen spectrum, 0.28–0.40 m below the sea bed.*

Species	Pollen (% of total land pollen)
Trees	
Betula	14.77
Pinus	0.26
Ulmus	0.26
Quercus	7.25
Alnus	*
Corylus	63.99
Herbs	
Gramineae	6.48
Cyperaceae	4.92
Caryophyllaceae *Lychnis*	0.26
Chenopodiaceae	*
Lamium	*
Lotus	0.26
Galium	1.03
Rumex	0.26
Umbelliferae	0.26

*Present in sample but not encountered during counting.

The likely environment indicated by the pollen evidence is a birch-, oak- and hazel-dominated woodland with additional herb cover, mainly grasses and sedge. This suggests a former temperate, freshwater and forested environment in the area prior to marine transgression. The likely age of the peat is early Holocene, perhaps 9–8 ka BP.

Examination of a second sub-sample from the base of the peat sample, at 2.50 m to 2.77 m below the sea-bed, revealed sparse, poorly preserved pollen grains which precluded detailed analysis. However, pollen types present included *Betula* (tree form), *Pinus diploxylon*, *Quercus*, *Corylus*, Gramineae, Cyperaceae, and possibly *Caltha*. In addition, fragments of the moss *Hypnum cupressiforme* were detected together with remains of beetles. The biological evidence from the second sub-sample indicates former subaerial conditions in the study area, with an environment similar to that indicated from the near-sea-bed peat sub-sample.

Origin and age of the valley infill

The oldest sediments in the area underlie the terrace (Units H, I and J). These are dominated by a gravel-filled fluvial channel complex (Unit I) and fine-grained (possibly estuarine) channel infill (Unit J). These deposits possibly represent sedimentation on an earlier floodplain of the same river that eroded to the valley bottom (Figs 3, 4 and 5).

The majority of the valley infill is represented by Unit C, a fine-grained (dominantly estuarine or intertidal) deposit, which is continuous into the north of the area. In the north, Unit C underlies a fluvial sand and gravel unit (Unit F) and a fine-grained channel infill (Unit G) of likely estuarine and shallow sub-littoral origin (Figs 3 and 5). Truncated reflectors immediately beneath the sea-bed on seismic profiles (Figs 6 and 8) are interpreted as indicating that upper surface erosion of infill deposits occurred during and after marine submergence. This erosion was penecontemporaneous with the deposition of a gravelly and shelly sea-bed sediment veneer (Unit K, Figs 6 and 9).

The valley bottom probably acted as the main drainage line in this part of the inner continental shelf during the last cold stage (sea-level below -50 m) and was therefore at least partly infilled during the late Devensian and Holocene. However, just as sediment body formation occurred in multiple cut-and-fill events, so evolution of the terrace deposits, valley and valley infill is likely to have occurred during several climatic cycles.

Regional significance of the study findings

The submerged valley east of Owers Bank records a sedimentary history of cut-and-fill during one or more cycles of relative sea-level rise and fall. From the disposition of the submerged valleys shown in Fig. 1, it is suggested that the valley bottom in the study area was the former course of the River Arun during the last cold stage. It is also likely that the predominantly fluvial cut-and-fill history discussed in the preceding sections is of regional significance in that it highlights base-level changes common to the other river systems in the area, including the Solent, Adur, Ouse and Cuckmere (Fig. 1).

Infilled channels and valleys on the south coast of England

The bases of infilled channels beneath present valley bottoms have been detected at -29.5 m OD for the Sussex Ouse at Newhaven (Jones 1971) and at -33.5 m OD for the Arun near Arundel (Reid 1903). In addition, the bases of infilled channels of the former Solent River have been detected at -24 m OD in Southampton Water from borehole records (Hodson & West 1972) and, from seismic data, at -46 m OD to the southwest of Selsey Bill (Dyer 1975). Each of these infilled channels probably relates to times of lower-than-present sea-level, when the inner continental shelf in the region was subaerially exposed (Jones 1981). The former River Arun, together with the adjacent drainage systems, was probably confluent with a southwesterly flowing trunk stream, draining much of the inner shelf in the region (Gibbard 1988; Smith 1989; Fig. 10).

Since the rockhead in the valley east of Owers Bank occurs at over -56 m CD and as this valley joins the deeper Northern Palaeovalley to the south (Fig. 1), regional relative sea-level was at times below -60 m during erosion of the channels and valleys off the south coast of England. During these episodes of lower-than-present sea-level, Britain would have been linked to the continent by the subaerially exposed southern English and northern French inner continental shelves.

Valley infills in the region

The valley infill in the study area includes deposits of peat, mud, fine-grained sand and sandy gravel represented by Units B–K (Fig. 3). Parts of the valley stratigraphy bear similarity to sedimentary successions investigated elsewhere in the region. For example, the sedimentary sequence represented by Units B and C – a coarse-grained 'lag' deposit overlain by a relatively thick sequence of muds and fine-grained sands – is similar to successions proved by boreholes in the infilled channels of Southampton Water (Hodson & West 1972) and in the lower reaches and estuary of the River Ouse, West Sussex (Jones 1971). Peat is an important component of these channel infills and also occurs in the study area as part of Unit E. Southampton Water and the lower reaches of the River Ouse possibly represent analogues for the study area during different phases of marine transgression.

Extensive fluvial sand and gravel deposits represent an important part of the valley

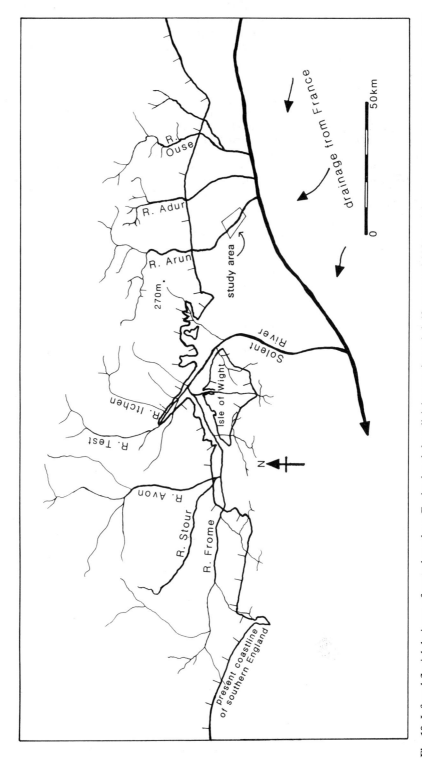

Fig. 10. Inferred fluvial drainage of central southern England and the adjoining continental shelf at around 20 ka BP (after Jones 1981; Gibbard 1988; Smith 1989 and Lawson & Hamblin 1989). The valley in the study area was possibly formed by the ancestral River Arun which was a tributary of a trunk stream flowing to the southwest in the Northern Palaeovalley. The Solent River probably had the largest drainage basin of the right bank tributaries in the area, including the reaches of rivers currently draining parts of Dorset and Wiltshire.

stratigraphy in the study area, both in the valley bottom and beneath the terrace (Units F and I, Fig. 3). The extension of rivers across the inner continental shelf, with consequent shelf valley development and sand and gravel aggradation at a time of lower-than-present sea-level, implies climatic conditions significantly colder than at present in the English Channel region. Indeed, conditions in drainage basins were such that large quantities of coarse-grained sediment (commonly flint) were released from bedrock (dominantly Chalk and poorly consolidated sandstones) and pre-existing flint-rich unconsolidated sediments to be deposited by rivers on the inner continental shelf. These conditions contrast with the relatively quiescent, temperate environment in southeast England at present.

Periglacial, particularly permafrost, environments potentially favour accelerated mass movement and transport of coarse-grained sediments, as catchment vegetation is relatively sparse, surface materials prone to gelifraction and slopes subject to gelifluction. Furthermore, river flow is commonly markedly unsteady and non-uniform, with the potential for spasmodic high-energy discharges induced by seasonal snow and ground-ice melt within catchments (Bryant 1983). With relative sea-level below -50 m and regional climate significantly cooler than at present, it is suggested that the deposition of the fluvial sands and gravels occurred in periglacial environments.

According to Dyer (1975), gravelly channel and terrace deposits are associated with the former channels of the Solent River immediately east of the Isle of Wight. These sediments might share the same origin as the fluvial sands and gravels infilling the valley east of Owers Bank, having been deposited in a periglacial environment by gravel-bed rivers draining southern England and the adjoining inner shelf during Quaternary cold stages.

Implications for evolution of the inner continental shelf off the south coast of England

The valley infill in the study area is interpreted as recording more than one cycle of gravel-bed river–estuarine/intertidal sedimentation. These cycles are represented in order of decreasing age by Units I and J, B and C and F and G (Fig. 3). It is therefore suggested that the formation of these deposits occurred through at least three phases of relative sea-level change of sufficient magnitude to lead to valley erosion to over -56 m, valley infill with over 20 m of sediments, and ultimate marine submergence during the Holocene transgression. The regional significance of such relative sea-level changes would not only be a possibly similar sequence of sedimentary events in the valleys adjacent to the study area, but also the linkage of Britain to the continent during times of relative sea-level below -50 m, with repeated isolation of Britain as an island during times of higher relative sea-level similar to the present day.

The eastern English Channel has therefore had a complex Quaternary history associated with repeated and large-scale changes of relative sea-level. It is considered that glacio-eustatic sea-level change is the likely dominant cause of large-scale relative sea-level fluctuation in the eastern English Channel, with the inner continental shelf submerged during interglacials and subaerial during cold stages.

A model of infilled valley evolution

The interpretations reached from the seismic and sample data from the submerged valley can be fitted into a broader context through modelling the development of inner-shelf sediment bodies in the eastern English Channel, by relating this to past changes on the present land surface (southeast England), with those towards the shelf break. An idealized long profile from the head of a drainage basin to the shelf break is shown in Fig. 11.

The model considerably simplifies the variability, fluctuations and graduation of Quaternary climatic and relative sea-level changes. Six phases of evolution are depicted.

(1) Temperate stage – cold stage transition

A glacio-eustatic sea-level fall leads to the gradual extension of rivers across the continental shelf. The climatic transition is characterized by a phase of gravel aggradation in the upper and middle reaches of rivers (Green & MacGregor 1980), with diachronous incision close to falling base-level. The influence of base-level rapidly declines upstream and rivers instead respond to the prevailing climate. The loss of river competence on the low-angle inner shelf results in the net aggradation of sands and gravels on river floodplains.

(2) Glacial maximum

A diachronous phase of fluvial incision close to the coast extends across the continental shelf with each large-scale fall of relative sea-level. At the glacial maximum, the coast is closest to the

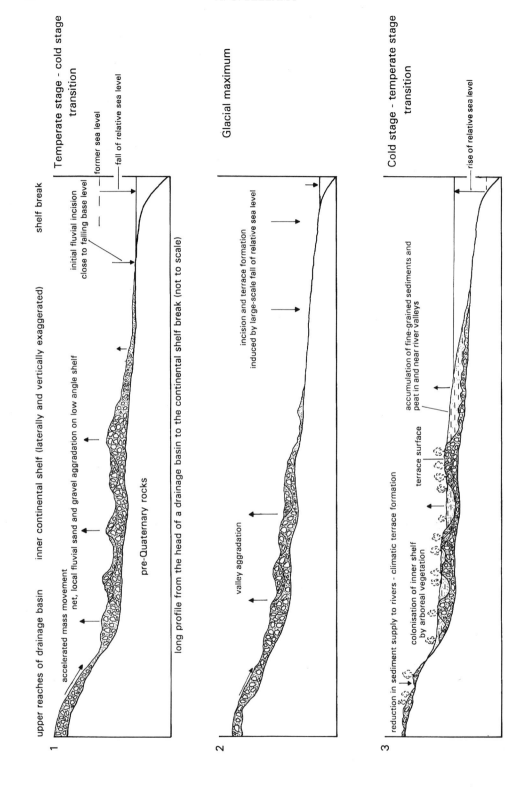

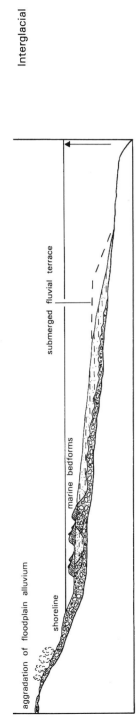
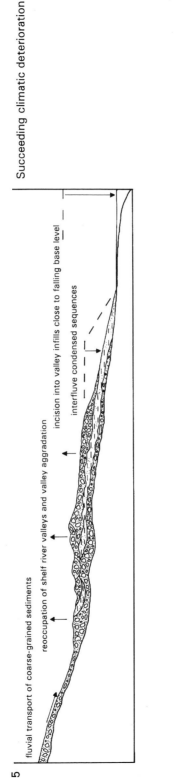
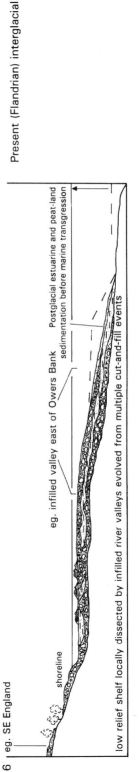

Fig. 11. Hypothetical stage model of infilled valley evolution in southern England and on the adjoining inner continental shelf. Contrasting sedimentary successions are suggested to form in response to repeated variations of climate and relative sea-level during the Quaternary.

shelf break. Fluvial incision is enhanced by episodic high-energy river discharges in a periglacial environment. Rivers are supplied primarily and seasonally by snow-patch and ground-ice melt which induces flooding, high rates of coarse-grained sediment transport and incision. Fluvial incision in a cold climate is suggested to be at least partly involved in the formation of the Northern Palaeovalley and the former River Arun valley east of Owers Bank. This cold, subaerial phase includes the reworking and decalcification of any earlier marine sediments, 'lag' gravel formation at the base of fluvial incisions, and coarse-grained sediment aggradation in the upper and middle reaches of rivers.

(3) Cold stage – temperate stage transition

Fluvial metamorphosis on the inner continental shelf occurs during the climatic transition with laterally and vertically unstable gravel-bed (possibly braided) rivers becoming relatively stable, single-channel rivers (Schumm 1977; Baker & Penteado-Orellana 1977, 1978). This change is due to denser vegetation colonization and succession on the shelf (as indicated by the peat sample from the infilled valley) and at the heads of drainage basins leading to slope stabilization, deceleration of mass movement and a reduction in coarse-grained sediment supply to rivers in southeast England. Penecontemporaneously, the aggradation of organic-rich, fine-grained sediments, including peat, occurs close to rising base-level in estuaries, from deposition on marshes and tidal flats.

Continuing fluvial incision occurs further upstream, due to the reduction in drainage basin sediment supply, and results in accelerated valley development and climatic terrace formation (Worssam 1973; Clayton 1977). This has possibly contributed to erosion of, for example, the Arun and Ouse Gaps, West Sussex, and the valley east of Owers Bank. Widespread marine submergence of the low-angle shelf subsequently occurs due to glacio-eustatic sea-level rise.

(4) Interglacial

Widespread shelf planation occurs during a marine transgression, with the erosion of landform highs and the development of ravinement surfaces by shoreface erosion (Swift 1968). Uppermost sediments are scattered into poorly sorted sea-bed veneers, about 0.2 m–2.5 m thick, and are locally concentrated to form diachronous, coarse-grained, retro-gradational shorelines away from estuaries. Subsequent local reworking of shelly sea-bed sediments into mobile bedforms occurs as water depth exceeds average wave-base and tides become dominant in current forcing.

(5) Succeeding climatic deterioration

This stage consists of a repetition of events in Stages 1 and 2: the reoccupation of continental-shelf river valleys occurs as relative sea-level falls. This includes incision into, and reworking of, marine sediments and pre-existing valley infills in fluvial environments. The dominant sedimentary process is cut-and-fill by laterally and vertically unstable channels within low-angle, coarse-grained, alluvial plains. River valleys are the depocentres on the inner continental shelf, with interfluves characterized by the development of condensed sequences.

(6) Holocene

Events described in Stages 3 and 4 are repeated during this stage. The present-day inner shelf is the product of repeated alternations of subaerial and marine conditions. Subaerial deposits, truncated during marine transgression, are unconformably overlain by the present-day marine sea-bed sediment veneer. Sea-bed sediments are reworked towards equilibrium with marine currents and sea-level. For example, gravelly sea-bed sediments are now immobile under present wave and tidal current regimes (Hamblin & Harrison 1989). As sea-level approaches the present level, river valleys along the coastline are infilled with thick sequences of fine-grained sediments in a continuum of temperate fluvial and estuarine environments, e.g. the valley of the River Ouse, West Sussex (Jones 1971).

Conclusions

Main study findings

Multiple cut-and-fill events have occurred in the evolution of the infilled valley east of Owers Bank. These events have included the action of gravel-bed rivers in a periglacial landscape, fine-grained sedimentation in estuarine environments, and peat accumulation in temperate climatic conditions. The valley has been submerged during at least one, and probably more, marine transgressions of likely regional significance. The other valleys off the south coast of England (Fig. 1) have possibly experienced similar environmental changes. Glacio-eustatic sea-level changes have probably repeatedly forced relative sea-levels in the eastern English

Channel to below −50 m, resulting in the linkage of Britain to the continent as a single landmass.

Relative importance of marine and subaerial environments in the region

Interpretation of the data from the study area indicates that the valley deposits were largely formed in subaerial environments and have been merely superficially modified by marine processes. It is therefore suggested that processes operative during marine (interglacial) stages have played a relatively minor part in the evolution of infilled valleys in the eastern English Channel.

Relative to the Quaternary as a whole, global ice volume is currently unusually low (Shackleton 1987), implying that eustatic sea-level is unusually high. Assuming the dominance of glacio-eustasy in determining large-scale relative sea-level fluctuations in the region, subaerial linkage of present-day southern England with the adjacent inner continental shelf (Fig. 11 (1), (2), (3) and (5)) has therefore persisted for most of the Quaternary. It is therefore suggested that Britain has been an island for only a relatively small part of the period. With the eastern English Channel subaerially exposed during cold stages, the British Isles and adjacent southern continental shelves have formed an extension of the northwest European landmass for most of Quaternary time.

I thank Professor Peter Worsley, Dr Ian Selby and Dr Peter Balson for advice in preparing the paper. P. E. F. Collins performed the pollen analysis in Table 1. The research was funded jointly by United Marine Dredging Limited and the Natural Environment Research Council through the Extractive Industries Partnership Scheme. The seismic and sample data are published courtesy of United Marine Dredging Limited. PRIS contribution number 409.

References

BAKER, V. R. & PENTEADO-ORELLANA, M. M. 1977. Adjustment to Quaternary climatic change by the Colorado River in central Texas. *Journal of Geology*, **85**, 395–422.
—— & —— 1978. Fluvial sedimentation conditioned by Quaternary climatic change in central Texas. *Journal of Sedimentary Petrology*, **48**, 433–451.
BRYANT, I. D. 1983. The utilisation of Arctic river analogue studies in the interpretation of periglacial river sediments from southern Britain. *In:* GREGORY, K. J. (ed.) *Background to Palaeohydrology*, Wiley, London, 413–431.
CLAYTON, K. M. 1977. River terraces. *In:* SHOTTON, F. W. (ed.) *British Quaternary Studies – Recent Advances*. Clarendon Press, Oxford 153–167.
DAWSON, M. 1987. Sedimentological aspects of periglacial terrace aggradations: a case study from the English Midlands. *In:* BOARDMAN, J. (ed.) *Periglacial Processes and Landforms in Britain and Ireland*. Cambridge University Press, Cambridge, 265–275.
DINGWALL, R. G. 1975. Sub-bottom infilled channels in an area of the eastern English Channel. *Philosophical Transactions of the Royal Society of London, A* **279**, 233–241.
DYER, K. R. 1975. The buried channels of the 'Solent River,' southern England. *Proceedings of the Geologists' Association*, **86**, 239–245.
GIBBARD, P. L. 1988. The history of the great northwest European rivers during the past three million years. *Philosophical Transactions of the Royal Society of London B*, **318**, 559–602.
GREEN, C. P. & MacGREGOR, D. F. M. 1980. Quaternary evolution of the River Thames. *In:* JONES, D. K. C. (ed.) *The Shaping of Southern England*. Institute of British Geographers, Special Publication, **11**, 177–202.
HAMBLIN, R. J. O. & HARRISON, D. J. 1989. *Marine Aggregate Survey Phase 2: South Coast*. British Geological Survey Marine Report, **88/31**.
HODSON, F. & WEST, I. M. 1972. Holocene deposits of Fawley, Hampshire and the development of Southampton Water. *Proceedings of the Geologists' Association*, **83**, 421–441.
JONES, D. K. C. 1971. The Vale of the Brooks. *In:* WILLIAMS, R. B. G. (ed.) *Guide to Sussex Excursions*. Institute of British Geographers Conference, Falmer, 43–46.
—— 1981 *South-east and Southern England*, Methuen, London.
LAWSON, M. J. & HAMBLIN, R. J. O. 1989. *Wight (50°N 02°W). Sea Bed Sediments and Quaternary Geology*. 1:250 000 Series Map, British Geological Survey.
MIALL, A. D. 1977. The braided river depositional environment. *Earth Science Reviews*, **13**, 1–62.
MITCHUM, R. M. & VAIL, P. R. 1977. Seismic stratigraphy and global changes of sea level, Part 7: seismic stratigraphic interpretation procedure. *In:* PAYTON, C. E. (ed.) *Seismic Stratigraphy – Applications to Hydrocarbon Exploration*. American Association of Petroleum Geologists, Memoir, **26**, 135–143.
——, —— & SANGREE, J. B. 1977. Seismic stratigraphy and global changes of sea level, Part 6: stratigraphic interpretation of seismic reflection patterns in depositional sequences. *In:* PAYTON, C. E. (ed.) *Seismic Stratigraphy – Applications to Hydrocarbon Exploration*. American Association of Petroleum Geologists, Memoir, **26**, 117–133.
REID, C. 1903. *The Geology of the Country around Chichester*. Memoir of the Geological Survey, UK.
SCHUMM, S. A. 1977. *The Fluvial System*. Wiley, Chichester.
SHACKLETON, N. J. 1987. Oxygen isotopes, ice volume

and sea level. *Quaternary Science Reviews*, **6**, 183–190.

SMITH, A. J. 1985. A catastrophic origin for the palaeovalley system of the eastern English Channel. *Marine Geology*, **64**, 65–75.

—— 1989. The English Channel – by geological design or catastrophic accident? *Proceedings of the Geologists' Association*, **100**, 325–333.

SWIFT, D. J. P. 1968. Coastal erosion and transgressive stratigraphy. *Journal of Geology*, **76**, 444–456.

WORSSAM, B. C. 1973. *A New Look at River Capture and at the Denudation History of the Weald*. Report of the Institute of Geological Sciences, **73/17**.

Raised beaches and sea-levels in the English Channel in the Middle and Late Pleistocene: problems of interpretation and implications for the isolation of the British Isles

D. H. KEEN

Division of Geography, Coventry University, Priory Street, Coventry CV1 5FB, UK

Abstract: This paper discusses the problems of the interpretation of the evidence for pre-Holocene high sea-level events and their age in the Strait of Dover and adjacent areas. This review suggests that there is evidence for the isolation of Britain from the continent in Oxygen Isotope Stages 7 and 5. The evidence for isolation in Stage 9 is less certain, and in Stage 11 is rather poor. In Stage 5, at least, isolation may have lasted for as long as 60 000 years.

The major control on the isolation of the British Isles is the level of the sea in the Strait of Dover and the southern North Sea. The present limiting water depth is c. 50 m in the Strait and c. 40 m in the area between Norfolk and the Rhine Delta. Below these levels, the British Isles would be connected to the European mainland allowing free passage for terrestrial organisms to pass to and from Britain to the rest of Europe. Sea-levels above −40 m would cause Britain to be severed from Europe and restrict direct biological links for terrestrial taxa. Theoretically, this situation has persisted since the initial breach in the Weald–Artois Chalk ridge occurred in the Middle Pleistocene, although the uncertainty of the pattern of bathymetry prior to the Holocene renders interpretations of water depth hazardous.

Further complications result from the lack of detail on the ages of the phases of separation. The number of sites in or near to the Strait which give unequivocal evidence for the age of any high sea-level event is very limited. Although the general number of high sea-level events of the Channel may be determined from sites remote from the Strait, the details of the interaction of sea-level and the depth of incision of the Chalk connection between Britain and the rest of the continent, which control the isolation of Britain, remain poorly known.

A number of factors related to the problem of the isolation of Britain will be examined in this paper, the age of the first cutting of the Chalk barrier to below modern sea-level; the sea-levels after this event and their identification and dating; and the length of time represented by each phase of isolation. In recent years it has become apparent, largely from evidence from deep sea cores (Shackleton & Opdyke 1973), that climatic fluctuations during the Pleistocene were more complex than can be accommodated by the sequence of named stages recorded from Britain (Mitchell et al. 1973). Because of uncertainties in identifying stages on land and the possibility that additional stages exist (Bowen et al. 1989; Jones & Keen 1993; Bridgland 1994; Sutcliffe 1995), oxygen isotope terminology from the deep sea will be adopted.

Age of the initial breach of the Chalk barrier

Evidence from Lower Pleistocene sediments from both shores of the North Sea, noted by Funnell (1961), Zagwijn (1974) and Norton (1977), suggests that the Channel and North Sea were separate bodies of water at least during some of this time. Norton (1977), in particular, cites the restricted pattern of marine mollusc faunas in the Lower Pleistocene of East Anglia, which he believed was due to the colonization of a North Sea 'bay' via a narrow strait between Norway and Scotland, and from the morphology of the bivalve species *Spisula solida* and *Spisula elliptica* which he considered developed in isolation and perhaps under conditions of low salinity in this North Sea 'bay'. Thus, the Weald–Artois Chalk remained intact, and mainland Britain was a large peninsula extension of western Europe. However, Meijer & Preece (1995) suggest that the marine molluscan faunas from the Dutch Maassluis Formation of the Middle Tiglian which is diverse and contain high proportions of southern taxa, are not indicative of a closed Strait during this time. Because of this uncertainty of its efficacy as a barrier to the free interchange of terrestrial taxa between

Fig. 1. Localities mentioned in the text. Key to initials:
Ab – Côte des Abers; Av – Aveley; Bem – Bembridge; Br – Brighton; Bx – Boxgrove; E – Earnley; H – Herzeele; Ip – Ipswich; Je – Jersey; P – Portland; Pen – Pennington; Pu – Purfleet; S – Sangatte; Sel – Selsey; Sh – Shoeburyness; St – Stone; Tb – Torbay; TV – Tourville; TancV – Tancarville; W – Woodston; Wi – Wissant; WT – West Thurrock.

Britain and the rest of the continent, or marine taxa between the Channel and the North Sea, the exact age of the severing of the Chalk ridge is a matter for dispute. Although the mechanism of separation is thought by most authors to be meltwater from an ice-sheet, or a proglacial lake ponded between an ice-sheet in the Southern Bight of the North Sea and the Chalk ridge (Roep et al. 1975; Gibbard 1988, 1995), the exact age of this breach is uncertain. Most authorities favour an Anglian age and thus a date within in Oxygen Isotope Stage 12 (Bowen et al. 1986; Gibbard 1988).

However, the probable persistence of a unified river system between the Rhine and the Thames in the Hoxnian (?Stage 11) that allowed the 'Rhenish' molluscan fauna (Kennard 1942) to immigrate into Britain late in the interglacial, suggests that if the Chalk barrier was broken in Stage 12, the depth of incision was not yet enough to isolate the British Isles. In contrast to this evidence from Kent, the deposits at Herzeele, NW France (Fig. 1), which Sommé et al. (1978) and Vanhoorne & Denys (1987) describe as exhibiting three separate marine levels attributed to the Cromerian and Holsteinian on the basis of pollen evidence, are thought to have been deposited when the Strait was open, because the morphological flat on which they rest can be traced into the northern end of the Strait at Wissant. Sommé et al. (1978) therefore conclude that by the Cromerian the general geomorphology of the area was similar to that of the present. T. Meijer (pers. comm., 1995), however, regards the Herzeele sediments as the product of a single interglacial, the Holsteinian, and discounts the possibility of more than one marine phase being represented. As further evidence he points to the presence of the freshwater gastropod *Theodoxus danubialis* (= *serratiliniformis*) at both Herzeele and Swanscombe, Kent, which suggests both a Holsteinian age and a continuing land connection between Britain and Flanders (Meijer & Preece 1995). Despite the varied evidence there is then no conclusive proof of the existence of the Dover Strait before Stage 12.

Post-breach sea-levels

Sites where details of sea-levels subsequent to the breach can be determined are rare for areas immediately adjacent to the Strait. Further away, on the coasts of the central Channel, the Lower Thames, East Anglia, Normandy, the Channel Islands and on the Flanders coastal plain (Fig. 1), there are raised beaches and estuarine deposits which give a record of sea-level history.

The development of a sea-level sequence is complicated by a number of factors. Firstly, the present height of raised beaches may not be a guide to the former height of the sea. Previous assumptions of a stable land affected only by the eustatic rise and fall of sea-level (e.g. Keen 1978) are now thought to be highly simplistic, with a number of complicating factors which may control the actual height of Pleistocene marine deposits in relation to former sea-levels. The displacement of marine deposits may be due to a number of effects, none of which is easily quantifiable and which may operate at different rates and rhythms.

Long-term tectonic effects have been known to affect the North Sea basin at least since Mesozoic times, and have caused progressive downwarping of the Dutch delta and the southern basin of the North Sea during the Pleistocene (Cameron *et al.* 1992). The effect of this axis of depression would have been to have lowered the present altitude of marine deposits below their height of deposition and thus below contemporary sea-level, as has occurred with the estuarine deposits of the Lower Thames (Bridgland 1994).

Other less easily recognizable effects such as hydro-isostasy and glacio-isostatic forebulge effects may also have a role to play in raising marine deposits above modern sea-level. Hydro-isostatic effects resulting from the transgression and regression of the sea over the shallow shelf of the Channel during the climatic cycles of the Pleistocene caused depression of the crust during loading in temperate stages, and uplift when the loading was removed by retreat of the sea during cold stages. This may have allowed the subsequent uplift of beaches deposited under conditions of crustal loading in interglacials.

General isostatic uplift of the land due to erosional unloading (Bridgland 1994) may also allow the progressive uplift of the land through time. The development of 'stair-cases' of marine terraces in areas like the Sussex coastal plain and the eastern Solent (Preece *et al.* 1990), where the highest terrace is invariably the oldest and the deposits become progressively younger as altitude declines, is seemingly, evidence that such effects have occurred.

Although most of the area from the Strait to the western part of the Channel has remained unglaciated and so free of direct glacio-isostatic rebound, the effect of a forebulge from ice limits of the Anglian (Stage 12) glaciation would extend well out into the English Channel, perhaps up to 400 km from the ice front (see Walcott 1972; Hails 1983). Such a forebulge would uplift the land during cold stages, then progressively lower land levels as isostatic equilibrium began to return. It is possible that raised beaches deposited on coasts affected by a forebulge would be downwarped in relation to their original height as the forebulge collapsed.

The detailed evaluation of the effects of isostatic deformation of raised beaches, such as has been undertaken for Scotland (Gray 1983), is impossible in the Channel area in the Middle Pleistocene given the uncertainties of dating of the raised beach remnants (see below), and the fragmentary nature of the marine deposits compared to the continuous shorelines of other areas (Hails 1983). Although not easily quantifiable, these effects cannot be discounted when considering former sea-levels.

A final control on the height of former marine deposits is their sedimentology allied to exposure and tidal regime. In areas of high tidal range (macro-tidal coasts) such as the Channel Islands shelf, modern littoral deposits extend up to 7 m above mean sea-level (Jones *et al.* 1990). It must be assumed that former tidal regimes would have had the same height relationship to mean sea-level for raised beaches as that pertaining to modern beaches, but this assumption is difficult to test, except by palaeotidal modelling (Scourse & Austin 1995).

On those coasts of the present Channel that are exposed to the maximum fetch from the Atlantic, waves may build marine deposits well above Ordnance Datum (OD). The classic example, Chesil Beach, Dorset, has a maximum crest height of about 14 m OD (Carr & Blackley 1973), and although such constructional features are not well known in the Pleistocene record, lesser elevations of raised beach sediments above

OD may be caused by this effect. As with the tidal effects, former wave regimes are difficult to determine and changes in direction and pattern of wave approach may have considerable effect on raised beach elevation. The example of the Bembridge Raised Beach, Isle of Wight, described by Preece et al. (1990) illustrates this point. The deposits at this site occupy what is now a sheltered locality on the eastern end of the island, but one which was open to far greater wave energy in the Pleistocene, as the deposits form a cuspate foreland which would have been unlikely to have developed under modern conditions of wave action.

Thus the determination of the heights of former sea-levels from raised beach remnants is very difficult. In the current state of knowledge probably all that can be firmly stated is that temperate stage sea-levels were generally at or slightly above OD, but that greater elevations than this (see below) owe their height to a combination of tectonic and isostatic effects which are difficult to either quantify or disentangle.

The second complicating factor is the problem of dating raised beaches. Although a number of dating methods are applicable to marine deposits and their included faunas, virtually all have short-comings. A review of the problems of dating techniques used in the Channel region may be found in Smart & Frances (1991), so only short comments will be included here on the quality and accuracy of the various dating methods.

Perhaps most widely applied as a dating technique in raised beach contexts is amino acid racemization. This has been extensively used for relative dating and, given certain assumptions, allows reasonable correlation of beaches containing shells. The technique has proved particularly useful in distinguishing between beaches formed during Stages 5e and 7, where shells are often abundant and analytical precision is high (Keen et al. 1981; Bowen et al. 1985; Davies & Keen 1985; Miller & Mangerud 1985; Bowen & Sykes 1988). In deposits older than these, amino acid ratios appear to give less clear-cut results and in sites thought to be of Stage 10 age or older there is considerable disagreement between the amino acid chronologies and those obtained by other methods (cf. Bowen & Sykes 1994; Roberts 1994). Problems of comparability can also occur between early pretreatment and analytical methods, and those adopted post-1985 (Bowen et al. 1985).

Radiometric age determinations on raised beaches in the Channel area have largely been attempted by U-series methods. The technique has been applied to molluscan shell (Sarnthein et al. 1986; Barabbas et al. 1988) and cave stalagmite (Keen et al. 1981; Proctor & Smart 1991). Of the two materials, shell appears to give the more problematic results because of the potential uptake of daughter material from the environment, while dates on stalagmite, which in general appear to operate as a closed system, seem more reliable but may be more difficult to relate to the actual raised beach. As with amino acid data, most dates obtained are of Stage 5 or 7 age, but reliability and also the number of datable sites declines before Stage 8.

Molluscan shell has also been used as the material for ESR dates (Barabbas et al. 1988; Schwarcz & Grün 1988), but considerable controversy exists over the validity of ages obtained by this method as they seem greatly at variance with ages determined by thermoluminescence or uranium series (see below).

Thermoluminescence (TL) dating has been applied to the French sites of Herzeele and Sangatte (Balescu & Lamothe 1991, 1993; Balescu et al. 1991, 1992) but there is considerable disagreement with the dates obtained by this method and those determined by others such as amino acid racemization or uranium series (see below). As with a number of other dating methods, the most reliable determinations derived from TL appear to centre around Stages 5 and 7.

Other age estimates have been obtained from biostratigraphy. The principal technique has been pollen biostratigraphy, which has also been one of the main pillars of the general Pleistocene stratigraphy of Britain (Mitchell et al. 1973; Jones & Keen 1993). The application of pollen analysis to marine deposits is difficult. Raised beaches are generally composed of sediments too coarse to preserve pollen, and even fine-grained marine deposits are usually poor in pollen of land plants. The pollen in estuarine deposits, despite being more abundant than in fully marine sediments, is often difficult to interpret for taphonomic reasons (Preece & Scourse 1987).

Perhaps a more serious objection to the use of pollen biostratigraphy is the fact that the vegetational histories of different stages can be very similar. The conventional stratigraphy (Mitchell et al. 1973) recognizes only two fully temperate stages between the Anglian and the present – the Hoxnian and Ipswichian. By contrast, the Oxygen Isotope record from the deep oceans shows four temperate phases during this time span, Stages 11, 9, 7 and 5e (Shackleton & Opdyke 1973). Clearly this more complete and complicated oceanic succession cannot easily be reconciled with the terrestrial record based largely on pollen stratigraphy. Increasingly, the more complicated stratigraphy derived from the oceans seems to fit the observed facts better than the more restricted pollen-based scheme (Jones & Keen 1993; Bridgland 1994; Sutcliffe 1995), but by doing so it complicates the interpretation of the raised beach record.

Evaluation of sea-levels using the methodologies noted above is best attempted in temporal sequence. Evidence for pre-Stage 12 sea-levels is sparse. Problems of dating are most acute during this time and there are few sites for comparison. Three sites, Boxgrove, Sussex (Roberts 1986; Roberts et al. 1994); the Steyne Wood Clay, Bembridge, Isle of Wight (Holyoak & Preece 1983; Preece et al. 1990) and Herzeele, NW France (Sommé et al. 1978) are sites that may provide evidence for early sea-level records.

The Boxgrove sequence consists of raised beach gravels resting on a bevelled surface of the chalk at c. 40 m OD, above which are estuarine and intertidal sands (the Slindon Sand). Terrestrial deposits yielding interglacial land snails, vertebrates and Acheulian artefacts (Roberts et al. 1994) immediately overlie the Slindon Sand. These, in turn, are buried by periglacial deposits. The small mammals (principally *Arvicola cantiana, Sorex savini* and *Pliomys episcopalis*) suggest a pre-Anglian age, possibly Stage 13 (Roberts 1986; Roberts et al. 1994). However, amino acid ratios from shells of the marine gastropods *Nucella lapillus* and *Littorina saxatilis* taken from the underlying Slindon Sand provide mean ratios of 0.29 ± 0.025 suggesting a Stage 11 age (Bowen & Sykes 1994).

The Steyne Wood Clay also lies at c. 40 m OD and consists of estuarine muds with pollen, Mollusca and other microfossils. The fossil content of the clay is undiagnostic as to age, but amino acid ratios from shells of the bivalve *Macoma balthica* gave mean values of 0.32 ± 0.04, which by comparison with the European sites reviewed by Miller & Mangerud (1985) suggest an age older than the Holsteinian (?Stage 11), but younger than the type Cromerian. Preece et al. (1990) suggest that the similar altitude to Boxgrove suggests that the two sites are probably contemporary, despite differences in the amino acid ratios.

The sea-level history in the Solent is further complicated by the site at Earnley, Sussex. This site lies at an altitude of c. 1 m OD, but has yielded fossil evidence of intertidal conditions and pollen of Middle Pleistocene character (West et al. 1984). The exact age of this site and its relationship to Boxgrove and the Steyne Wood Clay is as yet difficult to determine, although it would seem to be younger (Preece et al. 1990).

The major early Middle Pleistocene site on the northern French coast that occurs at Herzeele (Sommé et al. 1978; Vanhoorne & Denys 1987), has its base at 8 m NGF (French Mean Sea-level). Sommé et al. (1978) recognized three marine or tidal-flat deposits separated by peat. The oldest of the marine phases (Series I) is believed to be of Cromerian (Stage 13 or older) age while Series II and III, the upper two levels, are held to be of Holsteinian (?Stage 11) age on the basis of their contained pollen.

Later work by A. W. Burger & T. Meijer (pers. comm., 1995) suggests that these deposits represent only one temperate stage, the Holsteinian (?Stage 11), and that the complicated sequence of tidal flat sediments can be accommodated within one temperate phase. Burger & Meijer suggest that the three marine horizons of Sommé et al. (1978) can be explained as being due to reworking in channels in the dynamic conditions of a tidal or estuarine environment.

Attempts to date the Herzeele deposits by radiometric means have used a number of methods which have yielded a wide spread of ages. Sarnthein et al. (1986) and Barabbas et al. (1988) used the uranium series method on shells from the clays and obtained dates between 300 and 350 ka BP. Schwarcz & Grün (1988) obtained an age of 260 ka BP by electron spin resonance also on shells. Balescu & Lamothe (1991, 1993), using thermoluminescence, obtained ages of 228 ± 30 (corrected to 271 ± 36) ka BP for feldspar sand grains from Series III, and 239 ± 38 corrected to 289 ± 34) ka for Series I. Amino acid determinations on shell from Series III by Miller & Mangerud (1985), indicated a 'Holsteinian' age (which they tentatively equated with Stage 7) for the Herzeele deposits when compared to ratios obtained from other marine Holsteinian sites in Germany and Denmark. Amino acid ratios from shells from Series III, by contrast, were indicative of a Stage 9 age (Bowen & Sykes 1988). Of these dates, 350 ka BP is in the middle of Oxygen Isotope Stage 10, 300 ka is in the middle of Stage 9, and dates between 260 and 248 ka fall within Stage 8. The spread of these dates makes interpretation and confident age attribution difficult.

Deposits of Stage 9 age are also rare. If the possible Stage 9 attribution of Herzeele indicated by some dating methods (see above) is discounted, marine deposits possibly deposited during Stage 9 are known only from the Lower Thames, in the enlarged Wash embayment, and perhaps at Sangatte, France.

The Lower Thames sites reviewed by Bridgland (1994) show evidence of marine or estuarine sedimentation only a little above modern sea-level during Stage 9. At Purfleet (Fig. 1) some of the interglacial deposits accumulated under brackish conditions, as indicated by the presence of the mollusc *Hydrobia ventrosa* and the ostracod *Cyprideis torosa*. These deposits reach a maximum height of 15 m OD, and are regarded as being of Stage 9 age on the basis of their position in the Lower Thames terrace sequence (see Gibbard (1994) for alternative interpretation). Correlative Stage 9 deposits at Shoeburyness, near the current mouth of the Thames in Essex, occur below present sea-level, so placing little confidence in estimations of sea-level from this terrace information. Shoeburyness is sufficiently far east to have been affected by downwarping in the southern North Sea and therefore at a lower elevation than the level at which it was deposited (Bridgland 1994).

In the Wash embayment, Horton *et al.* (1992) describe an estuarine sequence which reaches heights of between 11 and 14 m OD and which lies below the Third Terrace of the River Nene. Amino acid ratios from the freshwater gastropod *Bithynia tentaculata* and the land gastropods *Cepaea* and *Trichia hispida* gave ratios of 0.244 ± 0.030 and 0.249 ± 0.028, 0.244 ± 0.005 and 0.253 ± 0.014, 0.233 ± 0.025 and 0.239 ± 0.022, respectively. These values suggest a Stage 9 age for these deposits and provide a similar indication of age and sea-level to that from the Lower Thames.

The Sangatte Raised Beach is the closest site to the Strait to provide potential evidence of former sea-levels. Despite this geographical proximity, the age of this beach is also uncertain. The sequence consists of beach sand between 8–10 m NGF overlain by head and loess with palaeosols developed in these terrestrial deposits (Balescu & Haesaerts 1984; Antoine 1989). Recent attempts to date the beach have used considerations of regional stratigraphy (Balescu & Haesaerts 1984), the development of palaeosols linked to general palaeosol stratigraphies for NW France (Antoine 1989), and thermoluminescence (Balescu *et al.* 1991, 1992). Balescu & Haesaerts (1984) suggested an age in Stage 7 for the marine levels because they are overlain by two complete periglacial cycles which they attribute to two cold stages (?2/4 and 6). TL dates obtained by Balescu *et al.* (1991, 1992) gave ages between 229 and 201 ka BP (with a mean of 210 ka BP), firmly placing the deposit in Stage 7 and supporting Balescu & Haesaerts' earlier suggestion based on stratigraphic considerations. However, Antoine (1989) preferred a Stage 9 age for the Sangatte marine level from a recognition of two palaeosols of interglacial type, probably of Stage 7 and 5 age, occurring above the marine horizon. The sparse evidence for Stage 9 sites makes clear comment on sea-levels of this time difficult.

Evidence for Stage 7 sea-levels can be obtained from the Lower Thames and from around the south coast of England, where both raised beaches and estuarine sites in the lower reaches of the Solent River provide data. At Aveley and West Thurrock, Essex, (Fig. 1), sites attributed by Bridgland (1994) to Stage 7 on the basis of the terrace stratigraphy, occur at 10 m OD, and show evidence of estuarine conditions.

Other evidence for the height of the sea in Stage 7 cannot be found close to the Strait on the British side, and although Sangatte on the French side is close to the Strait, its age is equivocal (see above). However, possible Stage 7 sites in the central and western Channel at Brighton and Selsey in Sussex, Stone and Pennington in Hampshire, Portland in Dorset and Torbay in Devon have marine and estuarine sediments up to 14 m OD.

At Brighton, the well known Black Rock Raised Beach has a base height at *c*. 8.5 m OD and consists of cobbles and pebbles of flint and Chalk which extend up to 11.9 m OD. Previously regarded as Ipswichian (Sub-stage 5e) in age (Mottershead 1977), amino acid ratios on molluscan shell from the beach (Davies 1984) gave ratios indicative of a Stage 7 age. The beach sediments are overlain by 20–25 m of coombe rock (chalky solifluction) which can be divided into two units by the radically different depositional dips (10° for the upper unit and 20° for the lower one). At the contact between the two heads a reddened horizon occurs which may be a palaeosol (D. H. Keen, unpublished data). The sequence is therefore strikingly similar to that of Sangatte and may be contemporaneous with a marine unit of Stage 7 age succeeded by terrestrial deposition in Stages 6 and 4/2 and with Stage 5 represented by the palaeosol.

The fossiliferous deposits at Selsey are freshwater muds which lie at or just below OD, although evidence of brackish conditions can first be seen at -1.76 m OD. These muds are overlain by 2–3 m of raised beach gravel, devoid of fossils and reaching a height of 7 m OD (West & Sparks 1960). The initial correlation of the Selsey sequence with the Ipswichian was on the basis of pollen biostratigraphy (West & Sparks 1960). This correlation becomes problematic when sites within the estuary of the former Solent River in Hampshire are considered. Two sites, Stone to the east of Lymington (West &

Sparks 1960; Brown et al. 1975), and Pennington, to the west (Allen et al. 1995) have provided fossil data. At Stone, estuarine muds, extending to at least 2 m OD, have yielded pollen and Mollusca indicative of a brackish environment. The pollen record was consistent with an Ipswichian age. However, brickearth overlying terrace gravel which caps the brackish muds, was divided into two by a palaeosol of interglacial type which Reynolds (1987) believed must have formed during the Ipswichian. If correct, the underlying estuarine muds must be of Stage 7 age or older. At Pennington, freshwater sediments with pollen and Mollusca suggestive of an Ipswichian (Sub-stage 5e) age occur between −3.9 and −5.3 m OD, below a lower terrace of the Solent River (Allen et al. 1995). Although the Stone and Pennington deposits have yielded pollen of different sub-stages, their difference in height and relative position within the terrace sequence of the Solent River System suggests attribution to different temperate stages. The suggested Ipswichian age of the Selsey site may also require reappraisal in view of faunal and amino acid evidence (Allen et al. 1995; Sutcliffe 1995).

In Dorset and Devon, evidence of former sea-levels comes from raised beaches on open coasts. The best preserved beaches are those with stalagmite cement and a high molluscan content, which make them suitable for dating by various methods; these occur on the Mesozoic and Palaeozoic limestones of that part of the Channel coast. At Portland, a complex of deposits, both terrestrial and marine, occurs across the tip of Portland Bill (Davies & Keen 1985; Keen 1985). The deposits on the west side of Portland Bill consist of 2–3 m of heavily cemented, well-rounded beach gravel which extends from 14 to 17 m OD. The beach is overlain by 1.7 m of 'loam', which is decalcified and of uncertain origin, and up to 4.6 m of limestone head. Amino acid ratios obtained from shells of the gastropods *Nucella lapillus* and *Littorina littorea* gave mean ratios of 0.18 and 0.183 ± 0.004 which suggest a Stage 7 age for the marine deposits. Work currently in progress (B. van Vliet-Lanoë and D.H. Keen) suggests that a palaeosol of interglacial type occurs between the 'loam' unit and the head suggesting that the loam may have formed during Stage 7 or 6, the palaeosol during Stage 5 and the head after Stage 5.

At Torbay, a raised beach between 9.1 and 12.1 m OD on Hopes' Nose, north of Torquay, consists of gravel and sand rich in marine Mollusca. This is overlain by blown sand and head with conspicuous palaeosol horizons (Mottershead et al. 1987). Amino acid ratios from the marine gastropods *Patella vulgata, Littorina littorea, L. obtusata* and *Nucella lapillus* provided mean values from 0.195 to 0.24 which were thought to be indicative of a Stage 7 age.

Across Torbay at Berry Head, a series of sea caves with stalagmite floors seal marine sediments and allow dating of the sea-levels which deposited them. Stalagmite and associated marine deposits at a base height of 7.2 m OD were dated using the uranium series method and dates of 210 + 34/−26 ka BP and 226 + 53/−76 ka BP obtained (Proctor & Smart 1991). These dates are indicative of a Stage 7 age.

Sediments dating from the one major pre-Holocene temperate stage after Stage 7, Stage 5e, are also poorly represented near the Strait. Former ideas that the Sangatte Raised Beach was of this age (see discussion in Balescu & Haesaerts 1984 and Antoine 1989) have now been revised and a Stage 7 or 9 age preferred (see above). Direct evidence of sea-level history during Sub-stage 5e is, however, obtainable from the Thames basin, southern East Anglia, the Solent, Dorset, Devon and the Channel Islands and Normandy.

In the Thames terrace sequence, Bridgland (1994) suggested that Sub-stage 5e deposits in the Kempton Park/East Tilbury Marshes Gravel lie at or just below modern sea-level. The Ipswichian (Stage 5e) stratotype at the Belstead Brook, Ipswich also occurs at this level (West 1957) but is not marine. The Bembridge Raised Beach (Fig.1), dated by Preece et al. (1990) to the Ipswichian, lies between 5 and 18 m OD, although the anamolous heights result from the fact that it forms part of a cuspate shingle foreland. Other sites in the Solent River estuary formerly thought to be of Ipswichian age on the basis of their pollen record – Selsey and Stone – are now believed to be of probable Stage 7 age (see above). At Pennington, also described above, sediment thought to be of Ipswichian age shows no evidence of marine conditions between −3.9 m and −5.3 m OD.

At Portland and Torbay, open coastline raised beaches and cave fills occur, similar to those described above for these localities in Stage 7. At Portland, cobbles, pebbles and sand of the Portland East Raised Beach occur on the east side of the peninsula between 6.95 and 10.75 m OD. Specimens of *Patella vulgata, Littorina littorea, L. obtusata* and *Nucella lapillus* have yielded mean amino acid ratios of 0.134 ± 0.006, 0.124 ± 0.01, 0.123 ± 0.01 and 0.13 ± 0.01 respectively. These ratios are regarded as indicative of an age in Sub-stage 5e (Davies & Keen 1985).

At Thatcher Rock, Torquay, Mottershead *et al.* (1987) describe marine sands and gravels between 7.8 and 10.1 m OD. Shells of *Patella vulgata* from the deposits yielded amino acid ratios of 0.12 held to be indicative of a Stage 5e age. At Berry Head, stalagmite at levels *c.* 3 m OD within Corbridge Cave gave dates between 145 + 10/−9 and 116 ± 9 ka BP, and were thought to have formed during Stage 5 (Proctor & Smart 1991). The occurrence of temperate marine Mollusca, Foraminfera and Ostracoda in the beaches at Torquay and Portland East led Proctor & Smart (1991) to suggest accumulation during the warmest part of Stage 5, Sub-stage 5e.

On the basis of amino acid ratios from marine Mollusca, obtained by modified analytical methods, Bowen *et al.* (1985), suggested two alternative ages for these beaches – either that they were of Sub-stage 5e age as suggested by Davies & Keen (1985) and Mottershead *et al.* (1987), or that they were deposited during a high sea-level episode in a later sub-stage of Stage 5, either Sub-stage 5c or 5a. For the later age to be possible, sea-temperatures and sea-levels would have to have been high during two phases in Stage 5, a possibility not readily accommodated by the Oceanic Oxygen Isotope curves. The uranium series dates of Proctor & Smart (1991) from Torbay also tend to confirm a Sub-stage 5e date for the raised beach at Thatcher Rock and thus also for Portland East, because the uranium series ages obtained fit best with a 5e age and not with ages as young as those suggested for Sub-stages 5c or 5a, which would be expected to be younger than 100 ka BP.

In the Lower Seine at Tancarville and Tourville (Lautridou 1982) estuarine sediments within the low terrace of the river range from 0 m up to 10 m NGF. The dating of these marine levels is dependent on considerations of the regional stratigraphy of overlying head and loess sequences and suggests an age within Stage 5.

In Jersey, the most prominent raised beach (the 8 m beach of Keen (1978)) extends from *c.* 3–10 m above mean sea-level (Keen 1993). As in Normandy, it is generally only datable by reference to the overlying stratigraphy. At Belcroute and Portelet in the SW of the island overlying head and loess successions suggest that beaches of Stage 7 and 5 ages approached the same level and were within the 3–10 m level envelope (Keen *et al.* 1993). This occurrence of more than one sea-level close to the present level and close to each other, complicates the interpretation of sea-levels for Stages 7 and 5.

On the north coast of Jersey, the calcite cement of the raised beach which reaches 7 m above mean sea-level at the Belle Hougue Cave, has been dated by uranium series to 121 + 14/−12 ka BP, suggesting a Sub-stage 5e age (Keen *et al.* 1981). Amino acid ratios from the shells of the mollusc *Patella vulgata* are also compatible with beaches dated to Sub-stage 5e on the north coasts of the Channel, although Hollin *et al.* (1993) suggest that the different thermal environment within the cave may render comparison of ratios obtained from open sites invalid. Despite these objections, the occurrence at Belle Hougue of the gastropod *Astralium rugosum*, which has its current northern limit at the Ile de Ré, north of Bordeaux, suggests a warmer sea temperature than the present, which would fit best with Sub-stage 5e.

Effect of sea-levels on isolation

The effect of the varying sea-levels noted above on the isolation of the British Isles from the rest of the continent is difficult to evaluate. It might be expected that after the breach of the Chalk barrier linking Kent and Artois, each temperate stage would produce sea-levels high enough to isolate Britain, with re-connection only taking place as the sea-level fell at the onset of glacial stages.

The situation is likely to have been more complicated especially in the light of evidence for a continuing land connection during at least part of the immediate post-Anglian temperate stage (Meijer & Preece 1995). The uncertainty of the dating of high sea-level events discussed above also complicates interpretation. Although the shortest distance from Britain to the continent is across the Strait, the shallowest water depths currently are between Norfolk and the Netherlands. During the Holocene sea-level rise the Strait was flooded while a land connection remained to the north (de Jong 1967; Jelgersma 1979). This shallow water area is thought to have occurred due to the existence of glacial deposits laid down in Late Pleistocene times (Bridgland & D'Olier 1995). A similar situation in earlier temperate phases may have allowed a continuous connection even as the deep channel of the Strait was flooded. In Stages 7 and 5 such a link was probably only temporary, as the sea-level rose from glacial low levels, and complete isolation occurred only in the warmest part of the interglacials as sea-level rose to its present height or above.

The former level of the sea at any one stage is very difficult to determine. It is highly likely that sea-levels reached or perhaps even slightly exceeded that of the present during each of the temperate stages 11, 9, 7 and 5, which probably post-date the breach in the Weald–Artois Chalk ridge. But the exact level of the sea in any of these stages cannot yet be determined. Sites like Boxgrove or Steyne Wood, perhaps older than Stage 12, owe their present elevation largely to tectonic or isostatic effects, and not to purely eustatic components. The same is probably equally true of the Stage 9 sites and Stage 7 sites, such as Portland West or Hopes' Nose, Torquay. Even the Sub-stage 5e sites at Portland or in Jersey, despite their elevation just above modern sea-level, may owe their present position to a degree of tectonic elevation. The occurrence of estuarine sites in eastern England or the Thames dated to Sub-stage 5e below modern mean sea-level may be explained by tectonic depression, but this explanation is more difficult to invoke for sites such as Pennington, where the evidence suggests that other sites nearby in the Solent of similar age have been affected by uplift.

The length of time that the British Isles was isolated is, like much of the detail surrounding this topic, uncertain. If Stage 5, the best known of the temperate stages of the Middle and Upper Pleistocene, is representative, maximum sea-level heights were attained in the warmest part of Sub-stage 5e, (cf. Keen *et al.* (1981)). The rest of Stage 5 is marked by fluctuating sea-levels. Sub-stages 5d and 5b were characterized by cold climate, and almost certainly had low sea-levels. How far the sea retreated during these phases and to what level it returned during the intervening temperate Sub-stages 5c and 5a, is a matter for conjecture. There is no evidence in the Strait, but to the south and west on the Channel Islands shelf, and along the coast of Brittany, evidence is available which gives some idea of the pattern of sea-level fluctuation in the later parts of Stage 5.

In Jersey, Lister (1989, 1993, 1995), considers that the sea fell to at least 15 m below current mean sea-level immediately after the maximum of Sub-stage 5e. This water depth at present would allow the re-connection of Jersey to the Cotentin. At Portelet and Belcroute, blown sand from beach sources was deposited in two separate phases on top of the Sub-stage 5e raised beach and before the deposition of the great thicknesses of periglacial head of early Weichselian/Devensian (Stages 4 and 2) times (Keen *et al.* 1993). Between these two episodes of aeolian transport was at least one phase of pedogenesis which occurred as sand-blowing lessened, probably as a result of the sea rising to near the present level. At Côte des Abers (Finistère) Hallégouët & van Vliet-Lanoë (1986) recognized a similar pattern of sea-level movement in the temperate sub-stages of the latter parts of Stage 5, with the sea returning to levels close to those of the present.

In the Strait, a fall in sea-level of 15 m, the minimum required to connect Jersey with the mainland, would still leave a considerable depth of water separating Britain from the rest of the continent if submarine contours were similar to the present in Stage 5. The current limiting depth in the Strait of −40 m would probably have been exceeded by the sea in the warm Sub-stages of 5c and 5a at least. As, in the western Channel, sand blowing continued through Sub-stages 5d and 5b, it is probable that a source of sand close to modern low water mark was being blown up into dunes, perhaps suggesting that sea-levels fell no lower than −25 m OD even in cold phases. If this is the case, the limiting water depths in the Strait of Dover and southern North Sea of −40 m would mean that Britain was isolated for most, or all of Stage 5, perhaps for the whole period between 130 and 70 ka BP, the length of this stage. Similar patterns of isolation might theoretically be expected in earlier temperate phases.

Conclusions

Sea-levels in the Strait of Dover are not certain for much of the Middle and Upper Pleistocene, but evidence from marine sequences elsewhere along the shores of the English Channel, and from the bathymetry of the Strait and the North Sea floor, suggest that the isolation of Britain from the Continent occurred certainly in Stages 5 and 7 and perhaps also in Stage 9. Isolation in Stage 11 is still uncertain, although if the cutting of the Strait was achieved by Anglian meltwater, any link with the rest of Europe was perhaps to the north of the Strait in the area of the combined Thames/Rhine delta. Because of the multiple uncertainties governing the height of the sea, altitudinal values for sea-level cannot be accurately determined for any of the temperate stages noted above. However, it is probable that sea-level never exceeded levels more than 5–10 m OD during the Middle and Upper Pleistocene (Shackleton 1987).

The duration of isolation in temperate stages may have been as long as the 60 ka time-span of the whole of Stage 5 on the evidence of the sea-levels determined in the western Channel.

These conclusions must be regarded as tentative. The possibility that neotectonics and/or isostatic effects causing a variation of land levels, and the possibility of sedimentation causing sea-bed level changes could greatly change the limiting water depth which controls isolation.

The problems of accurately dating the various sea-level events for which there is a record, further compound the complications. More conclusive evidence of isolation or connection must be sought through the various lines of fossil evidence which are covered in other papers in this volume.

I am grateful to Mr T. Meijer (Rijks Geologische Dienst, Haarlem) for providing unpublished data on the site of Herzeele and for comments on an early draft of this paper. Thanks are also due to Drs D. R. Bridgland, R. C. Preece, H. M. Roe and J. D. Scourse for discussion of the contents of this paper and for comments on an early draft of it.

References

ALLEN, L. G., GIBBARD, P. L., PETTIT, M. E., PREECE, R. C. & ROBINSON, J. E. 1995. Late Pleistocene interglacial deposits at Pennington Marshes, Hampshire, southern England. *Proceedings of the Geologists' Association*, **106**, in press.

ANTOINE, P. 1989. Stratigraphie des formations pléistocènes de Sangatte (Pas-de-Calais), d'après les premières travaux du tunnel sous la Manche. *Bulletin de l'Association Française pour l'étude du Quaternaire*, 1989–1, 5–18.

BALESCU, S. & HAESAERTS, P. 1984. The Sangatte raised beach and the age of the opening of the Straits of Dover. *Geologie en Mijnbouw*, **63**, 355–362.

—— & LAMOTHE, M. 1991. The blue emission of k-feldspar coarse grains and its potential for overcoming TL age underestimation. *Quaternary Science Reviews*, **11**, 45–51.

—— & —— 1993. Thermoluminescence dating of the Holsteinian marine formation of Herzeele (northern France). *Journal of Quaternary Science*, **8**, 117–124.

——, PACKMAN, S. C. & WINTLE, A. G. 1991. Chronological separation of interglacial raised beaches from northwestern Europe using thermoluminescence. *Quaternary Research*, **35**, 91–102.

——, ——, —— & GRÜN, R. 1992. Thermoluminesence dating of the Middle Pleistocene Raised Beach of Sangatte (Northern France). *Quaternary Research*, **37**, 390–396.

BARABBAS, M., MANGINI, A., SARNTHEIN, M. & STREMME, H. E. 1988. The age of the Holstein Interglaciation: a reply. *Quaternary Research*, **29**, 80–84.

BOWEN, D. Q. & SYKES, G. A. 1988. Correlation of marine events and glaciations on the northeast Atlantic margin. *Philosophical Transactions of the Royal Society of London*, **B318**, 619–635.

—— & —— 1994. How old is Boxgrove man? *Nature*, **371**, 751.

——, HUGHES, S., SYKES, G. A. & MILLER, G. H. 1989. Land–sea correlations in the Pleistocene based on isoleucine epimerization in non-marine molluscs. *Nature*, **340**, 49–51.

——, ROSE, J., MCCABE, A. M. & SUTHERLAND, D. G. 1986. Correlation of Quaternary glaciations in England, Ireland, Scotland and Wales. *Quaternary Science Reviews*, **5**, 299–340.

——, SYKES, G. A., REEVES, A., MILLER, G. H., ANDREWS, J. T., BREW, J. S. & HARE, P. E. 1985. Amino acid geochronology of raised beaches in south west Britain. *Quaternary Science Reviews*, **5**, 299–340.

BRIDGLAND, D. R. 1994. *Quaternary of the Thames*. London, Joint Nature Conservation Committee and Chapman & Hall.

—— & D'OLIER, B. 1995. The Pleistocene evolution of the Thames and Rhine drainage systems in the southern North Sea. *This volume*.

BROWN, R. C., GILBERTSON, D. D., GREEN, C. P. & KEEN, D. H. 1975. Stratigraphy and environmental significance of Pleistocene deposits at Stone, Hampshire. *Proceedings of the Geologists' Association*, **86**, 349–363.

CAMERON, T. D. J., CROSBY, A., BALSON, P. S., JEFFERY, D. H., LOTT, G. K., BULAT, J. & HARRISON, D. J. 1992. *The Geology of the Southern North Sea*. British Geological Survey United Kingdom Offshore Regional Report Series, HMSO, London.

CARR, A. P. & BLACKLEY, M. W. L. 1973. Investigations bearing on the age and development of Chesil Beach, Dorset and the associated area. *Transactions and papers, Institute of British Geographers*, **58**, 99–112.

DAVIES, K. H. 1984. *The Aminostratigraphy of British Pleistocene beach deposits*. PhD thesis, University of Wales, Aberystwyth.

—— & KEEN, D. H. 1985. The age of the Pleistocene marine deposits at Portland, Dorset. *Proceedings of the Geologists' Association*, **96**, 217–225.

FUNNELL, B. M. 1961. The Palaeogene and Early Pleistocene of Norfolk. *Transactions of the Norfolk and Norwich Naturalists' Society*, **19**, 340–364.

GIBBARD, P. L. 1988. The history of the great northwest European rivers during the past three million years. *Philosophical Transactions of the Royal Society of London*, **B318**, 559–602.

—— 1994. *Pleistocene History of the Lower Thames Valley*. Cambridge University Press, Cambridge.

—— 1995. The formation of the Strait of Dover. *This volume*.

GRAY, J. M. 1983. The measurement of shoreline altitudes in areas affected by glacio-isostasy, with particular reference to Scotland. *In:* SMITH, D. E. & DAWSON, A. G. (eds) *Shorelines and Isostasy*. Institute of British Geographers Special Publication No. 16. Academic Press, London.

HAILS, J. R. 1983. Coastal processes, relict shorelines and changes in sea level on selected mid- and low-latitude coasts. *In:* SMITH, D. E. & DAWSON, A. G. (eds.) *Shorelines and Isostasy*. Institute of British Geographers Special Pubication No. 16. Academic Press, London.

HALLÉGOUËT, B. & VAN VLIET-LANOË, B. 1986. Les oscillations climatiques entre 125000 ans et le maximum glaciaire d'après l'ètude des formations marines, dunaires et periglaciaires de la côte des Abers (Finistère). *Bulletin de l'Association Française pour l'Etude du Quaternaire*, **1–2**, 127–138.

HOLLIN, J. T., SMITH, F. L, RENOUF, J. T. & JENKINS, D. G. 1993. Sea-cave temperature measurements and amino acid geochronology of British Late Pleistocene sea stands. *Journal of Quaternary Science*, **8**, 359–364.

HOLYOAK, D. T. & PREECE, R. C. 1983. Evidence of a high Middle Pleistocene sea-level from estuarine deposits at Bembridge, Isle of Wight, England. *Proceedings of the Geologists' Association*, **94**, 231–244.

HORTON, A., KEEN, D. H., FIELD, M. H., ROBINSON, J. E., COOPE, G. R., CURRANT, A. P., GRAHAM, D. K., GREEN, C. P. & PHILLIPS, L. M. 1992. The Hoxnian Interglacial deposits at Woodston, Peterborough. *Philosophical Transactions of the Royal Society of London*, **B338**, 131–164.

JELGERSMA, S. 1979. Sea-level changes in the North Sea basin. *In:* OELE, E., SCHUTTENHELM, R. T. E. & WIGGERS, A. J. (eds.) *The Quaternary History of the North Sea. Acta Universitatis Upsaliensis Symposia Universitatis Upsaliensis Annum Quingentesimum Celebrantis*, **2**, 233–248.

JONES, R. L. & KEEN, D. H. 1993. *Pleistocene Environments in the British Isles*. Chapman & Hall, London.

——, ——, BIRNIE, J. F. & WATON, P. V. 1990. *Past Landscapes of Jersey: environmental changes during the last ten thousand years*. Société Jersiaise, St Helier.

DE JONG, J. D. 1967. The Quaternary of the Netherlands. *In:* RANKAMA. K. (ed.) *The Quaternary* (volume 2). Interscience, New York, 301–426.

KEEN, D. H. 1978. *The Pleistocene deposits of the Channel Isles. Report of the Institute of Geological Sciences*, 78/26.

—— 1985. Late Pleistocene deposits and Mollusca from Portland, Dorset. *Geological Magazine*, **122**, 181–186.

—— (ed.) 1993. *Quaternary of Jersey: Field Guide*. Quaternary Research Association, Cambridge.

——, HARMON, R. S. & ANDREWS, J. T. 1981. U-series and amino-acid dates from Jersey. *Nature*, **289**, 162–164.

——, VAN VLIET-LANOË, B. & LAUTRIDOU, J-P. 1993. Chronostratigraphy and interpretation: Belcroute and Portelet. *In:* KEEN, D. H. (ed.) *Quaternary of Jersey: Field Guide*. Quaternary Research Association, Cambridge.

KENNARD, A. S 1942. Pleistocene chronology (in discussion). *Proceedings of the Geologists' Association*, **53**, 24–25.

LAUTRIDOU, J-P. 1982. *The Quaternary of Normandy*. Quaternary Research Association, Cambridge.

LISTER, A. M. 1989. Rapid dwarfing of red deer on Jersey in the Last Interglacial. *Nature*, **342**, 539–542.

—— 1993. The dwarf red deer of Belle Hougue Cave. *In:* KEEN, D. H. (ed.) *Quaternary of Jersey: Field Guide*. Quaternary Research Association, Cambridge.

—— 1995. Sea-levels and the evolution of island endemics: the dwarf red deer of Jersey. *This volume*.

MEIJER, T. & PREECE, R. C. 1995. Malacological evidence relating to the insularity of the British Isles during the Quaternary. *This volume*.

MILLER, G. H. & MANGERUD, J. 1985. Aminostratigraphy of European marine interglacial deposits. *Quaternary Science Reviews*, **4**, 215–278.

MITCHELL, G. F., PENNY, L. F., SHOTTON, F. W. & WEST, R. G. 1973. *A correlation of Quaternary deposits in the British Isles*. Geological Society, London, Special Report No 4.

MOTTERSHEAD, D. M. 1977. The Quaternary evolution of the south coast of England. *In:* KIDSON, C. & TOOLEY, M. J. (eds.) *The Quaternary History of the Irish Sea*. Seel House Press, Liverpool.

——, GILBERTSON, D. D. & KEEN, D. H. 1987. The raised beaches and shore platforms of Torbay: a re-appraisal. *Proceedings of the Geologists' Association*, **98**, 241–257.

NORTON, P. E. P. 1977. Marine Mollusca in the East Anglian pre-glacial Pleistocene. *In:* SHOTTON, F. W. (ed.) *British Quaternary Studies: Recent Advances*. Clarendon Press, Oxford.

PREECE, R. C. & SCOURSE, J. D. 1987. Pleistocene sea-level history of the Bembridge area of the Isle of Wight. *In:* BARBER, K. E. (ed.) *Wessex and the Isle of Wight: Field Guide*. Quaternary Research Association, Cambridge.

——, ——, HOUGHTON, S. D., KNUDSEN, K. L. & PENNEY, D. N. 1990. The Pleistocene sea-level and neotectonic history of the eastern Solent, southern England. *Philosophical Transactions of the Royal Society*, **B328**, 425–477.

PROCTOR, C. J. & SMART, P. L. 1991. A dated cave sediment record of Pleistocene transgressions on Berry Head, Southwest England. *Journal of Quaternary Science* **6**(3), 233–244.

REYNOLDS, P. J. 1987. Lepe Cliff: the evidence for a pre-Devensian brickearth. *In:* BARBER, K. E. (ed.) *Wessex and the Isle of Wight: Field Guide*. Quaternary Research Association, Cambridge.

ROBERTS, M. B. 1986. Excavation of the Lower Palaeolithic site at Amey's Eartham Pit, Boxgrove, West Sussex: a preliminary report. *Proceedings of the Prehistoric Society*, **52**, 215–245.

—— 1994. How old is Boxgrove man? (reply to a discussion by D. Q. Bowen & G. A. Sykes). *Nature*, **371**, 751.

——, STRINGER, C. B. & PARFITT, S. A. 1994. A hominid tibia from Middle Pleistocene sediments at Boxgrove, UK. *Nature*, **369**, 311–313.

ROEP, Th. B., HOLST, H., VISSERS, R. L. M., PAGNIER, H. & POSTMA, D. 1975. Deposits of south-flowing, Pleistocene rivers in the Channel region near Wissant, NW France. *Palaeogeography, Palaeoclimatology, Palaeoecology*, **17**, 289–308.

SARNTHEIN, M., STREMME, H. E. & MANGINI, A. 1986. The Holstein interglaciation: time-stratigraphic position and correlation to stable-isotope stratigraphy of deep-sea sediments. *Quaternary Research*, **29**, 75–79.

SCHWARCZ, H. P. & GRÜN, R. 1988. Comment on M.

Sarnthein, H. E. Stremme and A. Mangini: The Holstein interglaciation: time-stratigraphic position and correlation to stable isotope stratigraphy of deep-sea sediments. *Quaternary Research*, **29**, 75–79.

Scourse, J. D. & Austin, R. M. 1995. Palaeotidal modelling of continental shelves: marine implications of a land-bridge in the Strait of Dover during the Holocene and Middle Pleistocene. *This volume*.

Shackleton, N. J. 1987. Oxygen isotopes, ice volume and sea-level. *Quaternary Science Reviews*, **6**, 183–190.

—— & Opdyke, N. D. 1973. Oxygen isotope and palaeomagnetic stratigraphy of Equatorial Pacific core V28-238: oxygen isotope temperatures and ice volumes on a 10^5 year and 10^6 year scale. *Quaternary Research*, **3**, 39–55.

Smart, P. L. & Frances, P. D. 1991. *Quaternary dating methods – a users guide*. Quaternary Research Association Technical Guide No 4, Quaternary Research Association, Cambridge.

Sommé, J., Paepe, R., Baeteman, C., Beyens, L., Cunat, N., Geeraerts, R., Hardy, R., Hus, A. F., Juvigné, E., Mathieu, L., Thorez, J. & Vanhoorne, R. 1978. La Formation de Herzeele: Un nouveau stratotype du Pléistocène moyen marin de la Mer du Nord. *Bulletin de l'Association Française pour l'Etude du Quaternaire*, 54–56, 81–149.

Sutcliffe, A. J. 1995. Insularity of the British Isles 250 000–30 000 years ago: the mammalian, including human, evidence. *This volume*.

Vanhoorne, R. & Denys, L. 1987. Further palaeobotanical data on the Herzeele Formation (Northern France). *Bulletin de l'Association Française pour l'Etude du Quaternaire*, 1987-1, 7–18.

Walcott, R. I. 1972. Past sea-levels, eustasy and deformation of the Earth. *Quaternary Research*, **2**, 1–14.

West, R. G. 1957. Interglacial deposits at Bobbitshole, Ipswich. *Philosophical Transactions of the Royal Society of London*, **B241**, 1–31.

—— & Sparks, B. W. 1960. Coastal interglacial deposits of the English Channel. *Philosophical Transactions of the Royal Society of London*, **B243**, 95–133.

——, Devoy, R. J. N., Funnell, B. M. & Robinson, J. E. 1984. Pleistocene deposits at Earnley, Bracklesham Bay, Sussex. *Philosophical Transactions of the Royal Society of London*, **B306**, 137–157.

Zagwijn, W. H. 1974. The palaeogeographic evolution of the Netherlands during the Quaternary. *Geologie en Mijnbouw*, **53**, 369–385.

Palaeotidal modelling of continental shelves: marine implications of a land-bridge in the Strait of Dover during the Holocene and Middle Pleistocene

J. D. SCOURSE[1] & R. M. AUSTIN[1,2]

[1] School of Ocean Sciences, University College of North Wales, Menai Bridge, Gwynedd, LL59 5EY, UK
[2] Present address: Institute of Hydrology, Wallingford, OX10 8BB, UK

Abstract: Numerical models of the tide on the NW European continental shelf are useful in predicting basic tidal dynamics (amplitude, current, bed stress, mixing) from bathymetry, coastline configuration and a known ocean tide at the shelf edge. Using the M_2 constituent of the ocean tide, a numerical model has been used to investigate the tidal regime of the English Channel/southern North Sea at successive stages during the Holocene transgression, and also during the maximal sea-levels of Middle Pleistocene temperate stages, by incorporating palaeogeographical interpretations of geological data from these time-slices. Prior to the breaching of the Strait of Dover in the early Holocene, the Southern Bight of the North Sea was a quiet, shallow embayment of low M_2 tidal amplitude (<0.5 m); following breaching it attained the present state of vigorous tidal action with mean amplitudes of around 2 m. Changes in sediment transport paths and shelf-sea fronts are also indicated. Models of the Middle Pleistocene interglacial English Channel incorporate modified palaeobathymetries to compensate for tectonic uplift in the central Channel area over the past 400 000 years. These models indicate increased seasonal stratification in the Channel compared with the present mixed condition. This is in agreement with independent evidence from coccolith assemblages contained within sediments of the same age from the central Channel area.

The presence or absence of land-bridges connecting Britain with the remainder of continental Europe during the Quaternary has as many implications for the marine realm as for the terrestrial. A land-bridge in the Strait of Dover area may provide a conduit for the migration of terrestrial organisms into Britain from the continent (Huntley & Birks 1983; Meijer & Preece 1995) but, in contrast, this same land-bridge will constitute a barrier for the transfer of energy and materials from the English Channel into the North Sea. Land-bridges in other critical areas, such as the Irish Sea (Devoy 1985, 1995), will have had equally significant effects. Quaternary shelf sequences contain empirical, particularly biological, data which provide evidence for the opening and closure of marine channels through which water masses, marine sediments and organisms can move (e.g. Norton & Spaink 1973). Though empirical data, both terrestrial and marine, provide the fundamental evidence for the interpretation of past marine channels and land-bridges, numerical models that incorporate these empirical data as major input variables, particularly concerning changing sea-levels and palaeobathymetries, can provide insights into the implications of such changing palaeogeographies which may not be recorded in the stratigraphic record or which may have been overlooked in that record. The outputs of numerical models therefore constitute hypotheses for future field testing, and as such they have an important role in the reconstruction of Quaternary environments.

Palaeotidal modelling

The most important numerical models in the context of insularity are those which reconstruct ancient tides in shelf seas. These are significant because major tidal parameters, such as elevation amplitude and amphidromal position, change in response to changing bathymetry and coastline palaeogeography as relative sea-level fluctuates. In addition, the changes in tidal parameters have important implications for other tide-dependent effects such as tidal currents, bed-sediment transport paths (peak bed stress vectors), and the position of biologically-significant tide-generated shelf-sea fronts.

Though numerical modelling of shelf-sea tidal amplitudes and depth-mean currents is a well established and successful technique (Heaps 1969; Flather 1976), there have been relatively few applications of such models to tides in ancient seas. However, provided reasonable estimates of bathymetry and coastline configuration for any chosen time-slice can be supplied as inputs, there is no reason why tidal models which predict modern tides in a particular area cannot be adapted to reconstruct the tides of the past.

Belderson et al. (1986) used a numerical model of the M_2 (lunar semi-diurnal) tide to investigate the origin of the large but currently moribund linear tidal sandbanks of the Celtic Sea (Stride 1963; Bouysse et al. 1976; Pantin & Evans 1984). This model generated tidal currents about twice as strong as those at present, with sea-level lowered by 100 m consistent with estimates for sea-level early in the Late Devensian–Holocene transgression. These tidal streams are interpreted by Belderson et al. (1986) as sufficient to generate and maintain the sandbanks which have subsequently been 'fossilized' by the reduction in tidal current velocities in this area as water depth has increased.

Proctor & Carter (1989) also used a model of the M_2 tide to investigate provenance of sediment in relation to the opening and closure of the Cook Strait, New Zealand, and Scott & Greenberg (1983) used a palaeotidal modelling approach to examine the development of the Bay of Fundy tidal system with rising sea-level. Hinton (1992a, b) has used a nested series of models, from oceanic to estuarine in scale, to investigate tidal changes in The Wash and Morecambe Bay during the Holocene.

Palaeotidal implications of opening and closure of the Straits of Dover during the Holocene

The palaeotidal modelling studies by Austin (1988, 1991) of the NW European continental shelf during the Holocene are of direct relevance to the opening and closure of the Strait of Dover. In his 1991 study, uniform depth reductions of between 5 m and 30 m were applied to an existing M_2 model of the NW European shelf in order to approximate bathymetric changes between about 9000 and 5000 years BP enabling an estimation of changes in the tidal dynamics of this area on scales of 20–1000 km. Eustatic sea-level curves were used to give a loose chronological interpretation of a predetermined set of bathymetric changes. The sea-level curve used in the 1991 study was derived from tabulated data in Shennan (1987), with a few points interpolated from Mörner's (1980) eustatic sea-level curve (Table 1). At the time of the study, these data were the most up-to-date available synthesis of previous work which had claims for regional validity. The effects of glacio-isostasy, hydro-isostasy and sedimentation/erosion on bathymetry and coastline changes were ignored in order to simplify the palaeogeographic basis for the models. The use of the lunar semi-diurnal (M_2) constituent only is justified in that it is by far the largest constituent of the ocean tide (the solar semi-diurnal constituent, S_2, is usually the next largest at around a third of the amplitude of M_2). The shelf-edge boundary of the model lies in deep water (300–3000 m) just off the shelf break, where the tide is largely oceanic in character, and it was assumed that there had been no change in the ocean tide over the Holocene timescale used. The average roughness of the sea-bed was also assumed to have remained constant. The modelling assumptions are discussed in full by Austin (1991).

Table 1. *Approximate ages of the predetermined bathymetric changes employed in the model reconstructions.*

Present mean sea-level (pmsl)	post-5000 years BP
−15 m below pmsl	7500 years BP
−25 m below pmsl	9000 years BP

Based on eustatic sea-level data in Mörner (1980) and Shennan (1987).

The base model used was developed at the Proudman Oceanographic Laboratory (POL). It is a two-dimensional finite-difference model employing 201 by 175 nodes with a spacing of 7.5′ longitude and 5′ latitude to cover an area from 12°W 48°N to 13°E 62.5°N. POL also provided digitized bathymetry and M_2 tidal amplitudes along the model boundary. The periodic components of the modelled tides were found to converge within 15 M_2 tidal cycles from a still-water start. Further technical details may be found in Austin (1991).

For the depth reductions used in Austin's study, the only major land-bridge considered is the link in the vicinity of the Strait of Dover. This is an emergent land-bridge in the model runs greater than 25 m below present mean sea-level (pmsl) (c. 9000 years BP), but thereafter becomes an increasingly deep marine channel. This change from land-bridge to marine channel in the early Holocene enables the tide-dependent effects of this barrier to be assessed.

Figure 1 shows a comparison of the contours of difference in M_2 amplitude between model

PALAEOTIDAL MODELLING OF CONTINENTAL SHELVES

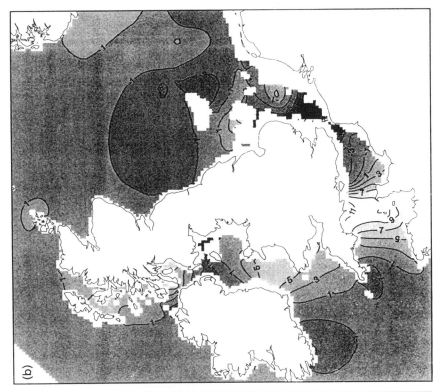

Fig. 1. Contours of difference in M_2 elevation amplitude between model pmsl (post-5000 years BP) and (**a**) model −15 m (c. 7500 years BP) and (**b**) model −25 m (c. 9000 years BP) in units of 10 cm. Positive values imply an amplitude greater than model pmsl. Land and external sea areas are shown unshaded. Reprinted with permission from Austin (1991) (*Terra Nova*, **3**, Fig.4, p. 281).

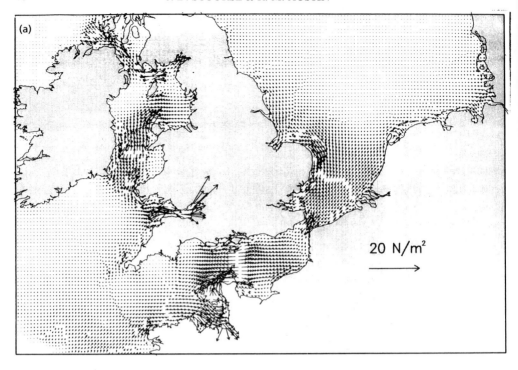

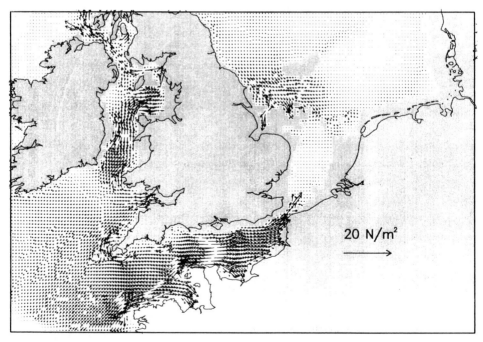

Fig. 2. Peak bed stress vectors from (**a**) model pmsl (post-5000 years BP) and (**b**) model −25 m (c. 9000 years BP). Shading indicates land and external sea areas. Reprinted with permission from Austin (1991) (*Terra Nova*, **3**, Figs 5(a) and 6(b), pp. 282–283).

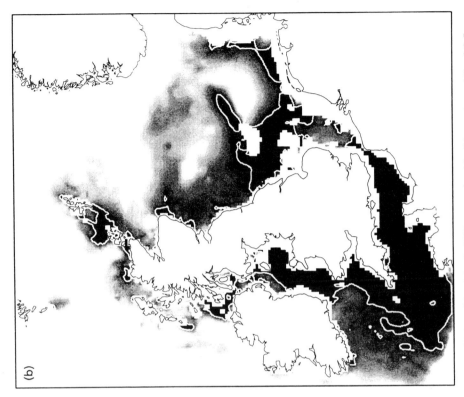
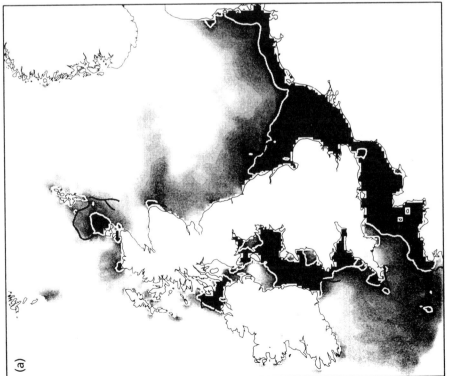

Fig. 3. Distribution of stratification parameter, S, from (a) model pmsl (post-5000 years BP) and (b) model -25 m (c. 9000 years BP), with white contour at $S = 1.5$. Lower values of S (darker areas) imply greater mixing. Black lines in (a) show positions of observed fronts reported by Pingree & Griffiths (1979). Land and external sea areas are shown unshaded. Reprinted with permission from Austin (1991) (*Terra Nova*, 3, Fig. 9, p. 286).

pmsl (post-5000 years BP) and (a) model −15 m (c. 7500 years BP) and (b) model −25m (c. 9000 years BP) in units of 10 cm. Positive values indicate amplitudes greater than model pmsl (effectively the present day) and *vice versa*. This indicates that with the Strait of Dover closed in the early Holocene, the Southern Bight of the North Sea and the eastern end of the 'English Channel' were both shallow embayments of low M_2 tidal amplitude (<0.5 m). Breaching of the land-bridge as sea-level increased caused a change towards the current state of vigorous tidal action, with increased amplitudes of around 2 m. This transition supports the idea that once breached during the Pleistocene, the Strait of Dover would be repeatedly vulnerable to tide-induced erosion with each successive high eustatic sea-level (Gibbard 1995). The modelled change in the Holocene tidal regime in the Southern Bight of the North Sea supports some similar M_2 model results by Franken (1987) using a different base model.

Calculations based on these model results suggest that tidal changes in mean high water of spring tides is nowhere greater than 8% of the total eustatic change (Austin 1991). These calculations assume that mean high water range is equal to M_2 range and that the typical value of 1.3 for the ratio of spring to mean range applies everywhere.

Following the work of Pingree & Griffiths (1979), Austin (1991) demonstrated that peak bed stress vectors derived from model pmsl correlate very closely with empirical data on sand transport paths around the British Isles at the present day. This suggests that peak bed stress vectors derived from the lowered sea-level models representing the Holocene past can be used with some confidence to approximate the changing patterns and strengths of sand transport through time on the NW European shelf. Errors in this approximation are most likely at local scales due to neglect of the evolution of bathymetric features such as sandbanks. On wider scales, these stress patterns arise largely from the interaction of the M_2 tide with its harmonic, M_4, which is generated by shallow-water effects and is hence chiefly dependent on depth. Changes in average sea-bed roughness are likely to influence the magnitude rather than the pattern of stresses.

Figure 2a shows the peak bed stress vectors for model pmsl which approximate very closely to the known sand transport paths around the British Isles at the present day. In contrast, Fig. 2b shows the peak bed stress vectors for model −25 m (c. 9000 years BP) with the Strait of Dover closed. The most noticeable features suggested are strong tidal movement of sediment within the eastern embayment of the 'English Channel' during the early Holocene and an increase in the strength of vectors into the Southern Bight of the North Sea as the tidal amplitude increases through time. The position of the major bedload parting in the central English Channel appears to have remained more or less static during the Holocene though with some diminution in the strength of vectors on either side of the parting.

Dissipated tidal energy on the shelf is the major cause of mixing over depth which may be opposed by thermal stratification causing the development of a thermocline due to the surface heat flux (Pingree & Griffiths 1978). In summer, well defined fronts are commonly found between areas of stratified and mixed water on the shelf; these fronts represent the surface expression of the thermocline. These break down in the autumn when the surface heat flux diminishes and tidal energy causes a return to mixing. Figure 3 shows the distribution of stratified and mixed water during the summer, based on derivation of the stratification parameter S (Pingree & Griffiths 1978) from model pmsl (post-5000 years BP) and −25 m (c. 9000 years BP). This derivation is based on the assumption of no change in the seasonal surface heat flux during the Holocene. Areas where $S<1$ are well mixed throughout the year, whereas areas where $S>2$ are generally stratified in summer. The results suggest that in the early Holocene, the Southern Bight of the North Sea was seasonally stratified, whereas at present it is mixed throughout the year. Shelf-sea fronts are zones of high organic productivity and may leave evidence in the biogenic component of shelf sediments accumulating on the bed (Houghton 1988; see below). The former positions of seasonal shelf-sea fronts, obtained from model reconstructions, therefore have implications for the biogenic content of shelf sediments and for sedimentation rates.

Palaeotidal implications of opening and closure of the Strait of Dover during the interglacial stages of the Middle Pleistocene

A recent study of tectonically uplifted Middle Pleistocene interglacial marine sediments (Preece *et al.* 1990) from central southern England has provided evidence on the palaeoceanography of the English Channel during such stages. This has in turn formed the empirical basis for a numerical modelling investigation of the tidal dynamics of the English Channel during Middle Pleistocene interglacial stages.

Table 2. *Coccolith assemblages from the marine interglacial deposits at Boxgrove (Slindon Sands) and Bembridge (Steyne Wood Clay).*

Taxa	Boxgrove	Bembridge
Gephyrocapsa oceanica Kamptner	A	A
Gephyrocapsa caribbeanica Boudreaux & Hay	A	C
Dictyococcites productus (Kamptner) Backman	C	A
Calcidiscus leptoporus (Murray & Blackman) Loeblich & Tappan	R	R
Coccolithus pelagicus (Wallich) Schiller	R	R
Reticulofenestra spp. (at least two species)	C	A
Syracosphaera pulchra Lohmann	R	R
Pontosphaera sp.	R	R
Braarudosphaera bigelowii (Gran & Braarud) Deflandre	R	R
Discosphaera tubifera (Murray & Blackman) Ostenfeld	R	R
Oolithus fragilis (Lohmann) Okada & McIntyre	R	–

Analysed by S. D. Houghton. Adapted from Preece *et al.* 1990.
A = abundant (over 10%), C = common (1–10%), R = rare (less than 1%), – = species absent.

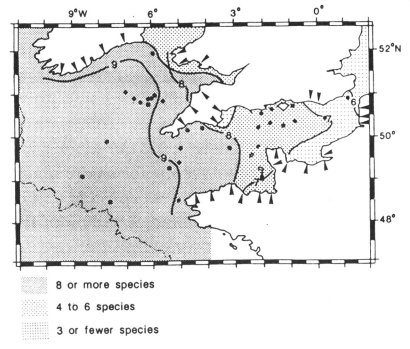

Fig. 4. Species diversity of coccolith assemblages and quantitative abundance of Recent coccoliths in surface sediments. Dots represent open-marine samples, arrows represent estuarine samples. Contours represent 10^n coccoliths/g of fine-grained silt (2–20 µm). Reprinted with permission from Houghton (1988) (*Marine Geology*, **83**, Fig. 2, p. 315).

The interglacial deposits at Bembridge (Steyne Wood Clay), eastern Isle of Wight (Holyoak & Preece 1983; Preece *et al.* 1990) and at Boxgrove (Slindon Sands) in Sussex (Roberts 1986) have been interpreted by Preece *et al.* (1990) as being of Middle Pleistocene marine origin on the basis of micropalaeontological, aminostratigraphic and palaeomagnetic data. They suggest accumulation of these sequences during an interglacial in the latter part of the 'Cromerian Complex', and probable correlation with oxygen isotope stages 9, 11 or 13; correlation with the 'Cromerian Complex' implies deposition prior to the Anglian cold stage that is usually thought to belong to stage 12. This is significant because Smith (1985, 1989) and Gibbard (1988, 1995)

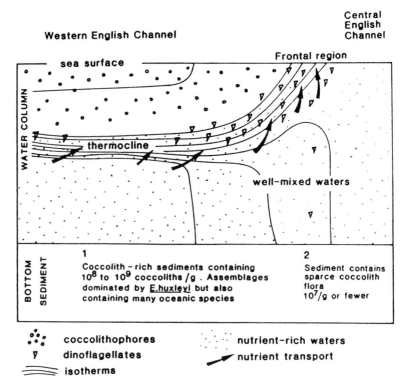

Fig. 5. Schematic representation of the spatial changes in coccolith abundance across the frontal region in the western English Channel. Water column conditions represent the summer period with thermocline development in the western English Channel. Blooms of coccolithophorids, dominated by *E. huxleyi*, occur in the nutrient-deficient, well illuminated waters above the thermocline. Coccolithophorids are scarce in the well mixed waters of the central English Channel. Reprinted with permission from Houghton (1988) (*Marine Geology*, **83**, Fig. 4, p. 318).

have suggested that the initial breaching of the Strait of Dover occurred during the Anglian glaciation as a result of the catastrophic drainage of a glacially impounded lake in the southern North Sea. This would therefore suggest that the Steyne Wood Clay and Slindon Sands were deposited in an interglacial stage predating this event. It is therefore possible to construct the hypothesis that prior to the Anglian, during Middle Pleistocene interglacial stages characterized by high eustatic sea-level, the English Channel would have formed a large embayment without any direct marine link to the southern North Sea. This is quite different from the Holocene pattern and possibly also other interglacial stages postdating the Anglian.

Calcareous nannofossils preserved in the Steyne Wood Clay and Slindon Sands (Preece *et al.* 1990) have shed some light on the environment of deposition and palaeoceanography of the Middle Pleistocene English Channel. The occurrence of diverse nannofossil assemblages in these sediments, comprising 11 and 12 species respectively (Table 2), including abundant gephyrocapsids, with subdominant *Calcidiscus leptoporus*, *Syracosphaera pulchra* and *Discosphaera tubifera*, strongly suggests that seasonally stratified water was present in the English Channel at the time of deposition. These nannofossil assemblages are only found today in the Celtic Sea and western English Channel in sediments underlying a watermass with pronounced thermocline development (Fig. 3a; Houghton 1986, 1988). Recent sediments in the English Channel deposited from tidally-mixed waters contain sparse nannofossil assemblages limited to six or fewer species characterized by dominant *Emiliania huxleyi* and *Coccolithus pelagicus* and rare gephyrocapsids (Fig. 4). Recent sediments in the Solent area are even more neritic, being limited to four species dominated by *C. pelagicus* (over 95%). Figure 5 demonstrates that coccolith diversity and production are both high in seasonally stratified

PALAEOTIDAL MODELLING OF CONTINENTAL SHELVES

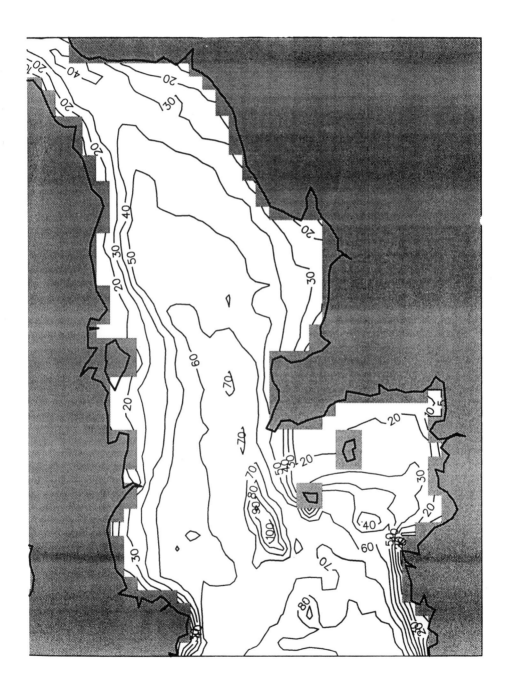

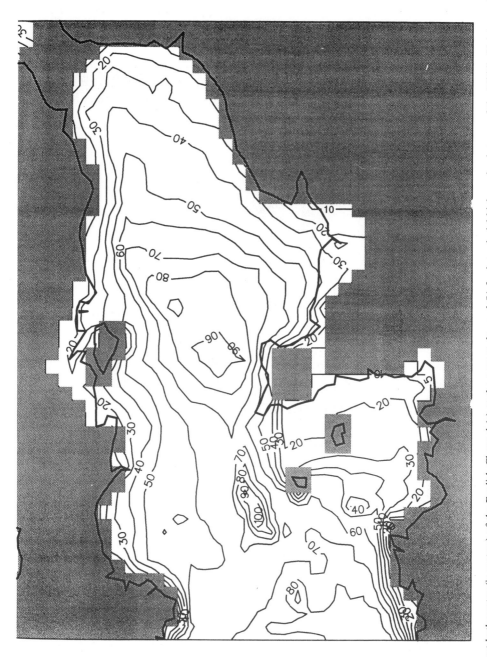

Fig. 6. Input bathymetry (in metres) of the English Channel (a) at the present day and (b) for interglacial high sea-level stages of the Middle Pleistocene incorporating tectonic uplift in the central Channel over the last 400 000 years. Note closure of the Strait of Dover.

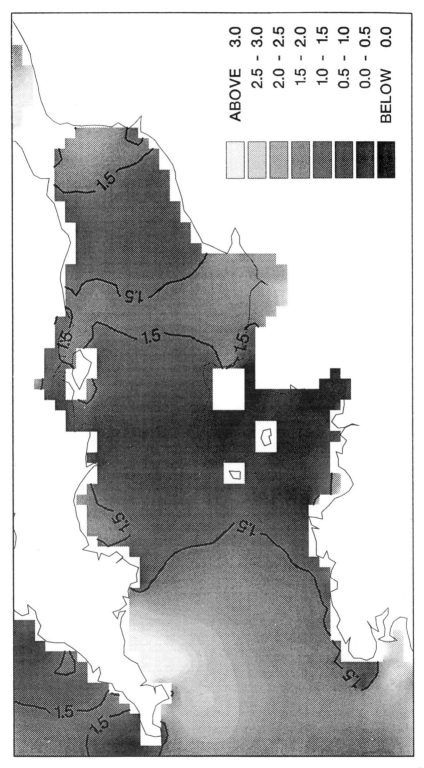

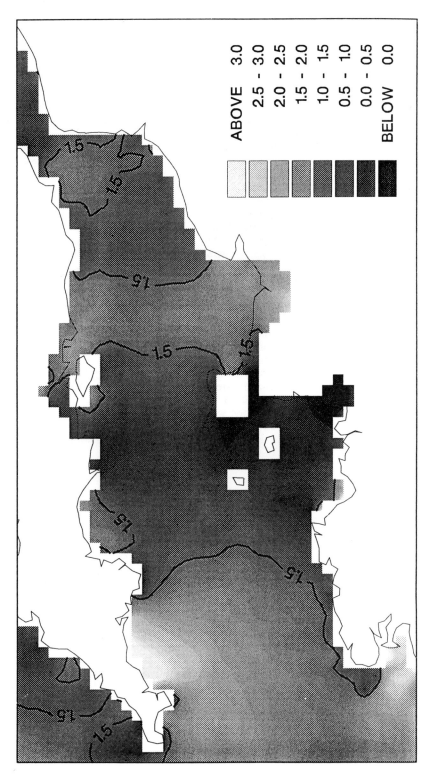

Fig. 7. Distribution of stratification parameter, S, for Middle Pleistocene English Channel with (**a**) Strait of Dover closed and (**b**) Strait open. Lower values of S (darker areas) imply greater mixing. Black contour at $S = 1.5$. Land areas are shown unshaded.

water above the thermocline and in stratified water, with low diversity/production in tidally mixed water. The absence of seasonally stratified water in the English Channel at the present day (Fig. 3a) explains the sparse nannofossil assemblages from recent sediments in this area.

Austin (1991) demonstrated, in the context of the Holocene, that through derivation of the stratification parameter S, palaeotidal models can provide information on the past distributions of stratified and mixed water, and on the positions of the shelf-sea fronts which form the boundaries between such areas. In order to test the hypothesis that the English Channel was seasonally stratified at the time of the deposition of these interglacial deposits, as suggested by the coccolith data, the present-day M_2 model with the same shelf-edge tide used by Austin (1991) for the Holocene was applied to this problem. In addition, the application of the model to two palaeogeographic situations, one with a closed English Channel, the other with a full open connection to the southern North Sea, provides a test for the hypothesis, based on the geological data, that these deposits pre-date the initial breaching of the Strait of Dover which probably occurred during the Anglian, as discussed above. No other palaeogeographic changes were applied to the present bathymetry in parts of the model outside the present English Channel area.

In addition to modification of the Strait of Dover area, a palaeobathymetric correction was applied to the English Channel in an attempt to compensate for the known tectonic uplift in the central Channel area. On the basis of the present elevations of the Steyne Wood Clay and Slindon Sands at around 40 m, the age and palaeoenvironment of these deposits, and comparison with the deep-sea oxygen isotope record, Preece et al. (1990) were able to document around 40 m of uplift over the last 400 000 years in this area. This they ascribed to compensatory isostatic uplift related to sediment loading in the southern North Sea. Figure 6a shows the bathymetry of the English Channel at the present day, and Fig. 6b the palaeobathymetry used in the model reconstructions with this tectonic component removed and with the Strait of Dover closed. The major differences shown for the palaeobathymetric simulation, compared with that for the present day, are the increased inundation of the Solent area and the Cotentin Peninsula in northern France, and the increased depth of the central Channel area.

Figure 3a shows the model-derived stratification parameter for the present English Channel and demonstrates the restriction of seasonally stratified water to the western Channel and South West Approaches. Figure 7a shows the same parameter for the Middle Pleistocene interglacial English Channel with the Strait of Dover closed, and Fig. 7b with the Strait open. Water depths for the open Strait of Dover are the same as at the present day. Both reconstructions in Fig. 7 incorporate the correction for tectonic uplift. They demonstrate that with a deeper and more extensive central Channel there is an enhancement of seasonal stratification, both in the central Channel and the eastern embayment in the vicinity of the Strait when compared with the present-day pattern. The area of stratification in the western Channel is also more extensive. The model results show that, given the assumptions, the tidal dynamics of the area are insensitive to the presence or absence of a channel on the scale of the present-day Strait and are therefore inconclusive as regards the timing of its initial breaching.

These palaeotidal reconstructions of an English Channel modified to take account of the palaeogeography of the Middle Pleistocene therefore corroborate the interpretations of the coccolith assemblages contained within deposits of this age. It appears that the English Channel, during the high sea-level interglacial stages of the Middle Pleistocene, was characterized by extensive seasonal stratification unlike the present state of vigorous mixing.

Conclusion

Palaeotidal models are able to provide informative insights into the marine effects of the changing palaeogeography of shelf seas through time. They are useful in hindcasting basic tidal dynamics such as amplitude, current, bed stress and mixing which have considerable implications for topics such as sediment transport paths, biological productivity and interpretation of the fossil record. Knowledge of past tidal ranges provides critical information for calibration of the sea-level record from established index points. In this way model reconstructions provide hypotheses for future field testing which can then be fed back as inputs into the models. This generates a process of fine-tuning which adds a new dimension to palaeoenvironmental reconstruction.

We would like to thank Roger Flather, Eric Jones and David Prandle of the Proudman Oceanographic Laboratory who developed the model on which this work is based, and who gave freely of their time and advice. Ed Hill kindly commented on an early draft of the paper and Anne Hinton's detailed comments as referee have been greatly appreciated.

References

Austin, R. M. 1988. *The Palaeotidal Regime on the North-West European Shelf.* MSc. thesis, University College of North Wales, Bangor.
—— 1991. Modelling Holocene tides on the NW European continental shelf. *Terra Nova*, **3**, 276–288.
Belderson, R. H., Pingree, R. D. & Griffiths, D. K. 1986. Low sea-level tidal origin of Celtic Sea sand banks – evidence from numerical modelling of M_2 tidal streams. *Marine Geology*, **73**, 99–108.
Bouysse, P., Horn, R., Lapierre, F. & Le Lann, F. 1976. Étude des grands bancs du sable du sud-est de la mer Celtique. *Marine Geology*, **20**, 251–275.
Devoy, R. J. N. 1985. The problem of a Late Quaternary landbridge between Britain and Ireland. *Quaternary Science Reviews*, **4**, 43–58.
—— 1995. Deglaciation, Earth crustal behaviour and sea-level changes in the determination of insularity: a perspective from Ireland. *This volume.*
Flather, R. A. 1976. A tidal model of the north-west European continental shelf. *Mémoires de la Société Royale des Sciences de Liège, 6 série*, **X**, 141–164.
Franken, A. F. 1987. *Rekonstruktie van het Paleo-Getijklimaat in de Noordzee.* Masters thesis, Delft Hydraulics Laboratory, The Netherlands.
Gibbard, P. L. 1988. The history of the great northwest European rivers during the past three million years. *Philosophical Transactions of the Royal Society of London*, **B318**, 559–602.
—— 1995. The formation of the Strait of Dover. *This volume.*
Heaps, N. S. 1969. A two-dimensional numerical sea model. *Philosophical Transactions of the Royal Society of London*, **A265**, 93–137.
Hinton, A. C. 1992a. *Modelling tidal changes within The Wash and Morecambe Bay during the Holocene.* PhD thesis, University of Durham.
—— 1992b. Palaeotidal changes within the area of the Wash during the Holocene. *Proceedings of the Geologists' Association*, **103**, 259–272.
Holyoak, D. T. & Preece, R. C. 1983. Evidence of a high Middle Pleistocene sea-level from estuarine deposits at Bembridge, Isle of Wight, England. *Proceedings of the Geologists' Association*, **94**, 231–244.
Houghton, S. D. 1986. *Coccolith Assemblages in Recent Marine and Estuarine Sediments from the Continental Shelf of Northwest Europe.* PhD thesis, University of Southampton.
—— 1988. Thermocline control on coccolith diversity and abundance in Recent sediments from the Celtic Sea and English Channel. *Marine Geology*, **83**, 313–319.
Huntley, B. & Birks, H. J. B. 1983. *An Atlas of Past and Present Pollen Maps for Europe: 0–13 000 Years Ago.* Cambridge University Press.
Meijer, T. & Preece, R. C. 1995. Malacological evidence relating to the insularity of the British Isles during the Quaternary. *This volume.*
Mörner, N-A. 1980. The northwest European 'sea-level laboratory' and regional Holocene eustasy. *Palaeogeography, Palaeoclimatology, Palaeoecology*, **29**, 281–300.
Norton, P. E. P. & Spaink, G. 1973. The earliest occurrence of *Macoma balthica* (L.) as a fossil in the North Sea deposits. *Malacologia*, **14**, 33–37.
Pantin, H. M. & Evans, C. D. R. 1984. The Quaternary history of the central and south-western Celtic Sea. *Marine Geology*, **57**, 259–293.
Pingree, R. D. & Griffiths, D. K. 1978. Tidal fronts on the shelf seas around the British Isles. *Journal of Geophysical Research*, **83**, 4615–4622.
—— & —— 1979. Sand transport paths around the British Isles resulting from M_2 and M_4 tidal interactions. *Journal of the Marine Biological Association of the UK*, **59**, 497–513.
Preece, R. C., Scourse, J. D., Houghton, S. D., Knudsen, K. L. & Penney, D. N. 1990. The Pleistocene sea-level and neotectonic history of the eastern Solent, southern England. *Philosophical Transactions of the Royal Society of London*, **B328**, 425–477.
Proctor, R. & Carter, L. 1989. Tidal and sedimentary response to the Late Quaternary closure and opening of Cook Strait, New Zealand: results from numerical modelling. *Paleoceanography*, **4**, 167–180.
Roberts, M. B. 1986. Excavation of the Lower Palaeolithic site at Amey's Eartham pit, Boxgrove, West Sussex: A preliminary report. *Proceedings of the Prehistoric Society*, **52**, 215–245.
Scott, D. B. & Greenberg, D. A. 1983. Relative sea-level rise and tidal development in the Fundy tidal system. *Canadian Journal of Earth Science*, **20**, 1554–1564.
Shennan, I. 1987. Holocene sea-level changes in the North Sea. *In:* Tooley, M. J. & Shennan, I. (eds) *Sea-Level Changes.* Blackwell, Oxford, 109–151.
Smith, A. J. 1985. A catastrophic origin for the palaeovalley system of the eastern English Channel. *Marine Geology*, **64**, 65–75.
—— 1989. The English Channel – by geological design or catastrophic accident? *Proceedings of the Geologists' Association*, **100**, 325–337.
Stride, A. H. 1963. North-east trending ridges of the Celtic Sea. *Proceedings of the Ussher Society*, **1**, 62–63.

Malacological evidence relating to the insularity of the British Isles during the Quaternary

T. MEIJER[1] & R. C. PREECE[2]

[1] *Rijks Geologische Dienst, PO Box 157, 2000 AD Haarlem, The Netherlands*
[2] *Department of Zoology, University of Cambridge, Downing Street, Cambridge CB2 3EJ, UK*

Abstract: The marine molluscan faunas from different temperate stages of the Pleistocene of the North Sea vary enormously both in terms of species richness and in the diversity of their biogeographical composition. The marine assemblages from the Middle Tiglian, Eemian and Holocene have all yielded *c*. 100 species or more, including many with southern or 'Lusitanian' affinities. It is thought that during these stages the Strait of Dover was open, allowing entry of these southern taxa into the southern North Sea. Conversely, the temperate stages from the Late Tiglian up to and including the Holsteinian have yielded relatively impoverished faunas (no more than about 40 species) virtually lacking any of the southern elements. During these stages it would appear that the Strait of Dover was closed, so preventing the spread of marine molluscs into the North Sea. Examination of the history of fluvial molluscs (particularly prosobranchs and larger bivalves) on either side of the English Channel supports this interpretation. Fluvial provinciality is recognized during the stages when Britain is thought to have become an island. Non-provinciality, pointing to fluvial exchange, occurs during the other stages.

The existence of a southern connection between the North Sea and the Atlantic Ocean during the Neogene and Quaternary has always been an important question in the NW European regional geology. A land-bridge in the region of the Strait of Dover may provide a conduit for the dispersal of terrestrial organisms into Britain from the continent, but at the same time would form a barrier preventing the spread of marine taxa into the southern North Sea. At other times, when the Strait was open, marine taxa would have had unimpeded access into the North Sea, but many non-marine organisms would have been excluded from Britain. Changing compositions of marine and non-marine faunas in the North Sea Basin are therefore assumed to be the result, in part, of the presence or absence of such a sea-way. There is much evidence to indicate that Britain remained a peninsula of Europe for much of the Quaternary, but there is no consensus about the periods when Britain became an island.

Molluscs are especially well suited to answer these kinds of palaeogeographical questions for the following reasons: (1) they occur on land, in freshwater and in the sea; (2) they generally disperse through their own habitat, have no powers of active flight and, except for those taxa with planktonic larvae, cannot be carried across large expanses of open water; (3) certain groups have narrow ranges in their distribution, both ecologically and geographically.

In this paper the fossil record of marine and freshwater molluscs from successive temperate stages of the Quaternary is examined. Faunal histories in the British Isles and mainland Europe, and in the English Channel and southern North Sea, are compared to ascertain what general conclusions can be drawn about the insularity of the British Isles during different stages. Knowledge is poor for the cold stages, but a land-bridge is assumed to have existed, since relative sea-level must have fallen below the critical point during all major stadials. Special attention has been focused on the Quaternary fossil record from The Netherlands, which is still unrivalled in NW Europe, particularly for the Lower and Middle Pleistocene. A map showing the locations of the main sites mentioned in the text is provided (Fig. 1).

Dispersal and migration of molluscs

Colonization of a suitable niche by a molluscan species may occur very rapidly, as the following three examples demonstrate.

The invasion of Europe by the ovoviviparous, parthenogenetic hydrobiid, *Potamopyrgus jen-*

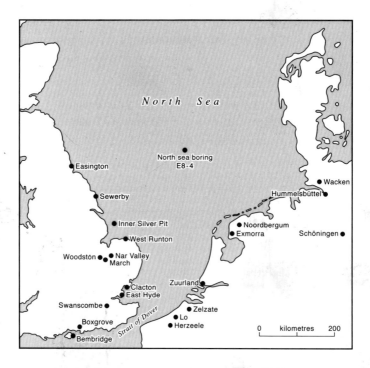

Fig. 1. Map showing the location of the more important sites mentioned in the text.

kinsi, is one of the most spectacular yet documented for any mollusc. It was first recorded in the River Thames in 1889, but was probably introduced from New Zealand, where it is known as *P. antipodarum*, as early as 1859 (Hubendick 1950, Ponder 1988). It reached the European mainland by about 1900 and is now widespread throughout NW Europe, and is the dominant snail in many freshwater and some brackish habitats.

The introduction of the American razor-shell (*Ensis americanus*) into the German Bight of the North Sea in 1979, and its westward spread along the Dutch sandy coast and northwards up the coast of Denmark within just 10 years, shows that marine species are also capable of rapid colonization (Cosel *et al.* 1982; Van Urk 1987; Swennen 1992). Unlike the European species of *Ensis*, it lives in brackish as well as marine conditions, and its rapid spread may be partly due to the filling of a niche not previously occupied. It certainly did not meet any serious barriers along the Dutch and Danish coasts.

Even this case has been surpassed by the rapid spread of two species of *Corbicula* into many Dutch rivers within only four years (bij de Vaate 1993). There has been some confusion about the species involved. They were originally recorded variously as *C. fluminalis* or *C. fluminea* with two forms (bij de Vaate & Greijdanus-Klaas 1990). However, Kinzelbach (1991) has revised the genus *Corbicula* in Europe and has shown that, at present, two species occur in the Rhine, namely *C. fluminea* and *C. fluviatilis*. Kinzelbach (1991) notes that it is very likely that both *Corbicula* species arrived in Europe in ships carrying ballast from southern North America, as there is no direct geographical contact with Asian *Corbicula* populations. Den Hartog *et al.* (1992) have discussed some of the biological attributes of *Corbicula* that have enabled them to be so successful as colonizers. However, these *Corbicula* have colonized only the River Rhine and its tributaries and have not yet been reported from the neighbouring River Scheldt, so the dividing watershed appears to form a barrier to dispersal.

The fact that watersheds are not crossed by all freshwater species is also illustrated by the recent immigration of *Lithoglyphus naticoides*. This prosobranch is a Danubian species that was absent in the Rhine system during the Quaternary, despite the fact that these are neighbouring river systems. After the first man-made connection between the two rivers was constructed in the last century, *L. naticoides* was able to

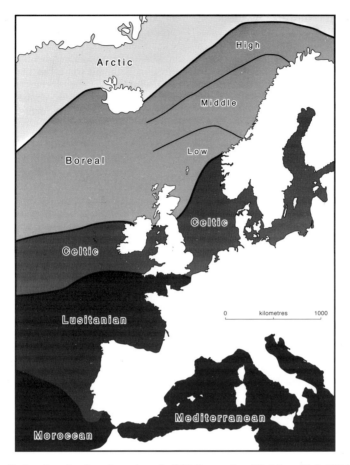

Fig. 2. Present limits of marine faunal provinces in NW Europe (modified from Hall (1964)).

Table 1. *Marine provinces according to Hall (1964).*

Province	Temperature conditions	Northern limit
Arctic		North of southern limit of winter ice
Boreal	< 4 months > 10°C	67° N. lat.
Celtic	6 months > 10°C	55° N. lat.
Lusitanian	4 months > 15°C, never below 10°C	48° N. lat.
Moroccan/Mediterranean	4 months > 20°C	36° N. lat.
West African	6 months > 20°C, never below 18°C	21° N. lat.

migrate into the Rhine, where it is now common, within only a few decades (Remy 1924). Today, after the construction of the new Rhine–Danube connecting canal, the proportion of all Danubian/Ponto-Caspian invertebrates in the Dutch part of the River Rhine is as high as 40% (G. van de Velde, pers. comm.).

Land therefore appears to be a barrier for some freshwater molluscs, but highly isolated pools are often readily colonized by other freshwater species. This is thought to be the result of passive dispersal by birds, mammals or insects (Kew 1893; Rees 1965). The assemblages found in such isolated pools consist mostly of pulmonates and small bivalves with very few species of prosobranchs (Reid 1892; Boycott 1919; Økland 1990). The former species are apparently better able to effect this kind of dispersal than prosobranchs and large bivalve species.

Most marine molluscs disperse by means of planktonic larval stages. This kind of dispersal may lead to rapid colonization, as is shown in the case of *Ensis americanus*. Several kinds of marine habitats, such as inland saline lakes, can presumably only be populated with the help of birds. All marine species living in supra-tidal conditions are likely to disperse in this way. This would include taxa belonging to the Littorinacea, Hydrobiacea, Rissoacea, Ellobiacea, Mytilacea, Cardiacea, Tellinacea and others. The distribution of some of these taxa may consequently depend on the routes of bird migration, which may in turn be controlled by the general coastal topography and the biomass of the tidal flats.

For land Mollusca, the main barrier to dispersal is water, especially sea-water. Only a few amphibious species are able to penetrate into freshwater habitats, but the potential for dispersal by doing so is very limited. Large, open expanses of water act as serious barriers for species restricted to closed woodland. Here, passive dispersal must also occur, but for species confined to more restricted habitats, such dispersal is exceptional.

Marine biogeography of the North Sea

The geographical ranges of marine taxa in the shelf seas around the British Isles are closely related to temperature. Consequently, the distributional limits of many species often coincide with certain isotherms and such faunal discontinuities form the basis of a series of molluscan provinces (Hall 1964) which is adopted here (Table 1, Fig. 2). Other factors, such as salinity and the nature of the substrate, are also important. Thus many southern species are prevented from entering the English Channel because of its sharp west–east thermal gradient. Some southern species are able to extend northwards along the western coasts of the British Isles, taking advantage of the Lusitanian current and North Atlantic drift. However, there is another major range boundary at the Shetland–Orkney channel, reflecting the presence of colder water. Those taxa that are able to cross this barrier will encounter further problems as they disperse southward into the North Sea. Not only does the water off the Northumberland coast become too turbid for many taxa, but south of Flamborough Head, the sediments become too sandy (Vincent 1990).

The modern marine fauna living on the Dutch coast is characterized by high numbers of Celtic species. Although the northernmost extension of the Lusitanian Province lies south of the Strait of Dover, a substantial number of Lusitanian species are still able to survive in the southern North Sea. Since many of these species are unlikely to be able to survive the cold waters in northern Scotland, or cross the other barriers mentioned above, their occurrence implies the existence of a southern connection with the Channel. Similarly, the occurrence of such southern taxa in the southern North Sea during older temperate stages is also taken to imply that the Strait of Dover must have been open.

Fluvial provinciality

The modern molluscan faunas of the major European rivers are relatively well known and only minor differences exist between them. Nevertheless, these minor differences may be enough to characterize particular river systems. These differences may be demonstrated with simple analyses of the faunas. Relative frequencies of prosobranch, pulmonate and bivalve species present in the catchments of selected major European river systems have been calculated. For taxonomic reasons, those inhabiting subterranean and phreatic waters have been omitted. The results (Table 2) show a two-fold sub-division; those from rivers discharging into the North Sea, and those from generally southward flowing rivers. The differences are most pronounced in the frequencies of the prosobranchs, which are significantly lower in the 'North Sea river group' than in the remaining rivers. Pulmonate gastropods and small bivalves, like *Pisidium*, show only minor differences between the selected systems. Almost all prosobranch species in the 'North Sea river group' are also present in the other rivers. However, besides these catholic prosobranchs, additional species inhabit the southward flowing rivers, but may be confined to a few or possibly only single systems, where they appear to be en-demic. This applies principally to prosobranchs, but endemism is also found to a lesser extent among larger bivalves, such as *Unio*, *Anodonta* and *Sphaerium*.

The differences between the two main groups do not result from latitude. The upper course of the River Dnepr is further north than the area occupied by the 'North Sea river group'. It is assumed that the fauna of the 'North Sea river group' has become impoverished as a result of the environmental stresses imposed during the successive Quaternary glaciations. During harsh periods, the catchment area of these rivers was situated entirely within a (peri)glacial environment, leaving few niches for many fluvial species. Obviously this was not the case with the southward flowing rivers, where these

Table 2. *Composition of the living freshwater molluscan faunas known from the catchment areas of a selection of modern European river systems. Subterranean, phreatic, brackish and introduced species are not included.*

River system	Pros	%	Pulm	%	Biv	%	T
Thames	8	15.6	21	41.1	22	43.1	51
Rhine	10	15.5	29	45.3	25	39.1	64
Scheldt	9	16.7	25	46.3	20	37.0	54
Meuse	8	15.1	25	47.2	20	37.7	53
Ebro	9	24.3	16	43.2	12	32.4	37
Po	12	23.1	22	42.3	18	34.6	52
Danube	23	29.1	30	38.0	26	32.9	79
Southern Bug	13	24.5	22	41.5	18	34.0	53
Dnepr	14	22.6	27	43.5	21	33.9	62

Data compiled from: Adam (1960); Alzona (1971); Ehrmann (1956); Ellis (1962, 1969); Fretter & Graham (1978); Girod et al. (1980); Giusti & Pezzoli (1980); Haas (1929); Illies (1978); Zhadin (1952); Zilch & Jaeckel (1962).
Pros = Prosobranchia; Pulm = Pulmonata; Biv = Bivalvia; T = Total

Table 3. *Composition of the fossil freshwater molluscan faunas known from The Netherlands.*

Temperate stage	Pros	%	Pulm	%	Biv	%	T
Middle Tiglian	14	40.0	13	29.0	14	31.0	41
Late Tiglian	16	30.8	20	34.8	17	34.6	53
Waalian	11	26.8	18	34.8	20	38.4	49
Bavel	15	23.4	25	39.1	24	37.5	64
Leerdam	5	14.8	14	51.9	9	33.3	28
Noordbergum Interglacial	6	–	2	–	4	–	12
Holsteinian	7	20.5	17	43.6	14	35.9	38
Belvédère Interglacial	7	17.1	19	46.3	15	36.6	41
Eemian	9	24.0	22	41.0	19	35.0	50
Holocene	11	21.8	24	36.0	27	42.2	62
Modern	10	22.1	25	36.9	28	41.0	63

Modified from Meijer (1990a, b)
Pros = Prosobranchia; Pulm = Pulmonata; Biv = Bivalvia; T = Total

species could survive in extra-glacial parts of the system. If this is the explanation, one should expect a higher relative frequency of prosobranch species in the older Quaternary deposits of a river system. This is precisely the observed pattern in the Quaternary fluvial assemblages of the River Rhine in The Netherlands (Table 3). During the Early Quaternary the composition of the Rhenish fauna was very similar to that now inhabiting southward flowing rivers.

Endemic fluvial species are important biogeographical indicators, since they may characterize particular river systems. Provinciality exists between two river systems when different endemic species occur in each of them. This probably means that the systems have been separated for a long time by an important watershed. The opposite case, that of shared endemic species by two systems, points to a fluvial connection between them, today or in the past. In the latter case, this implies the long-term persistence of the correct ecological niches in both systems since the connection ended.

Based on biogeographical and fossil data, as well as on recent observations upon dispersal, the following conclusions may be drawn:

1. Stratigraphically and palaeogeographically significant freshwater species are found mainly among fluvial prosobranch gastropods and, to a lesser extent among the larger bivalves.
2. Certain fluvial species characterize one, or only a few, river system(s). Provinciality exists when a species is restricted to only one particular river system. The first appearance of such species in other river systems is taken to indicate a physical connection with the source system, rather than passive migration across the watershed. It is important to note that provinciality may be expressed in only a very minor part (possibly by only one species) of the total fauna.

The critical species

The question of the existence of a Strait of Dover may now be answered as follows.

A sea-way is thought to have existed if:
(a) temperate marine species form a substantial part of the North Sea fauna;
(b) fluvial provinciality occurs on either side of the North Sea.

A land connection is assumed if:
(a) temperate marine species are absent or form only a minor part of the North Sea fauna;
(b) Fluvial exchange between the British Isles and the continent exists.

The primary evidence comes from thermophilous marine and fluvial species which occur in these latitudes only during temperate climates when relative sea-levels were high. Therefore, insularity of the British Isles is likely to be shown only during these stages.

Stratigraphical framework

For the main stratigraphical framework we follow Zagwijn (1985) and for correlations between England and the continent in the Early to Middle Pleistocene we follow Gibbard et al. (1991). For the late Middle and Late Pleistocene, we accept the broad correlation of the British Hoxnian and Ipswichian with the continental Holsteinian and Eemian, respectively, but realize that the stratigraphy between these stages has not yet been satisfactorily resolved. At several sites in Britain, as well as on the continent, interglacial deposits have been assigned to warm phases between these two interglacials. Palynological data are either lacking or are inconclusive, but palaeozoological and other data point to one or more additional stages. For example, in The Netherlands the Belvédère Interglacial was recently recognized in an intra-Saalian context (van Kolfschoten et al. 1993). This interglacial may correlate with similar deposits at Fransche Kamp (Ruegg 1991) and appears to equate with the Hoogeveen Interstadial (Zagwijn 1973). In England, many localities apparently belong to a similar stage, which is thought to correlate with stage 7 of the oxygen isotope record (cf. Jones & Keen 1993). Other authors (e.g. Bridgland 1994) think that further stages are also represented in the British late Middle Pleistocene. Litt & Turner (1993) have recently proposed a new scheme for the sub-division of the 'Saalian Complex', in which the temperate deposits at the German localities of Dömnitz, Schöningen and Wacken were thought to represent a warm stage between Holsteinian and Eemian. They correlated this intra-Saalian temperate stage with the Hoogeveen Interstadial.

The fossil marine record

Late Pliocene

Starting in the Late Pliocene, the marine molluscan assemblages indicate moderately deep to shallow, fully marine conditions. The faunas are of high diversity and include many temperate species. In The Netherlands these faunas represent assemblage zone C (Spaink 1975), which has yielded a total of 160 molluscan species. This zone most probably correlates with the Red Crag (Waltonian) assemblages of East Anglia and with the molluscan faunas of the Belgian Kruisschans and Merksem Members. The British and Belgian assemblages likewise show high faunal diversity with many temperate species.

Early Pleistocene

In the North Sea Basin, marine Early Pleistocene deposits are known only from England and The Netherlands. The oldest known temperate shell-bearing deposits are of Ludhamian (= Early Tiglian) age (Norton 1967). The assemblages show considerable reworking and are therefore not discussed further.

The first well dated Pleistocene marine assemblages are from the Middle Tiglian. In large areas of The Netherlands, a marked change to shallow marine conditions with lowered salinities is indicated by the molluscan assemblages, which point to intertidal conditions resembling a wadden environment. The assemblages are impoverished when compared with the Late Pliocene, but southern species are still present and faunal diversity remains high (140 species).

Biostratigraphically these faunas are assigned to assemblage zone A (Spaink 1975), and correlated with the faunas of the Norwich Crag Formation in East Anglia, which are of Bramertonian/Antian age. The environmental conditions reflected by the Norwich Crag assemblages closely resemble those of Dutch Middle Tiglian. The number of species and the frequency of southern species are also comparable.

In The Netherlands, most marine deposits of Late Tiglian age are decalcified, and shell-bearing deposits have been encountered only in the Zuurland borehole at Brielle (Meijer 1988a). Although the faunas may be assigned to assemblage zone A, many species characteristic of this zone are missing. The species point to a tidal environment with a strong fluvial influence. The total number of marine species is low (14) and southern species are virtually absent.

In East Anglia, the faunas of the Weybourne Crag (Pre-Pastonian a) are thought to be of similar age to those of the Late Tiglian (Gibbard et al. 1991). The marine fauna (with *Macoma balthica*) is impoverished when compared with that from the Norwich Crag.

Marine deposits of supposedly Waalian age are known from only a few boreholes in The Netherlands. The marine molluscan assemblages are very similar to the Dutch Late Tiglian faunas, but are even more impoverished (Meijer 1987). No other Early Pleistocene marine deposits are known after the Waalian.

It is clear that the marine molluscan assemblages of the Early Pleistocene generally indicate shallow sub-tidal to tidal conditions and can be sub-divided into two main groups. Assemblages with a large number of species, a high diversity and presence of temperate species occur in the Middle Tiglian. After this stage the assemblages are impoverished, of low diversity and with a virtual absence of southern species.

Middle Pleistocene

In the early Middle Pleistocene, the '*Leda myalis* Bed', which occurs above the type Cromerian sequence at West Runton, Norfolk and in neighbouring sections, contains a boreal marine assemblage of low diversity lacking temperate species (Norton 1980; Gibbard et al. 1991; Meijer & Preece 1995). *Yoldia* and *Mya* are important species. In view of the boreal character of this assemblage, it is possible that this unit belongs to the early Anglian, rather than the 'Cromerian' *sensu stricto* (cf. West 1980). There is no counterpart of this deposit in The Netherlands.

During the latter part of the 'Cromerian Complex', interglacial shallow marine deposits, reflecting an estuarine environment, were laid down in the northern part of The Netherlands (Tesch 1942). These belong to the Noordbergum Interglacial, which has been designated the stratotype of 'Interglacial IV' of the Dutch sequence. The marine assemblages have very low diversity, rarely exceeding five species, and completely lack southern species. An overall total of 17 marine species is known from this stage (Meijer & Preece 1995), which seems to have no equivalent elsewhere in the North Sea Basin.

On the English Channel coast, high-level marine deposits occur at elevations of about 42 m OD at Bembridge, Isle of Wight (Steyne Wood Clay) and at Boxgrove, Sussex (Holyoak & Preece 1983). Both these deposits are thought to be early Middle Pleistocene in age and to represent an interglacial at the end of the 'Cromerian Complex' (Preece et al. 1990). A typical estuarine fauna was recovered from Bembridge (Holyoak & Preece 1983) and a rocky shore fauna, with *Nucella lapillus* and *Littorina saxatilis*, was present at Boxgrove (R. C. Preece & M. R. Bates unpublished data). The only really noteworthy marine species at Boxgrove is the sinistral whelk *Neptunea contraria*, which today lives along the coast of Portugal and does not reach the English Channel. This species is distinct from the Red Crag species usually known by that name (Nelson & Pain 1986).

Marine deposits of Holsteinian age are more widely distributed in The Netherlands than those of preceding interglacials. In addition to the northern region, marine deposits are also found in the west. The faunal composition is almost exactly the same as that of the Noordbergum Interglacial (Gibbard et al. 1991).

Outside The Netherlands, the oldest late Middle Pleistocene temperate marine shell-bearing deposits are from the Holsteinian/Hoxnian. In England, assemblages from several localities have been attributed to this interglacial, including the Nar Valley, Woodston, Clacton-on-Sea and elsewhere in eastern Essex. With the exception of Woodston, no modern malacological research has been undertaken on the marine molluscs from any of these sites. Seventeen marine species have been reported from Clacton (Kennard & Woodward 1923; Warren 1923, 1955; Baden-Powell 1955; Turner & Kerney 1971), but most of these were already mentioned in very early work carried out in the last century and have simply been repeated by later authors without confirmation. New collections need to be made from each of the various channels at Clacton, as it seems likely that they do not all belong to the same stage.

The marine molluscan assemblages from the Nar Valley also require reappraisal, as the records are based on very old data. Baden-Powell (1967) produced a composite list of 37 species, of which about 15 may be considered temperate elements. The identifications of several species are questionable (e.g. *Parvicardium exiguum*) and require re-examination.

The Woodston marine assemblage is very small and points to an intertidal and estuarine environment (Horton et al. 1992). The brackish water hydrobiids *Semisalsa stagnorum* and *Mercuria confusa* appear to be the only temperate elements.

Table 4. *Marine and brackish molluscs known from Hoxnian deposits in England compared with Holsteinian and Eemian deposits on the continent.*

Species	1	2	3	4
Mytilus edulis	x	x	x	H,E
Monia patelliformis	-	-	x	E
Ostrea edulis	x	x	x	H,E
Chlamys varia	x	-	x	E
Mysella bidentata	x	-	x	H,E
Kellia suborbicularis	-	-	x	E
Acanthocardia echinata	-	-	x	E
Cerastoderma edule s.l.	x	x	x	H,E
Parvicardium exiguum	x	-	-	E
Venerupis decussata	x	-	x	E
Spisula elliptica	-	x	x	?H,E
Spisula solida	-	-	x	E
Spisula subtruncata	-	-	x	H,E
Macoma balthica	x	-	x	H,E
Macoma calcarea	-	-	x	H,E
Solecurtus scopula	-	-	x	?
Abra alba	-	-	x	E
Scrobicularia plana	x	x	x	H,E
Corbula gibba	-	-	x	E
? *Mya arenaria*	-	-	x	-
Mya truncata	-	-	x	H,E
Pholadidae	x	-	-	?
Gibbula cineraria	-	-	x	E
Littorina littorea	x	x	x	H,E
Littorina obtusata s.l.	-	-	x	H,E
Littorina rudis s.l.	x	-	x	H,E
Hydrobia ulvae	x	x	x	H,E
Hydrobia ventrosa	x	x	-	H,E
Semisalsa stagnorum	x	x	-	E
Paladilhia radigueli auct.	x	-	-	-
Mercuria confusa	x	x	-	-
Turritella communis	-	-	x	H,E
Turritella pliorecens	-	-	x	-
Epitonium clathrus	-	-	x	E
Euspira poliana	-	-	x	H,E
Bittium reticulatum	-	-	x	E
Aporrhais pespelicani	-	-	x	H,E
Buccinum undatum	-	-	x	H,E
? *Hinia incrassata*	-	-	x	E
Hinia pygmaea	-	-	x	E
Hinia reticulata	-	-	x	?H,E
Haedropleura septangularis	-	-	x	E

Locality: 1, Clacton-on-Sea; 2, Woodston; 3, Nar Valley; 4, occurrence elsewhere in the North Sea Basin: H, Holsteinian, E, Eemian, x, Present. Further explanation in the text.

The British marine species occurring in the Hoxnian are summarized in Table 4. It should be remembered that the data are based mainly upon old literature records. Their occurrence outside England in Holsteinian as well as in Eemian deposits is also indicated. It appears that 16 species out of 41 have not been found in Holsteinian deposits on the continent, and are known only from the Eemian. These are all temperate species. Of the remainder, 20 have been found in Holsteinian *and* in Eemian deposits, while five are unknown from either interglacial outside England. Most of the species known from both interglacials are from the Celtic province, whereas two are transitionally Lusitanian/Celtic.

It is uncertain how the differences between the English Hoxnian and the continental Holsteinian marine faunas may be explained. One possible explanation is that the English and continental assemblages do not belong to the same interglacial. It is also possible, although

Table 5. *Molluscs known from the Herzeele Formation (NW France and SW Belgium).*

Species	1	2	3	4	5	6	7
Littorina saxatilis	-	-	-	-	-	-	x
Hydrobia ulvae	x	x	x	x	x	x	x
Hydrobia ventrosa	x	x	x	x	x	x	x
Semisalsa stagnorum	x	x	x	x	x	x	x
Rissoa membranacea	-	-	-	-	-	-	x
Nucella lapillus	x	-	x	x	x	x	x
Oenopota turricula	-	-	x	-	x	-	-
Chrysallida spiralis	-	-	-	-	-	-	x
Retusa obtusa	-	x	x	-	-	x	x
Ovatella myosotis	x	-	-	-	-	-	x
Mytilus edulis	x	x	-	x	x	x	x
Mytilaster minima	-	-	-	-	-	-	x
Cerastoderma edule	T	T	T	T	T	T	x
Cerastoderma glaucum	?	?	?	?	?	?	x
Macoma balthica	x	x	x	x	x	x	x
Scrobicularia plana	x	x	x	x	x	x	x
Abra tenuis	x	-	-	-	-	x	x
Barnea candida	-	-	-	-	-	-	x
**Theodoxus danubialis*	x	-	x	x	x	x	x
Bithynia tentaculata (opercula)	-	-	-	-	-	-	x
Lymnaea truncatula	-	-	x	-	-	-	-
Limacidae	-	-	-	-	-	-	x
Aegopinella nitidula	-	-	-	-	-	x	-

Locality: 1, Vinkem (Belgium), Izenberge Crag; 2, Lo, near road to Kaaskerke (Belgium), Izenberge Crag; 3, Izenberge, Rijckeboer farm (Belgium), Izenberge Crag; 4, Beveren, road to Eikhoek (Belgium), Izenberge Crag; 5, Beveren, De Hand farm (Belgium), Izenberge Crag; 6, Gijverinckhove, Degraeve farm (Belgium), Izenberge Crag; 7, Pit Heem near Herzeele (France), Herzeele Formation, Unit 4 (According to Sommé et al. 1978)

X, present; T, according to Tavernier & de Heinzelin (1962) (specimens not seen); ?, presence uncertain; *, *Bela plicifera* in Tavernier & de Heinzelin (1962); **, *Theodoxus fluviatilis* in Tavernier & de Heinzelin (1962).

less likely from palynological evidence, that the English sites represent more than one interglacial. Indeed, Bowen et al. (1989) have obtained higher amino acid ratios from the Nar Valley (and Swanscombe) than from Hoxne itself, perhaps suggesting different ages. Ratios from Woodston, however, were lower and did broadly match those obtained from Hoxne (Bowen et al. 1989; Horton et al. 1992). Lower ratios have also been obtained from foraminifera from the Nar Valley Clay at Tottenhill compared with those of the same species from the Sand Hole Formation of the Inner Silver Pit (J. D. Scourse, pers. comm.), which was originally thought to be an offshore extension of the Nar Valley sequence (Ansari 1992). Moreover, evidence from Schöningen near Helmstedt, Germany, has recently revealed three interglacial horizons between the Elsterian and Saalian tills. In the light of such evidence, it would be wise to exercise caution in the attribution of incomplete late Middle Pleistocene interglacial sequences to either the Hoxnian or Holsteinian.

Holsteinian marine assemblages from Germany were reviewed by Grahle (1936). He listed a total of 55 species from 55 localities, but included assemblages from underlying late Elsterian deposits. Only a small number of species are known (mean, seven species) from each of the localities, but the most diverse assemblages reach 18 species. The Holsteinian type locality at Hummelsbüttel, near Hamburg, has yielded a very impoverished intertidal assemblage, lacking temperate species. The assemblage from Wacken again lacks temperate taxa and includes several cool species. This assemblage, comprising just 15 species, indicates tidal to shallow sub-tidal conditions. Absolute dating techniques have provisionally suggested an intra-Saalian age, but the molluscan assemblage is indistinguishable from those found in other German Middle Pleistocene localities. Clearly, further work is needed.

Hinsch (1993) also analysed the German Middle Pleistocene assemblages, drawing heavily on the work of Grahle, and listed 51 species from purely Holsteinian deposits. Many of these species suggest deeper water than that indicated by the Dutch assemblages, which may perhaps account for the greater diversity. Most species today live in the Boreal and Celtic provinces; only a few (*Aporrhais pespelicani*, *Hinia reticulata* and *Ostrea edulis*) are transitional and these may be considered the warmest elements in these assemblages.

In NW France, the Herzeele Formation was defined by Sommé *et al.* (1978). The shallow marine/estuarine deposits extend into western Belgium, where they are known as the '*Cardium* sands of Lo' (Tavernier & de Heinzelin 1962). The fauna of Herzeele (Table 5) has been studied, together with a re-examination of the faunas from the Belgian localities (Meijer 1988*b*). The assemblages are of low diversity and point to lowered salinities in a shallow marine sub-tidal environment. They include several temperate elements such as *Semisalsa stagnorum*, *Rissoa membranacea*, *Ovatella myosotis* and *Mytilaster minima*. In the North Sea basin, *M. minima*, today an intertidal species from the Mediterranean/Lusitanian province, is known only from the Holsteinian deposits at Herzeele. The other species occur in brackish water and/or supra-tidal conditions.

The Middle Pleistocene marine assemblages are all poor in numbers of species, are of low diversity and virtually lack warm temperate elements. The temperate species that do occur are small and possibly prone to passive dispersal by birds (*Mercuria confusa*, '*Paladilhia radigueli*', *Semisalsa stagnorum*, *Rissoa membranacea*, *Ovatella myosotis*, *Mytilaster minima*).

Eemian

Marine Eemian deposits from The Netherlands yield rich, highly diverse assemblages with many Lusitanian and Mediterranean species. The total number of marine species known from the Dutch Eemian is 150 (Spaink 1958). The assemblages are indicative of a variety of habitats ranging from supra-tidal to water depths of several dozens of metres. Most sub-tidal assemblages point to quiet, euryhaline conditions with clear water and rich submerged vegetation, such as *Zostera* meadows. There are many epibionts that would have lived on hard substrates. These may initially have lived on glacial boulders and subsequently on extensive oyster beds. Temperate species typical of the Dutch Eemian include *Mytilaster lineatus*, *Flexopecten flexuosus*, *Lucinella divaricata*, *Acanthocardia tuberculata*, *Sphaerocardium paucicostatum*, *Venerupis aurea*, *Abra ovata*, *Gastrana fragilis*, *Angulus distortus*, *Gastrochaena dubia*, *Gibbula magus*, *Turboella radiata*, *Bittium reticulatum*, *Ocenebra erinacea*, *Hinia pygmaea*, *Haedropleura septangularis*, *Ebala nitidissima* and *Haminea navicula* (Harting 1852, 1875; Brouwer 1941, 1943*a*, *b*; Spaink 1958, 1973; Doeksen 1975, 1991; Hordijk & Janse 1987; Meijer 1988*a*).

In England, fossiliferous marine Ipswichian (= Eemian) deposits are not widespread. Along the North Sea coast, shell-bearing deposits that have been attributed to this stage occur around the Wash (March Gravels), in east Yorkshire (Sewerby and lower part of the Speeton Shell Bed) and Durham (Easington Raised Beach). Of these, only the Raised Beach at Sewerby with *Hippopotamus* can be ascribed to the Ipswichian with any confidence (Catt & Penny 1966). Shells from the Easington Raised Beach, which occurs on a rock platform at about 32 m OD, have yielded amino acid ratios that would seem to indicate a pre-Ipswichian age (Bowen *et al.* 1991). The March Gravels have yielded moderately diverse assemblages (Baden-Powell 1934; Keen *et al.* 1990) including species such as *Monia patelliformis*, *Pholas dactylus*, *Gibbula cineraria*, *Ocenebra erinacea* and *Hinia pygmaea*, which occur in Eemian deposits in The Netherlands and elsewhere. However, recent work has suggested that these interglacial marine shells are older, since they have not only yielded amino acid ratios suggestive of a pre-Ipswichian age, but they also occur in terrace deposits above those in which Ipswichian organic sediments have been found (Bridgland *et al.* 1991). It must be remembered that many of these marine shells appear to have been reworked into fluvial gravels of a subsequent cold stage (West 1987). Other English deposits of Ipswichian age accumulated in predominantly freshwater environments subject to minor tidal influence (e.g. Holyoak & Preece 1985).

Marine shell-bearing Eemian deposits are also known from Denmark, Germany and Belgium (Dittmer 1941; Hinsch 1985, 1993; Janssen 1975; Nolf 1974; de Moor & de Breuck 1973; Nordmann 1908, 1928). These assemblages include high proportions of thermophilous sub-tidal species, and are similar to those from the Eemian type area in The Netherlands.

In conclusion, marine assemblages of the Eemian are widespread, but not in Britain, and are rich in species, have high diversities and contain many thermophilous sub-tidal elements.

Holocene and present

Holocene marine molluscan assemblages from

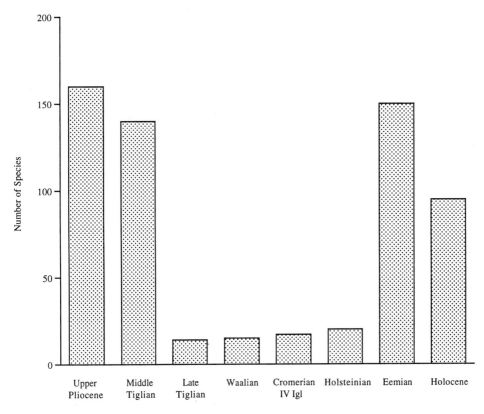

Fig. 3. Histograms showing the number of marine molluscs known from various temperate stages in The Netherlands.

The Netherlands are rich in species, have a high diversity and include many thermophilous sub-tidal taxa (e.g. Janssen 1981; Meijer 1984, 1988a; Raven 1979, 1982, 1983). A total of 95 marine species have been recorded from the Dutch Holocene, the composition being virtually indistinguishable from the modern molluscan fauna from the Dutch coast, where 104 species are known. A high proportion of southern species also occurs in the recent fauna (Janssen 1975). Significantly, the proportion of warm species in the allochthonous fauna (dead shells, washed ashore) is much higher than in the autochthonous fauna. For example, ten 'Mediterranean' taxa are known only from dead shells on the Dutch coast, and much higher frequencies of Lusitanian taxa (134 as opposed to 40 known live) are represented as dead sshells.

Outside The Netherlands, little information has been published on Holocene marine molluscan faunas. In Belgium, the boreholes of the De Panne area have yielded assemblages rich in species with many thermophiles (Spaink & Sliggers 1983). The same is true for Danish localities (Petersen 1994).

The Holocene and modern marine assemblages essentially have close affinities to those of the Eemian. There are a large number of species and the assemblages have a high diversity, but there is a lower proportion of thermophiles in the Holocene.

Species richness and biogeographical composition of the total marine fauna known from the temperate stages of the Dutch Quaternary are given in Figs 3 and 4. From these data a threefold sub-division emerges:

1. Late Pliocene–Middle Tiglian: high number of species, low proportion of Celtic species, many warm sub-tidal taxa.
2. Late Tiglian–Holsteinian: low number of species, high proportion of Celtic species, very few warm sub-tidal taxa.
3. Eemian–Holocene: high number of species, low proportion of Celtic species, many warm sub-tidal species.

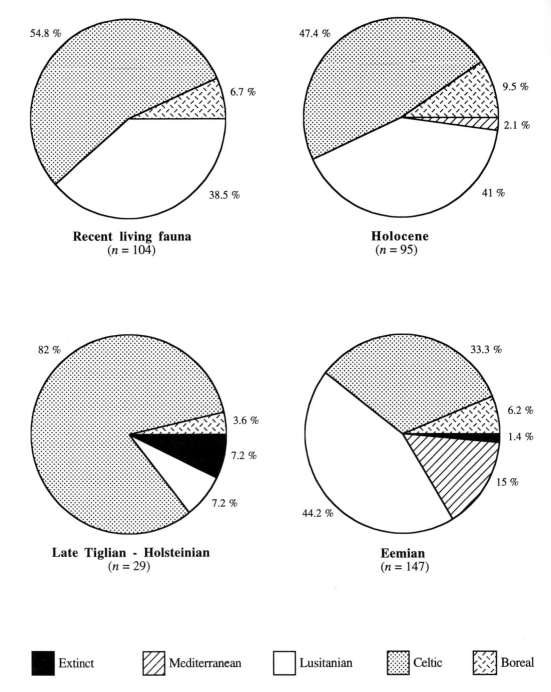

Fig. 4. Biogeographical composition of marine molluscan assemblages from various temperate stages in The Netherlands. Data for the modern fauna are taken from Janssen (1975).

A comparison with marine Quaternary assemblages elsewhere in the North Sea shows that this is the common pattern and we therefore consider it to be generally valid for the whole basin.

The fossil fluvial record

Late Pliocene

Deposition in the North Sea basin during the Late Pliocene is predominantly marine and virtually no significant fluvial molluscs are known.

Lower part of the Early Pleistocene

The earliest significant records of fluvial molluscs come from the Middle Tiglian. In The Netherlands, fluvial species are found in the Tegelen Formation (Rhine deposits) and in the interdigitating shallow marine Maassluis Formation. Stratigraphically, the most important fluvial species are: *Theodoxus danubialis* (= *serratiliniformis*), *Viviparus glacialis*, *Valvata goldfussiana*, *Lithoglyphus jahni*, *Tournouerina belnensis* and *Sphaerium rosmalense* (Meijer 1990a). In the contemporaneous deposits on the British side of the basin, these species are unknown. There, in the shallow marine Norwich Crag Formation of Bramertonian/Antian age, fluvial species such as *Viviparus medius* and *Sphaerium icenicum* occur (Holyoak & Preece 1986; Gibbard *et al.* 1991). These, in turn, are unknown from equivalent deposits of Middle Tiglian age on the continent. The main river systems on either side of the basin, the Rhine and the Thames, therefore appear to exhibit provinciality during the Middle Tiglian.

In England, *Viviparus medius* and *Sphaerium icenicum* have not been found in deposits of the subsequent 'Pre-Pastonian a' and later stages. Instead, *Viviparus glacialis*, *Valvata goldfussiana* and *Lithoglyphus jahni* appear in the Weybourne Crag and the Bure Valley Beds, which are thought to equate with the Late Tiglian (Gibbard *et al.* 1991). In The Netherlands these species persist into the Late Tiglian (and in the case of *V. goldfussiana* into the Bavelian). This similarity in the fluvial prosobranch fauna points to an absence of provinciality during the Late Tiglian.

In conclusion, the lower part of the Early Pleistocene fluvial assemblages show a two-fold sub-division: provinciality during the Middle Tiglian and non-provinciality, pointing to fluvial exchange, during the Late Tiglian.

Upper part of the Early Pleistocene and Middle Pleistocene

In England, the type Cromerian at West Runton, and other Cromerian sites such as Sugworth near Oxford, have yielded assemblages with several characteristic fluvial species such as *Viviparus viviparus gibbus*, *Valvata goldfussiana* and *Tanousia runtoniana* (Meijer & Preece 1995). In The Netherlands, these species have their last appearances in late Early Pleistocene to early Middle Pleistocene deposits (Bavelian to early 'Cromerian Complex').

In borehole E8/4 (54°32'44" N, 3°24'16" E), situated in the middle of the North Sea (Fig. 1), *Valvata goldfussiana* occurs in a fluvial assemblage dating from the early 'Cromerian Complex' (Meijer & Preece 1995). This occurrence provides direct palaeontological evidence for interglacial fluvial activity in the middle of the North Sea during a period when deltas from the Rhine and other rivers extended far beyond present coastlines (Zagwijn 1979; Cameron *et al.* 1992).

The assemblages of the Bavel and Leerdam Interglacials (Bavelian), West Runton (Cromerian *sensu stricto*) and North Sea borehole E8/4 suggest an absence of provinciality between England and The Netherlands during the late Early and early Middle Pleistocene.

The late Middle Pleistocene fluvial deposits near Swanscombe, Kent, were clearly formed after the Thames was diverted to its present course during the Anglian, and are thought to be Hoxnian in age (but see comments above). A striking feature of the Swanscombe sequence is the change in character of the freshwater fauna which occurs between the Lower and the Middle Gravels. Assemblages from the upper part of the Swanscombe aggradation include a number of species (*Theodoxus danubialis* (= *serratiliniformis*), *Viviparus diluvianus*, *Valvata naticina*, *Belgrandia marginata*, *Corbicula fluminalis*, *Pisidium clessini* and others) absent from the lower deposits (Kerney 1971). Two of these species (*Theodoxus danubialis* and *Pisidium clessini*) first appear, very rarely, near the top of the Lower Loam. Since certain members of this assemblage today have central European affinities, Kennard (1942*a*, *b*) suggested that the Thames and Rhine river systems only became linked at a time after the deposition of the Lower Gravel.

A similar pattern occurs at Clacton-on-Sea, Essex, where the molluscan succession can be related, in part, to different sub-stages of the Hoxnian by means of pollen analysis. At

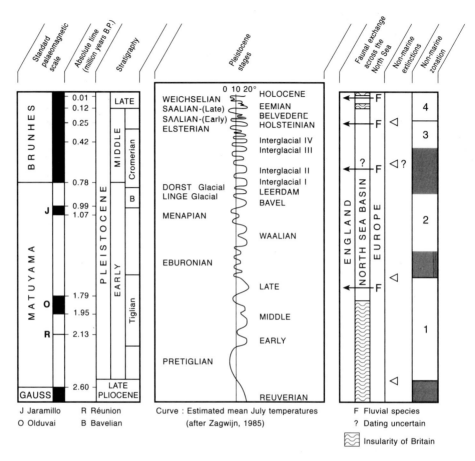

Fig. 5. Summary diagram showing the periods when Britain appears to have been isolated and episodes when fluvial exchange occurred (arrows). Major molluscan zones (Meijer 1986) are numbered and the more important extinction horizons are shown. Periods from which no molluscan data are known are shaded.

Clacton, freshwater deposits, yielding pollen spectra attributable to Ho IIb and Ho IIIa, are overlain by estuarine beds that began to accumulate during the latter part of Ho IIIa (Turner & Kerney 1971). Except for *Pisidium clessini*, members of the so-called 'Rhenish fauna' are absent from the lowermost freshwater beds. *Belgrandia marginata* first appears in their upper part and *Corbicula fluminalis*, *Viviparus diluvianus* and a few other taxa are unknown before the estuarine stage (Kerney 1971).

Kerney (1971) used the evidence from Clacton to suggest that the Lower Gravel and Lower Loam at Swanscombe formed during the early temperate sub-stage (Ho II), whereas accumulation of the Middle Gravels began not before the end of the late temperate sub-stage (Ho III), and probably extended into the early glacial sub-stage. The abrupt change in the freshwater fauna supports the view that there is a hiatus in the succession, as does the presence of a palaeosol at the top of the Lower Loam (Kemp 1985). Kerney suggested that the presence of *Corbicula* and a well developed 'Rhenish' fauna at the base of the Lower Middle Gravel indicated a correlation with Ho IIIb, rather than with Ho IIIa.

Several members of the 'Rhenish' fauna (*Theodoxus serratiliniformis* (= *danubialis*), *Viviparus diluvianus*, *Corbicula fluminalis* and *Pisidium clessini*) have recently been discovered, with occasional brackish species, in fluvial sediments that also accumulated in the late temperate sub-stage (Ho IIIb) at East Hyde near Tillingham in eastern Essex, 20 km SW of Clacton (Roe 1994; Roe & Preece 1995).

Elsewhere in NW Europe, several of these 'Rhenish' species have their last appearance during the Holsteinian, but ironically not all are known from the Rhine system and most are not confined to it. For example, *Theodoxus danubialis* is unknown from Rhine deposits of this age, although it has been found in deposits of equivalent age from the River Scheldt occurring at Herzeele (NW France) and Lo (SW Belgium) (Meijer 1988b). The other 'Rhenish' species are characteristic of many localities of the same age, such as Neede in The Netherlands (Meijer 1972). During the late Middle Pleistocene there is apparently also evidence for the absence of provinciality.

Support for a physical link between Britain and mainland Europe during the Holsteinian is also provided by terrestrial molluscan faunas, particularly those from calcareous tufas. A highly distinctive woodland fauna, composed of a diverse mixture of biogeographical elements, has been recognized from interglacial tufas in France, Germany and Britain (Kerney 1976; Rousseau 1987; Preece et al. 1991; Rousseau et al. 1992). Perhaps the most remarkable discovery in these assemblages is the occurrence of *Lyrodiscus*, a sub-genus of *Retinella*, which today is confined to the Canary Islands (Rousseau & Puisségur 1990). Most of these tufas are thought to be Holsteinian in age and have been correlated with oxygen isotope stage 11 of the marine record (Rousseau et al. 1992).

It may be concluded that during the late Early to Middle Pleistocene fluvial assemblages point to fluvial exchange between England and the continent. Terrestrial faunas on either side of the Channel also appear to have been similar.

Late Pleistocene to present

After the Holsteinian, the fluvial molluscan faunas of NW Europe suffered the last major episode of extinctions (Fig. 5). Several species did not reappear in this region after this interglacial and others became totally extinct. Consequently, the fluvial component of freshwater assemblages in NW Europe declined sharply after the Holsteinian. However, one species, *Theodoxus fluviatilis*, appears for the first time in NW Europe after this interglacial (Steusslof 1953). It is unknown from the British Ipswichian, but is present in many Dutch Eemian sequences and occurs in the Holocene in many countries in NW Europe, including Britain (Kerney 1977).

In England, the unique occurrence of *T. fluviatilis* during the Holocene may perhaps be explained by a later and slower transgression, compared with the Eemian, combined with less distant refugia. Holocene estuarine assemblages, with temperate shallow marine and freshwater species including *T. fluviatilis*, are known from the centre of the North Sea (Meijer 1988b). It therefore seems likely that fluvial faunal exchange was still possible during the early Holocene. Comparable Eemian deposits from the central North Sea are unknown, which suggests a pre-interglacial separation of British and continental river systems.

Another fluvial species, *Margaritifera auricularia*, occurs in the British Ipswichian (Kerney 1958) and possibly slightly earlier (Preece 1988), but is unknown from the Dutch Eemian, the opposite situation to that of *T. fluviatilis*. *M. auricularia* is a large freshwater mussel characteristic of large lowland rivers, but is rare and possibly extinct today. It is known from the Holocene in The Netherlands, England and elsewhere (Preece et al. 1983) and occurred in the Rivers Rhine and Meuse in historical times (Kuijper 1988). Its disappearance from many European rivers in comparatively recent times has been attributed to anthropogenic causes. The different histories of *T. fluviatilis* and *M. auricularia* during the Late Pleistocene suggest fluvial provinciality during the Last Interglacial (Eemian/Ipswichian) but not during the Holocene. This may perhaps be explained by the different patterns and timings of the transgressions.

Another example that may possibly indicate fluvial provinciality during the Last Interglacial is the case of *Corbicula fluminalis*. This freshwater bivalve is known from many Dutch Eemian sites (e.g. Meijer 1990b), but in Britain there is a growing consensus that it was absent during the Ipswichian (e.g. Jones & Keen 1993). It has been recorded from many sites that have been attributed to the Ipswichian (e.g. Sparks 1964) but several of these sites are now thought to date from an earlier interglacial. This reappraisal is based on various lines of evidence. It is well known that in Britain *Corbicula* and *Hippopotamus* (taken to be indicative of the Last Interglacial) have never been found in direct association. Horses (and evidence of human activity) were also apparently absent during the Last Interglacial but are not uncommon at sites with *Corbicula* (Sutcliffe 1995). Moreover, amino acid ratios from *Corbicula* sites are invariably higher than those from 'hippo' sites (Bowen et al. 1989). *Corbicula*, as discussed earlier, is an invasive colonizer and would have readily spread from the continent into Britain had any linkage existed. The fact that it appears not to have done so suggests that Britain had become isolated.

The following periods may now be distinguished using evidence from fluvial molluscan assemblages:

1. Middle Tiglian: fluvial provinciality of Rhine and Thames.
2. Late Tiglian–Holsteinian: fluvial exchange between England and the continent.
3. Eemian–Holocene: fluvial provinciality of Rhine and Thames.
4. Holocene: fluvial exchange between England and the continent during the early part of the interglacial.

Unresolved problems

The occurrence of some thermophilous sub-tidal marine species in several British localities (such as the Nar Valley and Clacton-on-Sea) remains problematic. As stated, the observations were made long ago and urgently need confirmation. Besides, it is clear that a multidisciplinary re-examination of these important sites is needed to elucidate new stratigraphical problems. This should clarify their value concerning the issue of insularity.

In the Middle and Late Pleistocene, species of the freshwater prosobranch genus *Theodoxus* provide important evidence that has a direct bearing on several palaeogeographical and stratigraphical questions. In The Netherlands, and elsewhere in NW Europe, *T. fluviatilis* is unknown from any pre-Eemian deposit. Several older records have proved to be errors for *T. danubialis*, which is a typical Middle Pleistocene species in NW Europe (Meijer 1969, 1988b). The occurrence of *T. fluviatilis* in Holsteinian deposits at Bad Oldesloe (Grahle 1936) likewise requires confirmation, as do the Belgian records in the Oostende Formation (de Moor & Heyse 1975). The presence of *T. danubialis* in the excavation at Zelzate in Belgium (Janssen 1965; Meijer 1969), however, does pose a problem. The estuarine deposits at this locality have been assigned to the Eemian. This attribution is based on the correlation of the associated soil with the Rocourt Soil (Paepe & Zagwijn 1972), and the occurrence of typical marine Eemian species such as *Bittium reticulatum*, *Hinia reticulata*, and *Venerupis aurea*. We consider the malacological data to be the most reliable. Amino acid epimerization data have recently been obtained from Zelzate, and from an adjacent borehole (Bakkersdam, 54A/34) in The Netherlands, in which *T. danubialis* was found in a similar stratigraphical position. Two specimens of *Macoma balthica* from Zelzate gave D/L ratios (total fraction) of 0.204 and 0.297 (mean 0.251 ± 0.066; ABER-1249), whereas three specimens from Bakkersdam gave D/L ratios of 0.259, 0.267 and 0.315 (mean 0.280 ± 0.030; ABER-1248). These ratios are indicative of a pre-Eemian age. If this is correct, then several species thought to typify the Eemian have been shown to occur in a pre-Eemian temperate stage. This stage, however, is not necessarily the Holsteinian but could be an intra-Saalian warm event.

If this is the case, it could explain the occurrence of these species in pre-Eemian deposits in England and Germany. Marine deposits of early Saalian age are known from Wacken in Germany (Grahle 1936), where extremely impoverished molluscan assemblages have been recovered, but inadequate sampling may well account for the apparent absence of warm sub-tidal species there.

The precise dating of the deposits at Zelzate and Bakkersdam does not affect our hypothesis. If these sites are of Middle Pleistocene age, then the occurrence of *T. danubialis* points to non-provinciality between England and the continental mainland. These localities can then be grouped with those at Herzeele, Swanscombe and East Hyde. If they are of Eemian age, then they suggest provinciality of the River Scheldt with respect to the British Isles. A similar conclusion is reached from the occurrence of *T. fluviatilis* in Rhine deposits of Eemian age.

Discussion

The above survey has revealed that the marine molluscan faunas from different temperate stages of the Pliocene and Pleistocene vary enormously both in terms of species richness and in the diversity of their biogeographical composition (Table 6, Fig. 5). The marine assemblages from the Middle Tiglian are extremely rich (140 species) and contain a relatively high proportion of species with 'Lusitanian' and Mediterranean affinities, although less than in the Late Pliocene. A southern seaway connecting the North Sea with the Atlantic Ocean is thought to have existed at this time, a conclusion supported by foraminiferal evidence (Funnell & West 1977; Funnell 1995). Moreover, freshwater molluscan faunas on either side of the North Sea show differences suggesting fluvial provinciality, although the British data are based on allochthonous shells washed into marine sediments (Norwich Crag Formation).

The character of the marine molluscan assemblages from the Late Tiglian is completely different. These are impoverished (< 20 species) and southern species are no longer present. In

Table 6. *Composition and significance of the marine molluscan faunas from temperate stages of the Pliocene and Pleistocene of the North Sea Basin.*

Temperate stages	A	B	C	D
Middle Tiglian	High	High	Present	Yes
Late Tiglian–Holsteinian	Low	Low	Absent	No
Eemian–Holocene	High	High	Present	Yes

A, marine species number; B, proportion of marine southern species; C, fluvial provinciality; D, existence of Strait of Dover

The Netherlands, similar depauperate marine assemblages are also known from the Waalian, temperate stages of the 'Cromerian Complex' and Holsteinian. British data from the late Early Pleistocene are sparse, although the marine faunas from the Weybourne Crag are likewise impoverished. Freshwater shells (e.g. *Viviparus glacialis*, *Lithoglyphus jahni*) again occur as allochthonous elements in the Weybourne Crag and Bure Valley Beds and have provided evidence for their correlation with the Late Tiglian (Gibbard et al. 1991). They also suggest that the fluvial provinciality of the Middle Tiglian has broken down, as these taxa are common in equivalent deposits in The Netherlands.

The marine molluscan faunas from the British Hoxnian are known from several localities in eastern England. Some of these sites, such as Woodston, have likewise yielded only restricted faunas of brackish or estuarine character. However, old records from the Nar Valley and from some of the channels at Clacton suggest the existence of rather more diverse assemblages. These pose a problem and suggest that correlation with the impoverished Dutch Holsteinian assemblages may not be sound. Moreover, amino acid data have suggested that not all of these British sites necessarily belong to one stage (e.g. Bowen et al. 1989), despite pollen-stratigraphical similarities. This suggestion requires further testing.

It is instructive to establish when the marine transgression occurred within the interglacial. At Woodston the earliest marine sediments occur within Ho IIc and lie between 11 and 14 m OD (Horton et al. 1992). In the Nar Valley they also occur within Ho IIc, but reach a maximum during Ho IIIb, by which time they have aggraded to 25 m OD (Ventris 1985). At Clacton, the first signs of the transgression appear later in Ho IIIa, but the 'Estuarine Beds' accumulated mostly during Ho IIIb, do not reach more than 11 m OD (Warren 1955). It is not clear whether the differences in the height of these marine sediments reflect differential movements between the sites or whether they can be explained by local physiographic factors.

If the marine transgression occurred at the end of Ho II (or beginning of Ho III), as it did in both Britain and The Netherlands (P. Cleveringa, pers. comm.), then it is difficult to explain the main influx of 'Rhenish' species into the Thames system at about the same time. One possibility is that the English and continental rivers were linked early in the interglacial, and the late appearance of 'Rhenish' elements might simply be due to retarded immigration reflecting more distant refugia. Examples of modern colonizations by freshwater molluscs suggest that it would not have taken long for several of these species to have spread throughout the Thames system once access had been achieved. Presumably this colonization must have occurred before the main marine transgression. The terrestrial molluscan fauna from the Lower Loam at Swanscombe (i.e. before the main influx of 'Rhenish' species) contains many forest species, including several with modern central European ranges (Kerney 1971). Likewise, the distinctive *Lyrodiscus* faunas from penecontemporaneous calcareous tufas suggest that Britain was joined to the continental mainland for much of the first half of the interglacial. This conclusion is also supported by vertebrate evidence (Stuart 1995).

Such a conclusion apparently conflicts with the widely held view that the chalk joining Britain and France was breached by an overflow from an impounded proglacial lake in the southern North Sea during the Anglian/Elsterian (cf. Gibbard 1995). However, there is likely to have been a lag between the breaching and the main eustatic rise of sea-level. Moreover, it is likely that the sea-level history following deglaciation of this major cold stage would have been extremely complex, involving interactions of glacio-isostasy, hydro-isostasy and related processes. These, and possible forebulge effects around previously isostatically depressed coastlines, may hold the key as to why southern marine molluscs are absent from continental

Holsteinian sequences and why terrestrial organisms such as snails and amphibians were able to enter Britain.

Interpretation of events in the late Middle and Late Pleistocene are similarly hampered by uncertainty regarding the correlation of deposits. Marine molluscan faunas from Dutch Eemian deposits are rich and biogeographically diverse and there seems little doubt that southern taxa reached the North Sea via the Strait of Dover. Evidence from freshwater molluscan assemblages, although not as strong, points to fluvial provinciality, also suggesting that Britain had become an island at this time. Amino acid data suggest that some assemblages that have been attributed to the Ipswichian/Eemian may belong to earlier interglacial stages. If correct, this suggests that Britain may also have become isolated during one or more interglacial stages before the Ipswichian/Eemian (Keen 1995).

The sea-level history of the Holocene has been intensively studied and is moderately well known. The marine molluscan faunas, like those of the Eemian, are rich in species and include southern elements. Radiocarbon dating of index points, coupled with a consideration of bathymetric data, suggest that Britain last became isolated about 8500 years ago or somewhat earlier (Jelgersma 1979). Radiocarbon dates of 9560 years BP have been obtained from cockle shells (*Cerastoderma edule*) which lived in the initially brackish waters of the Southern Bight (Eisma *et al.* 1981) but this area did not become fully saline until about 7000 years BP (Funnell 1995). Unlike the situation in the Eemian, fluvial provinciality is not exhibited during the Holocene. Presumably the transgression must have occurred relatively earlier and more rapidly in the Eemian than in the Holocene or the refugia of the critical taxa must have been more distant.

T. Meijer wishes to thank the Director of the Rijks Geologische Dienst for permission to publish the results of this research. We are grateful to Adri Burger, Piet Cleveringa, Robert-Jan van Leeuwen, David Bridgland, Philip Gibbard, Helen Roe and James Scourse for fruitful discussions and for critically reading earlier versions of the manuscript.

References

ADAM, W. 1960. *Mollusques terrestres et fluviatiles*. Faune de Belgique, Bruxelles.
ALZONA, C. 1971. Malacofauna Italica. *Atti della Società Italiana di Scienze Naturali e del Museo Civile di Storia Naturales, Milano*, **111**, 433.
ANSARI, M. H. 1992. *Stratigraphy and palaeobotany of Middle Pleistocene interglacial deposits in the North Sea*. PhD thesis, University of Wales, Bangor.
BADEN-POWELL, D. F. W. 1934. On the marine gravels at March, Cambridgeshire. *Geological Magazine*, **71**(5), 193–219.
—— 1955. Report on the marine fauna of the Clacton channels. *In:* WARREN, S. H. The Clacton (Essex) channel deposits. *Quarterly Journal of the Geological Society of London*, **111**, 283–307.
—— 1967. On the marine mollusca of the Nar Valley clay and their relation to the palaeolithic sequence. *Transactions of the Norfolk Norwich Naturalists' Association*, **21**, 32–42.
BIJ DE VAATE, A. 1993. Exotic aquatic macroinvertebrates in the Dutch part of the River Rhine: causes and effects. *In:* VAN DIJK, G. M. MARTEIJN, E. C. L. (eds) *Ecological rehabilitation of the River Rhine, the Netherlands Research Summary Report (1988–1992)*. Report of the project 'Ecological Rehabilitation of the rivers Rhine and Meuse', report number **50**, 27–29.
—— & GREIJDANUS-KLAAS, M. 1990. The Asiatic clam, *Corbicula fluminea* (Müller, 1774) (Pelecypoda, Corbiculidae), a new immigrant in the Netherlands. *Bulletin Zoölogisch Museum Universiteit van Amsterdam*, **12**, 173–178.
BOWEN, D. Q., HUGHES, S., SYKES, G. A. & MILLER, G. H. 1989. Land–sea correlations in the Pleistocene based on isoleucine epimerization in non-marine molluscs. *Nature*, **340**, 49–51.
——, SMITH, D. B. & SYKES, G. A. 1991. The age of the Easington Raised Beach, County Durham. *Proceedings of the Yorkshire Geological Society*, **48**, 415–420.
BOYCOTT, A. E. 1919. The freshwater Mollusca of the parish of Aldenham. *Transactions of the Hertfordshire Natural History Society and Field Club*, **17**, 153–200.
BRIDGLAND, D. R. 1994. *Quaternary of the Thames*. Geological Conservation Review Series, 7, Chapman & Hall, London.
——, DAVEY, N. D. W. & KEEN, D. H. 1991. Northam Pit, Eye, near Peterborough (TF 230036). *In:* LEWIS, S. G., WHITEMAN, C. A. & BRIDGLAND, D. R. (eds) *Central East Anglia and The Fen Basin*. Field Guide, Quaternary Research Association, 173–183.
BROUWER, J. 1941. Bijdrage tot de kennis van het hoogterras in mariene facies in de ondergrond van's Gravenhage en Wassenaar. *Geologie en Mijnbouw*, **3**, 73–84, 247–252.
—— 1943*a*. Procentgetallenonderzoek van de molluskenfauna der Nederlandsche eemlagen. *Geologie en Mijnbouw*, NS **5**, 49–60.
—— 1943*b*. Procentgetallenonderzoek van de molluskenfauna der Nederlandsche eemlagen, IV. Resultaten van het onderzoek van een aantal grondmonsters, verkregen uit boringen, gemaakt ten behoeve van de Zuiderzeewerken. *Basteria*, **8** (1/2), 20–25.
CAMERON, T. D. J., CROSBY, A., BALSON, P. S., JEFFERY, D. H., LOTT, G. K., BULAT, J. & HARRISON, D. J. 1992. *United Kingdom Offshore Regional Report: the Geology of the Southern North Sea*. HMSO, London.
CATT, J. A. & PENNY, L. F. 1966. The Pleistocene

deposits of Holderness, East Yorkshire. *Proceedings of the Yorkshire Geological Society*, **35**, 375–420.

DE MOOR, G. & DE BREUCK, W. 1973. Sedimentologie en stratigrafie van enkele pleistocene afzettingen in de Belgische kustvlakte. *Natuurwetenschappelijk Tijdschrift*, **55**, 3–96.

—— & HEYSE, I. 1975. Litostratigrafie van de kwartaire afzettingen in de overgangszone tussen de kustvlakte en de Vlaamse Vallei in Noordwest-België. *Natuurwetenschappelijk Tijdschrift*, **56**, 85–109.

DEN HARTOG, C., VAN DEN BRINK, F. W. B. & VAN DER VELDE, G. 1992. Why was the invasion of the river Rhine by *Corophium curvispinum* and *Corbicula* species so successful? *Journal of Natural History*, **26**, 1121–1129.

DITTMER, E. 1941. Das nordfriesische Eem. Ein Beitrag zur Geschichte der junginterglazialen Nordsee. *Kieler Meeresforschungen*, 1941, 169–199.

DOEKSEN, G. 1975. Eemfossielen uit waterboringen op Terschelling. *Correspondentieblad van de Nederlandse Malacologische Vereniging*, **163**, 366–371.

—— 1991. Eemfossielen uit waterboringen op Terschelling II. *Correspondentieblad van de Nederlandse Malacologische Vereniging*, **260**, 828–841.

EHRMANN, P. 1956. *Die Tierwelt Mitteleuropas*, **II**(1), *Weichtiere, Mollusca*. Quelle & Meyer, Leipzig.

EISMA, D., MOOK, W. G. & LABAN, C. 1981. An Early Holocene Tidal Flat in the Southern Bight. *International Association of Sedimentologists Special Publication*, **5**, 229–237.

ELLIS, A. E. 1962. *British Freshwater Bivalve Molluscs*. Synopsis of the British Fauna, **13**, Linnean Society of London.

—— 1969. *British Snails*. 2nd edn, Clarendon Press, Oxford.

FRETTER, V. & GRAHAM, A. 1978. The prosobranch molluscs of Britain and Denmark, 3. *Journal of Molluscan Studies, Supplement*, **5**, 101–152.

FUNNELL, B. M. 1995. Global sea-level and the (pen)-insularity of late Cenozoic Britain. *This volume*.

—— & WEST, R. G. 1977. Preglacial Pleistocene deposits of East Anglia. *In:* SHOTTON, F. W. (ed.) *British Quaternary Studies – Recent Advances*. Clarendon Press, Oxford.

GIBBARD, P. L. 1995. The formation of the Strait of Dover. *This volume*.

——, WEST, R. G., ZAGWIJN, W. H. ET AL. 1991. Early and Early Middle Pleistocene correlations in the southern North Sea Basin. *Quaternary Science Reviews*, **10**, 23–52.

GIROD, A., BIANCHI, I. & MARIANI, M. 1980. Gasteropodi, 1. *Guide per il Riconoscimento delle Specie Animali delle Acque Interne Italiane*, 7. Consiglio Nazionale delle Richerche AQ/1/47, Verona.

GIUSTI, F. PEZZOLI, E. 1980. Gasteropodi, 2. *Guide per il Riconoscimento delle Specie Animali delle Acque Interne Italiane*, **8**. Consiglio Nazionale delle Richerche AQ/1/47, Verona.

GRAHLE, H. O. 1936. Die Ablagerungen der Holstein-See (Mar. Interglaz. I.), ihre Verbreitung, Fossilführung und Schichtenfolge in Schleswig-Holstein. *Abhandlungen der Preussischen Geologischen Landesanstalt*, NF**172**.

HAAS, F. 1929. Fauna malacológica terrestre y de agua dulce de Cataluña. *Trabajos del Museo de Ciencias Naturales de Barcelona*, **13**.

HALL, C. A. 1964. Shallow-water marine climates and molluscan provinces. *Ecology*, **45** (2), 226–234.

HARTING, P. 1852. De bodem onder Amsterdam onderzocht en beschreven. *Verhandelingen der eerste klasse van het Koninklijk, Nederlandsche Instituut van Wetenschappen, Letterkunde en Schoove Kunsten*, 3e Reeks deel, **5**; 73–232.

—— 1875. Le système Eemien. *Archives Néerlandaises des Sciences Exactes et Naturelles de Société Hollandaise de Sciences à Harlem*, **10**, 443–454.

HINSCH, W. 1985. Die Molluskenfauna des Eem-Interglazials von Offenbüttel-Schnittlohe (Nord-Ostsee-Kanal, Westholstein). *Geologisches Jahrbuch*, **A86**, 49–62.

—— 1993. Marine Molluskenfaunen in Typusprofilen des Elster-Saale-Interglazials und des Elster-Spätglazials. *Geologisches Jahrbuch*, **A138**, 9–34.

HOLYOAK, D. T. & PREECE, R. C. 1983. Evidence of a high Middle Pleistocene sea-level from estuarine deposits at Bembridge, Isle of Wight, England. *Proceedings of the Geologists' Association*, **94**, 231–244.

—— & —— 1985. Late Pleistocene interglacial deposits at Tattershall, Lincolnshire. *Philosophical Transactions of the Royal Society of London B*, **311**, 193–236.

—— & —— 1986. An undescribed species of *Sphaerium* (Bivalvia: Sphaeriidae) from the Norwich Crag (Early Pleistocene) of East Anglia. *Journal of Conchology*, **32**, 195–197.

HORDIJK, L. & JANSE, A. 1987. Mollusken uit het Pleistoceen van het waterweggebied. *Afzettingen Werkgroep voor Tertiaire en Kwataire Geologie*, **8**, 68–74.

HORTON, A., KEEN, D. H., FIELD, M. H. ET AL. 1992. The Hoxnian Interglacial deposits at Woodston, Peterborough. *Philosophical Transactions of the Royal Society of London B*, **338**, 131–164.

HUBENDICK, B. 1950. The effectiveness of passive dispersal in *Hydrobia jenkinsi*. *Zoologiska Bidrag från Uppsala*, **28**, 493–504.

ILLIES, J. (ed.), 1978. *Limnofauna Europaea*. G. Fischer, Stuttgart.

JANSSEN, A. W. 1965. Mollusca uit de Pleistocene ontsluiting te Zelzate. *Mededelingen van de Werkgroep voor Tertiaire en Kwartaire Geologie*, **2**, 24–37.

—— 1975. Systematische lijst van Nederlandse recente en fossiele mollusken. *Mededelingen van de Werkgroep voor Tertiaire en Kwartaire Geologie*, **12** (4), 115–170.

—— 1981. A Holocene mollusc faunule from a temporary excavation near Standdaarbuiten (The Netherlands, Province of Noord Brabant). *Mededelingen van de Werkgroep voor Tertiaire en Kwartaire Geologie*, **18** (1), 29–36.

JELGERSMA, S. 1979. Sea-level changes in the North Sea basin. *In:* OELE, E., SCHUTTENHELM, R. T. E.

& WIGGERS, A. J. (eds) *The Quaternary history of the North Sea*. Acta Universitatis Upsaliensis Symposia Universitatis Upsaliensis Annum Quingentesimum Celebrantis, **2**, 233–248.

JONES, R. L. & KEEN, D. H. 1993. *Pleistocene Environments in the British Isles*. Chapman & Hall, London.

KEEN, D. H. 1995. Raised beaches and sea-levels in the English Channel in the Middle and Late Pleistocene: problems of interpretation and implications for the isolation of the British Isles. *This volume*.

——, ROBINSON, J. E., WEST, R. G., LOWRY, F., BRIDGLAND, D. R. & DAVEY, N. D. W. 1990. The fauna and flora of the March Gravels at Northam Pit, Eye, Cambridgeshire, England. *Geological Magazine*, **127** (5), 453–465.

KEMP, R. A. 1985. The decalcified Lower Loam at Swanscombe, Kent: a buried Quaternary soil. *Proceedings of the Geologists' Association*, **96**, 343–355.

KENNARD, A. S. 1942*a*. Discussion on Pleistocene Chronology. *Proceedings of the Geologists' Association*, **53**, 24–25.

—— 1942*b*. Faunas of the High Terrace at Swanscombe. *Proceedings of the Geologists' Association*, **53**, 105.

—— & WOODWARD, B. B. 1923. The non-marine mollusca of Clacton-on-Sea. *Quarterly Journal of the Geological Society of London*, **79**, 629–634.

KERNEY, M. P. 1958. On the occurrence of *Margaritifera auricularia* (Spengler) in the English Pleistocene. *Journal of Conchology*, **24**, 250.

—— 1971. Interglacial deposits in Barnfield Pit, Swanscombe, and their molluscan fauna. *Journal of the Geological Society, London*, **127**, 69–93.

—— 1976. Mollusca from an interglacial tufa in East Anglia, with the description of a new species of *Lyrodiscus* Pilsbry (Gastropoda: Zonitidae). *Journal of Conchology*, **29**, 47–50.

—— 1977. British Quaternary non-marine Mollusca: a brief review. *In:* SHOTTON, F. W. (ed.) *British Quaternary Studies – Recent Advances*. Clarendon Press, Oxford, 32–42.

KEW, H. W. 1893. *The Dispersal of Shells*. Kegan Paul, London.

KINZELBACH, R. 1991. Die Körbchenmuscheln *Corbicula fluminalis*, *Corbicula fluminea* und *Corbicula fluviatilis* in Europa (Bivalvia: Corbiculidae). *Mainzer Naturwissenschaftliches Archiv*, **29**, 215–228.

KUIJPER, W. J. 1988. Over het vroegere voorkomen van de rivierparelmossel *Margaritifera auricularia* in Nederland. *Basteria*, **52**, 133–137.

LITT, T. & TURNER, C. 1993. Arbeitsergebnisse der Subkommission für Europäische Quartärstratigraphie: Die Saalesequenz in der Typusregion (Berichte der SEQS 10). *Eiszeitalter und Gegenwart*, **43**, 125–128.

MEIJER, T. 1969. *Theodoxus danubialis* (Pfeiffer 1828) uit de Vlaamse Vallei te Zelzate (België). *Mededelingen van de Werkgroep voor Tertiaire en Kwartaire Geologie*, **6** (3), 53–54.

—— 1972. Enkele mollusken uit de Klei van Neede. *Mededelingen van de Werkgroep voor Tertiaire en Kwartaire Geologie*, **9** (3/4), 87–88.

—— 1984. Holocene molluskenfauna's uit de Stevenshofjespolder in Leiden. *Bodemonderzoek in Leiden 1983*, 134–151.

—— 1986. Non-marine biozonation of Quaternary deposits in the Netherlands. *Proceedings of the Eighth International Malacological Congress, Budapest, 1983*, 161–163.

—— 1987. De molluskenfauna van het Waalien in Nederland. *Correspondentieblad van de Nederlandse Malacologische Vereniging*, **236**, 276–279; **237**, 288–297.

—— 1988*a*. Mollusca from the borehole Zuurland-2 at Brielle, The Netherlands (an interim report). *Mededelingen van de Werkgroep voor Tertiaire en Kwartaire Geologie*, **25** (1), 49–60.

—— 1988*b*. Fossiele Zoetwaternerieten uit het Nederlandse Kwartair en enkele opmerkingen over het voorkomen van deze groep in het Kwartair van Noordwest Europa. *De Kreukel, Jubileumnummer*, 89–108.

—— 1990*a*. Notes on Quaternary freshwater Mollusca of The Netherlands, with descriptions of some new species. *Mededelingen van de Werkgroep voor Tertiaire en Kwartaire Geologie*, **26** (4) (1989), 145–181.

—— 1990*b*. Two new freshwater molluscan species from the early Quaternary of the Netherlands. *Contributions to Tertiary and Quaternary Geology*, **27** (4), 107–112.

—— & PREECE, R. C. 1995. Malacological evidence relating to the stratigraphical position of the Cromerian. *In:* TURNER, C. (ed.) *The Early Middle Pleistocene in Europe*. Balkema, Rotterdam, in press.

MILLER, G. H. & MANGERUD, J. 1985. Aminostratigraphy of European marine interglacial deposits. *Quaternary Science Reviews*, **96**, 217–225.

NELSON, C. M. & PAIN, T. 1986. Linnaeus' *Neptunea* (Mollusca: Gastropoda). *Zoological Journal of the Linnean Society*, **88**, 291–305.

NOLF, D. 1974. Mollusken uit het marien Kwartair te Meetkerke (West-Vlaanderen, België). *Natuurwetenschappelijk Tijdschrift*, **55** (1973), 97–120.

NORDMANN, V. 1908. *Molluskfaunaen i Cyprinaleret og Mellem – Europas andre Eem-aflejringer*. Studier over interglaciale aflejringer i Danmark, Holland og Nord-Tyskland, København.

—— 1928. Position stratigraphique des Dépôts d'Eem. *Danmarks geologiske Undersøgelse. II. Raekke*, **47**.

NORTON, P. E. P. 1967. Marine molluscan assemblages in the Early Pleistocene of Sidestrand, Bramerton and the Royal Society Borehole at Ludham. *Philosophical Transactions of the Royal Society of London B*, **253**, 161–200.

—— 1980. Marine mollusc faunas. *In:* WEST, R. G. *The pre-glacial Pleistocene of the Norfolk and Suffolk coasts*. Cambridge University Press, Cambridge, 125–127.

ØKLAND, J. 1990. *Lakes and snails*. Universal Book Services/Dr W. Backhuys, Oegstgeest.

PAEPE, R. & ZAGWIJN, W. H. 1972. Possibilités de corrélation des dépôts Weichseliens de la Belgique et des Pays Bas. *Bulletin de l'Association Française pour l'Étude du Quaternaire*, 1972 (1), 59–69.

PETERSEN, K. S. 1994. Environmental changes recorded in the Holocene molluscan faunas from Djursland, Denmark. *Scripta Geologica, Special Issue*, **2**, 359–369.

PONDER, W. F. 1988. *Potamopyrgus antipodarum* – a molluscan coloniser of Europe and Australia. *Journal of Molluscan Studies*, **54**, 271–285.

PREECE, R. C. 1988. A second British interglacial record of *Margaritifera auricularia*. *Journal of Conchology*, **33**, 50–51.

——, BURLEIGH, R., KERNEY, M. P. & JARZEMBOWSKI, E. A. 1983. Radiocarbon age determinations of fossil *Margaritifera auricularia* (Spengler) from the River Thames, West London. *Journal of Archaeological Science*, **10**, 249–257.

——, LEWIS, S. G., WYMER, J. J., BRIDGLAND, D. R. & PARFITT, S. 1991. Beeches Pit, West Stow, Suffolk (TL 798719). *In:* LEWIS, S. G, WHITEMAN, C. A. & BRIDGLAND, D. R. (eds) *Central East Anglia and The Fen Basin*. Field Guide, Quaternary Research Association, 94–104.

——, SCOURSE, J. D., HOUGHTON, S. D., KNUDSEN, K. L. & PENNEY, D. N. 1990. The Pleistocene sea-level and neotectonic history of the eastern Solent, southern England. *Philosophical Transactions of the Royal Society of London B*, **328**, 425–477.

RAVEN, J. G. M. 1979. The Subboreal coastal barriers at Leidschendam, with a description of the faunas (Province of Zuid Holland, The Netherlands). *Mededelingen van de Werkgroep voor Tertiaire en Kwartaire Geologie*, **16** (1), 17–54.

—— 1982. Changes in the macrofauna of a shallowing subtidal channel (Subatlantic, Holocene) in the mouth of the Oosterschelde (Province of Zeeland, The Netherlands). *Mededelingen van de Werkgroep voor Tertiaire en Kwartaire Geologie*, **19** (2), 59–78.

—— 1983. Subatlantic (Holocene) tidal flat and marsh deposits at Katwijk aan Zee (Province of Zuid Holland, The Netherlands). *Mededelingen van de Werkgroep voor Tertiaire en Kwartaire Geologie*, **20** (2), 51–66.

REES, W. J. 1965. The aerial dispersal of Mollusca. *Proceedings of the Malacological Society of London*, **36**, 269–282.

REID, C. 1892. On the natural history of isolated ponds. *Transactions of the Norfolk and Norwich Naturalists' Society*, **5**, 272–286.

REMY, P. 1924. Note sur la répartition géographique de *Lithoglyphus naticoides* de Férussac (Gastrop. Prosobr.). *Annales de Biologie Lacustre*, **13** (1/2), 83–91.

ROE, H. M. 1994. *Pleistocene Buried Channels in Eastern Essex*. PhD thesis, University of Cambridge.

—— & PREECE, R. C. 1995. A new discovery of the Middle Pleistocene 'Rhenish' fauna in Essex. *Journal of Conchology*, **35**, 272–273.

ROUSSEAU, D-D. 1987. Les associations malacologiques forestières des tufs 'Holsteiniens' de la France Septentrionale. Une application du concept de biome. *Bulletin du Centre de Géomorphologie du CNRS, Caen*, **32**, 9–18.

—— PUISSÉGUR, J-J. 1990. Phylogénèse et biogéographie de *Retinella* (*Lyrodiscus*) Pilsbry (Gasteropoda; Zonitidae). *Géobios*, **23**, 57–70.

——, —— & LECOLLE, F. 1992. West-European terrestrial molluscs assemblages of isotopic stage 11 (Middle Pleistocene): climatic implications. *Palaeogeography, Palaeoclimatology, Palaeoecology*, **92**, 15–29.

RUEGG, G. H. J. (ed.) 1991. Geology and archaeology of ice-pushed Pleistocene deposits near Wageningen (The Netherlands). *Mededelingen Rijks Geologische Dienst*, **46**.

SOMMÉ, J., PAEPE, R., BAETEMAN, C. *ET AL.* 1978. La Formation d'Herzeele: un nouveau stratotype du Pléistocène moyen marin de la Mer du Nord. *Bulletin de l'Association Française pour l'Etude du Quaternaire*, 1978, 81–149.

SPAINK, G. 1958. De Nederlandse Eemlagen. I Algemeen overzicht. *Wetenschappelijke Mededelingen Koninklijke Nederlandse Natuurhistorische Vereniging*, **29**.

—— 1973. Boringen ten behoeve van toekomstige zandopspuitingen. *De Kreukel*, **9** (1), 3–10.

—— 1975. Zonering van het mariene Onder-Pleistoceen en Plioceen op grond van molluskenfauna's. *In:* ZAGWIJN, W. H. & VAN STAALDUINEN, C. J. (eds) *Toelichting bij Geologische Overzichtskaarten van Nederland*. Rijks Geologische Dienst, Haarlem, 118–122.

—— & SLIGGERS, B. C. 1983. Mollusc investigation. *In:* DENYS, L., LEBBE, L., SLIGGERS, B. C., SPAINK, G., VAN STRYDONCK, M. & VERBRUGGEN, C. Litho- and biostratigraphical study of Quaternary deep marine deposits of the western Belgian coastal plain. *Bulletin de la Société belge de Géologie*, **92** (2), 125–154.

SPARKS, B. W. 1964. The distribution of non-marine Mollusca in the Last Interglacial in south-east England. *Proceedings of the Malacological Society of London*, **36**, 7–25.

STEUSSLOF, U. 1953. Wanderungen und Wandlungen der Süsswassermollusken Mitteleuropas während des Pleistozäns. *Archiv für Hydrobiologie*, **48**, 210–236.

STUART, A. J. 1995. Insularity and Quaternary vertebrate faunas in Britain and Ireland. *This volume*.

SUTCLIFFE. A. J. 1995. Insularity of the British Isles 250,000–30,000 years ago; the mammalian, including human, evidence. *This volume*.

SWENNEN, C. 1992. De groei van de Amerikaanse Zwaardschede (*Ensis americanus*) in de Waddenzee. *Het Zeepaard*, **52** (6), 129–131.

TAVERNIER, R. & DE HEINZELIN, J. 1962. De Cardiumlagen van West-Vlaanderen. *Natuurwetenschappelijk Tijdschrift*, **44**, 49–58.

TESCH, P. 1942. De Noordzee van historisch–geologisch standpunt. *Mededelingen van's Rijks Geologischen Dienst*, **A9**, 1–23.

TURNER, C. & KERNEY, M. P. 1971. A note on the age of the freshwater beds of the Clacton Channel. *Journal of the Geological Society, London*, **127**, 87–93.

VAN KOLFSCHOTEN, T., ROEBROEKS, W. & VANDENBERGHE, J. 1993. The Middle and Late Pleistocene sedimentary and climatic sequence at Maastricht-Belvédère: the type locality of the Belvédère Interglacial. *Mededelingen Rijks Geologische Dienst*, **47**, 81–90.

VAN URK, R. M. 1987. *Ensis americanus* (Binney) (syn. *E. directus* auct. non Conrad) a recent introduction from Atlantic North-America. *Journal of Conchology*, **32**, 329–333.

VENTRIS, P. A. 1985. *Pleistocene Environmental History of the Nar Valley, Norfolk*. PhD thesis, University of Cambridge.

VINCENT, P. 1990. *The Biogeography of the British Isles: An Introduction*. Routledge, London.

VON COSEL, R., DÖRJES, J. & MÜHLENHARDT-SIEGEL, U. 1982. Die amerikanische Schwertmuschel *Ensis directus* (Conrad) in der Deutschen Bucht. *Senckenbergiana maritima*, **14**, 147–173.

WARREN, S. H. 1923. The *Elephas-antiquus* Bed of Clacton-on-Sea (Essex) and its flora and fauna. *Quarterly Journal of the Geological Society of London*, **79**, 606–619.

—— 1955. The Clacton (Essex) channel deposits. *Quarterly Journal of the Geological Society of London*, **111**, 283–299.

WEST, R. G. 1980. *The pre-glacial Pleistocene of the Norfolk and Suffolk coasts*. Cambridge University Press, Cambridge.

—— 1987. A note on the March Gravels and Fenland sea levels. *Bulletin of the Geological Society of Norfolk*, **37**, 27–34.

ZAGWIJN, W. H. 1973. Pollen analytical studies of Holsteinian and Saalian Beds in the Northern Netherlands. *Mededelingen Rijks Geologische Dienst*, NS **24**, 139–156.

—— 1979. Early and Middle Pleistocene coastlines in the southern North Sea Basin. *In:* OELE, E., SCHÜTTENHELM, R. T. E. & WIGGERS, A. J. (eds) *The Quaternary History of the North Sea*. Acta Universitatis Upsaliensis Symposia Universitatis Upsaliensis Annum Quingentesimum Celebrantis, **2**, 31–42.

—— 1985. An outline of the Quaternary stratigraphy of the Netherlands. *Geologie en Mijnbouw*, **64**, 17–24.

ZHADIN, V. I. 1952. *Mollusks of Fresh and Brackish Waters of the USSR*. Moskva, Leningrad (English translation: Jerusalem, 1965).

ZILCH, A. & JAECKEL, S. G. H. 1962. *Mollusca. Tierwelt Mitteleuropas*. II(1), Ergänzung, Leipzig.

Insularity and Quaternary vertebrate faunas in Britain and Ireland

A. J. STUART

*Castle Museum, Norwich NR1 3JU, UK and
Department of Geology, University of Manchester, Manchester M13 9PL, UK*

Abstract: Terrestrial vertebrate faunas (mammals, reptiles and amphibians) from the Middle and Upper Pleistocene of Britain, Ireland and adjacent Continental Europe are compared. Problems arise from uncertainties of status and correlation of stages. Evidence of insularity is provided by reduced diversity in (a) Britain *versus* the Continent, and (b) Ireland *versus* Britain. British temperate/interglacial faunas older than the Last Interglacial are very similar to Continental faunas, notably the Cromerian faunas of West Runton, England, and Voigtstedt, Germany. The earliest indications of insularity are in Last Interglacial faunas, in which, for example, pine voles *Pitymys* spp., extinct rhinoceros *Stephanorhinus kirchbergensis*, horse *Equus ferus* and humans, present in equivalent Continental faunas, are absent from Britain. Local extinction of horses and humans is implied. The British Last Cold Stage (Devensian) faunas are very similar to those of the adjacent Continent, indicating unimpeded migration. However, Irish Last Cold Stage faunas are impoverished (absences include woolly rhinoceros *Coelodonta antiquitatis* and humans), indicating no connection from Ireland to Britain. Early Holocene separation of Britain is reflected in the absences of several small vertebrates found on the adjacent Continent. Irish Holocene faunas are much more impoverished, lacking, for example, voles, frogs and snakes, indicating continued isolation from Britain.

The present paper deals with the Middle and Late Pleistocene terrestrial vertebrate faunas (mammals, reptiles and amphibians) of Britain and Ireland. Evidence of insularity is sought in reduced diversity in (a) Britain *versus* adjacent Continental Europe, and (b) Ireland *versus* Britain.

The presence of a moderate water gap, especially if largely spanned by islands (even geologically ephemeral islands), does not prevent faunal interchange altogether, although amphibians are unlikely to cross as they are killed by salt-water. Such a gap acts as a filter allowing some species to cross and not others. Large mammals may cross by swimming, smaller mammals by swimming or on rafted vegetation. During cold stages, migration across frozen sea or on ice floes may have been possible. Exactly which animals successfully cross and which do not is to some extent a matter of chance. In this situation the 'island' fauna will be impoverished in comparison with that of the 'mainland', and is likely to be ecologically unbalanced.

Care is needed in making comparisons between faunas from different sites. Varying factors of taphonomy, habitat distribution and chance can distort our perception of what is significant in presences and absences of particular taxa. It is striking, for example, how the Cromerian mammal faunal list from Sugworth has come to resemble that of West Runton much more closely with continued collecting from the latter site (Stuart 1980, 1982, in press; Table 1). This effect results from the fact that some taxa that are common in the much smaller assemblage from Sugworth are very rare at West Runton.

Another factor to be borne in mind is that at any particular time, because of climatic and/or vegetational limitations, the range of a given species may not have extended to Britain or Ireland, whether or not the relevant land-bridges were in operation. This effect is more easily judged in the Holocene from the distribution patterns of extant species, and in some cases from the success (or failure) of human introductions.

Finding evidence of insularity before the Last Cold Stage in the British sequence (Irish records are confined to the Last Cold Stage and Holocene) is inevitably confused by current uncertainties of correlation both within the British Isles and from Britain to the Continent (see discussions in Jones & Keen (1993)). In particular, notwithstanding the author's earlier view (e.g. Stuart (1976); cf. Sutcliffe & Kowalski (1976)), the occurrence of one or more temperate stages between the Hoxnian and Ipswichian now appears very probable (Jones & Keen 1993). However, the attribution of many sites, the climatic nature of these episodes, the apparent

Table 1. *West Runton Freshwater Bed mammals compared with those from Voigtstedt.*
(West Runton list revised from Stuart (1992, in press); large cats revised by Turner (1995); Voigtstedt list revised from Kahlke (1965), Stuart (1981), Kurtén (1986)).
In a few cases where identifications in the two columns do not coincide but are probably the same taxon, they have been placed on the same line.

West Runton Freshwater Bed*	Voigtstedt
Insectivora	
1. *Erinaceus* cf. *europaeus*, hedgehog	
2. *Beremendia fissidens*, extinct large shrew	large shrew
3. *Sorex minutus*, pygmy shrew	cf. *Sorex minutus*, pygmy shrew
4. *Sorex runtonensis*, extinct shrew	*Sorex runtonensis*, extinct shrew
5. *Sorex savini*, extinct shrew	*Sorex savini*, extinct shrew
6. *Neomys newtoni*, extinct water shrew	*Neomys newtoni*, extinct water shrew
7. *Talpa europaea*, common mole	*Talpa europaea*, common mole
8. *Talpa minor*, extinct mole	*Talpa minor*, extinct mole
9. *Desmana moschata*, Russian desman	*Desmana moschata*, Russian desman
Primates	
10. *Macaca sylvanus*, Barbary macaque	*Macaca sylvanus*, Barbary macaque
Lagomorpha	
11. *Lepus* sp., a hare	*Lepus* sp., a hare
Rodentia	
12.	*Petauria voigtstedtensis*, extinct flying squirrel
13.	*Spermophilus dietrichi*, extinct ground squirrel
14. *Trogontherium cuvieri*, extinct beaver-like rodent	*Trogontherium cuvieri*, extinct beaver-like rodent
15. *Castor fiber*, beaver	*Castor fiber*, beaver
16. *Cricetus cricetus*, common hamster	*Cricetus cricetus*, common hamster
17. *Clethrionomys glareolus*, bank vole	*Clethrionomys glareolus*, bank vole
18. *Pliomys episcopalis*, extinct vole	
19. *Mimomys savini*, extinct water vole	*Mimomys savini*, extinct water vole
20. *Pitymys arvaloides*, extinct pine vole	
21. *Pitymys gregaloides*, extinct pine vole	
22. *Microtus* cf. *arvalis*, common vole	*Microtus* cf. *arvalis*, common vole
23. *Microtus oeconomus*, northern vole	*Microtus oeconomus*, northern vole
24. *Apodemus sylvaticus*, wood mouse	*Apodemus sylvaticus*, wood mouse
Carnivora	
25. *Canis lupus*, wolf (small)	*Canis lupus*, wolf (small)
26. *Ursus deningeri*, extinct bear	*Ursus deningeri*, extinct bear
27. *Mustela nivalis*, weasel	
28. *Mustela erminea*, stoat	
29. *Mustela putorius*, polecat	*Mustela* cf. *eversmanni*, steppe polecat
30. *Martes* sp., extinct marten	*Martes* sp., extinct marten
31.	*Meles meles*, badger
32. *Pannonictis* sp., extinct mustelid	
33. *Lutra simplicidens*, extinct otter	*Lutra simplicidens*, extinct otter
34. *Crocuta crocuta*, spotted hyaena	*Crocuta crocuta*, spotted hyaena
35. *Felis* sp., small cat	*Felis* sp., small cat
36. *Panthera* cf. *gombaszoegensis*, jaguar-sized cat	
37.	*Panthera* cf. *pardus*, leopard
38. cf. *Homotherium* sp., sabretooth	*Homotherium* sp. (*H. moravicum*) sabretooth
Proboscidea	
39. *Mammuthus trogontherii*, extinct elephant (mammoth)	*Mammuthus meridionalis* / *M. trogontherii*, extinct elephant (mammoth)

Table 1. Continued

West Runton Freshwater Bed*	Voigtstedt
Perissodactyla	
40. *Equus* cf. *altidens*, extinct small horse	*Equus* cf. *altidens*, extinct small horse
41. *Equus* sp., a horse	*Equus suessenbornensis*, extinct horse
42. *Stephanorhinus hundsheimensis*,† (*Dicerorhinus etruscus*) extinct rhinoceros	*Stephanorhinus hundsheimensis*, (*Dicerorhinus etruscus*) extinct rhinoceros
Artiodactyla	
43. *Sus scrofa*, wild boar	*Sus scrofa*, wild boar
44. *Megaloceros verticornis*, a giant deer	*Megaloceros verticornis*, a giant deer
45. *Megaloceros savini*, a giant deer	*Megaloceros savini*, a giant deer
46. *Dama dama*, fallow deer	
47. *Cervus elaphus*, red deer	*Cervus elaphus* (*Cervus acoronatus*), red deer
48. *Alces latifrons*, extinct elk (moose)	*Alces latifrons* (*Alces* sp.), extinct elk
49. *Capreolus capreolus*, roe	*Capreolus capreolus* (*Capreolus suessenbornensis*) roe
50. *Bison schoetensacki*, extinct bison	*Bison schoetensacki*, extinct bison

*In addition, noctule bat *Nyctalus noctula* is recorded from the West Runton Freshwater Bed (WRFB). An extinct squirrel *Sciurus whitei* (probably reworked) is recorded from the marine gravel above the WRFB, and a small hamster cf. *Cricetulus migratorius* occurs in a shelly freshwater silt (Bed g) immediately overlying the WRFB (see Stuart 1993)
† See Fortelius *et al.* (1993)

failure of pollen to distinguish them, and correlations with Continental sites, are all problems which need much further work.

Britain

Middle Pleistocene

Few cold stage assemblages are known from Britain prior to the Devensian (Stuart 1982). As would be expected with lowered sea-levels and resultant broad land connection to the Continent, as far as they go the British cold-stage faunas (see lists in Stuart (1982) and Lister & Brandon (1991)) are similar to those of approximately similar ages from Germany and elsewhere (Kolfschoten 1990; Turner 1991).

British temperate (interglacial) faunas are much better known. An exceptionally good fauna is available from a single site, the Cromerian stratotype at West Runton, Norfolk. At this, the most important site in the Cromer Forest Bed Formation, well over a century of collecting has resulted in a long list of mammalian taxa (Stuart 1981, 1982, 1992, 1993; Lister 1993), which continues to grow with additional discoveries. This diversity alone suggests free immigration from Continental Europe, presumably via the southern North Sea Basin and eastern English Channel. Moreover, the West Runton Freshwater Bed fauna (46 taxa) is strikingly similar to the more modest assemblage (41 taxa) from a clay pit at Voigtstedt, Thuringia, Germany (Kahlke 1965; Stuart 1981; Table 1).

Far fewer small mammal fossils are available from Voigtstedt in comparison with the abundant material from West Runton. Nevertheless, currently about 37 taxa appear to be common to both localities, and only four taxa occur solely at Voigtstedt. Of the latter the flying squirrel *Petauria voigtstedtensis* and ground squirrel *Spermophilus dietrichi* may have had an easterly continental distribution, not ranging as far west as Britain. The other two species, badger *Meles meles* and leopard *Panthera* cf. *pardus*, may well turn up in the future at West Runton.

Faunas from sites which appear to post-date the Cromerian but are older than the Anglian, notably Boxgrove, Sussex (Roberts 1986, 1990; Roberts *et al.* 1994), and Westbury-sub-Mendip, Somerset (Bishop 1982; Andrews 1990), show no obvious evidence of insularity when compared with the small number of sites of apparently similar age in Germany. The faunal lists from Boxgrove and Miesenheim 1 (Turner 1991) compare closely. The first appearance of humans in Britain at this time is also consistent with the existence of a land connection.

British Hoxnian faunas are known principally from Hoxne, Suffolk (Stuart 1982; Stuart *et al.* 1993), Swanscombe, Kent (Sutcliffe 1964; Stuart

Table 2. British Interglacial amphibian and reptile records (data from Holman 1993, 1994; Ashton et al. 1994).

	1	2	3	4	5	6
Amphibia						
Triturus vulgaris, smooth newt	WR	Bx	Cd,Bn		Sh	+
Triturus helveticus, palmate newt			Bn			+
Triturus cristatus, great crested newt			Cd,Bn			+
Rana arvalis, moor frog	WR,Sg,LO	Bx	Cd,Bn		SM,Sh	−
Rana temporaria, common frog	WR	Wb,Bx	Bn		It,SM,Sh	+
Rana esculenta/ridibunda, edible/marsh frog	WR		Cd		It	−
Rana lessonae, pool frog			Cd			−
Hyla sp., tree frog			Cd		It	−
Hyla arborea, common tree frog			Bn			
Pelodytes punctatus, parsley frog		Wb				−
Pelobates fuscus, common spadefoot		Bx				−
Bufo bufo, common toad	WR	Bx	Sw,Cd	Ss	It,SM,Sh	+
Bufo calamita, natterjack			Bn	Ss		+
Reptilia						
Emys orbicularis, European pond tortoise	LO	Wb	IV,Cd,Bn	Ss,St,ST	It,SM,Sh	*
Anguis fragilis, slow worm	WR	Wb,Bx	Cd,Bn			+
Lacerta vivipara, viviparous lizard			Cd		Sh	+
Lacerta agilis, sand lizard						+
Vipera berus, adder	WR	Wb	Cd			+
Elaphe longissima, aesculapian snake			Cd,Bn,BP			−
Natrix natrix, grass snake	WR,Sg,LO	Wb	IV,Cd,Bn	Ss	It,SM,Sh	+
Natrix maura/tessellata, viperine/dice snake			Cd		Sh	−
Coronella austriaca, smooth snake		Wb				+

1, Cromerian; 2, post-Cromerian, pre-Anglian; 3, Hoxnian; 4, post-Hoxnian, pre-Ipswichian; 5, Ipswichian/Last Interglacial; 6, Holocene.
Bn, Barnham, Suffolk; BP, Beeches Pit, Suffolk; Bx, Boxgrove, Sussex; Cd, Cudmore Grove, Essex; It, Itteringham, Norfolk; IV, Ingress Vale, Kent; LO, Little Oakley, Essex (NB may be post-Cromerian *sensu stricto* (Stuart in press); Sg, Sugworth, Berkshire; Sh, Shropham, Norfolk; SM, Swanton Morley, Norfolk; Ss, Selsey, Sussex; St, Stutton, Suffolk; ST, Stoke Tunnel, Suffolk; Sw, Swanscombe, Kent; Wb, Westbury-sub-Mendip, Somerset; WR, West Runton Freshwater Bed, Norfolk
*Present earlier in Holocene, but now extinct in British Isles

1982); Clacton, Essex (Sutcliffe 1964), Cudmore Grove, Essex (Currant, quoted in Holman et al. 1990), and Barnham, Suffolk (Ashton et al. 1994). Faunas from correlative Holsteinian sites on the adjacent Continent include Bilzingsleben, Germany (Kahlke & Mania 1994), Steinheim, Germany (Adam 1954; Sutcliffe 1964), Schöningen, Germany (Thieme et al. 1993), and Neede in The Netherlands (Kolfschoten 1990). From the available data, the British faunas closely resemble those from the Continent and provide little indication of insularity. Notable taxa in common include: macaque monkey *Macaca sylvanus*; extinct beaver *Trogontherium cuvieri*; pine vole *Pitymys subterraneus*; and extinct rhinos *Stephanorhinus kirchbergensis* (formerly *Dicerorhinus kirchbergensis*, see Fortelius *et al.* (1993)) and *S. hemitoechus*.

Possible significant absences from the British faunas are sabretooth *Homotherium latidens*, recorded from Steinheim, and extinct buffalo *Bubalus murrensis*, recorded from several Holsteinian localities in Germany (Koenigswald 1986; Kahlke & Mania 1994). However, *H. latidens* is very scarce at the German sites and may yet be found in Britain, and *B. murrensis* probably did not range further west than central Europe. The absence of spotted hyaena *Crocuta crocuta* from the British faunas is not attributable to insularity as it is also absent from Holsteinian faunas from the adjacent Continent.

The high diversity of the rich amphibian and reptile faunas (Table 2) from Cudmore Grove, Essex (14 taxa) (Holman *et al.* 1990), attributed on pollen evidence to Hoxnian substage IIIb, and from Barnham, Suffolk (11 taxa) (Ashton *et al.* 1994) also strongly suggests free immigration from the Continent.

Faunas from several British sites which were regarded as Ipswichian (e.g. Stuart 1976, 1982), including Stutton (Suffolk), Stoke Tunnel (Suffolk), Selsey (Sussex), Ilford (Essex) and Crayford (Kent), are probably post-Hoxnian, pre-Ipswichian (Jones & Keen 1993). Distinguishing characteristics include the occurrence at the older sites, in suitable fluvial deposits, of the distinctive freshwater bivalve *Corbicula fluminalis*, apparently absent from the Ipswichian/Last Interglacial. Conversely, hippopotamus *Hippopotamus amphibius*, common in Ipswichian faunas, does not occur at the earlier sites. Another important feature is that the molars of water vole *Arvicola terrestris cantiana* invariably have enamel thicker on the convex side of the angles (as in the ancestral *Mimomys savini*) whereas those from the Ipswichian are phylogenetically more advanced (e.g. Kolfschoten 1990; Stuart 1982). These pre-Ipswichian/Last Interglacial sites may be of the same or similar age to Maastricht Belvédère, The Netherlands (Kolfschoten 1985, 1990), and the Lower Travertine at Weimar-Ehringsdorf, Germany (Kahlke 1974; Heinrich 1981). With one possible exception (the absence of pine voles *Pitymys* spp. in the British faunas) there are no obvious features which indicate insularity.

Last Interglacial (Ipswichian)

With the removal from the list of several sites which appear to be earlier (see above), the faunas which can be definitely attributed to the Last Interglacial show convincing evidence of insularity. This situation contrasts strongly with previous stages.

The British faunas show reduced diversity compared with those of German sites, which are more or less reliably attributed to the Eemian/Last Interglacial (Table 3). However, comparisons are not altogether straightforward. The British faunal records are from fluvial deposits and from caves, whereas much of the German material comes from travertines. Our knowledge of British Ipswichian faunas is still inadequate, with especially poor records for insectivores and carnivores.

Not all the significant differences between the British and German faunas are likely to result from insularity. There is a strong impression that the east–west climatic gradient was steeper in the Last Interglacial than in say the Cromerian (perhaps due to ocean circulation around the British Isles with the opening of the Strait of Dover). More continental environments at the easternmost German sites are suggested by the presence of common hamster *Cricetus cricetus* and ground squirrel *Spermophilus citellus*, and the absence of hippopotamus *Hippopotamus amphibius*, which has not been recorded further east than the Rhine Basin in the Eemian (Koenigswald 1988).

Of the taxa that are recorded from Britain (Table 3), several, such as *Microtus agrestis*, *Microtus oeconomus*, *Arvicola terrestris cantiana*, *Panthera leo*, *Crocuta crocuta*, *Canis lupus*, *Cervus elaphus*, *Megaloceros giganteus* and *Bison priscus*, probably survived from the previous cold stage. Other species confined to southern Europe in the previous cold stage, such as hippopotamus *Hippopotamus amphibius*, straight-tusked elephant *Palaeoloxodon antiquus*, narrow-nosed rhinoceros *Stephanorhinus hemitoechus* and fallow deer *Dama dama* (Stuart 1991), may have arrived early in the interglacial or alternatively swum across if the water gap was moderate.

The decidedly thermophilous pond tortoise *Emys orbicularis* (Table 2), which requires mean July temperatures at least 2°C warmer than at present in order for its eggs to hatch (Stuart 1979, 1982), probably arrived early in the stage when temperatures were already high, but a land-bridge was still in existence. This parallels the situation in the Holocene (see below). Our knowledge of Ipswichian amphibians and reptiles is inadequate at present (Table 2), but the

Table 3. *Comparison of German and British mammal faunas from the middle part of the Last Interglacial (Eemian/Ipswichian)*

	Germany open sites	Britain fluvial deposits	Britain caves
Insectivora			
1 *Erinaceus europaeus*, hedgehog	Sf		
2 *Talpa europaea*, common mole	Eh,Bg,Ta,Sf		BH
3 *Sorex araneus*, common shrew	Bg,Ta,Sf	Sh,SM*	BH
4 *Sorex minutus*, pygmy shrew		Sh,SM	
5 *Crocidura russula/leucodon*, a large white-toothed shrew†	Eh		
6 *Crocidura* cf. *suaveolens* lesser white-toothed shrew	Sf		
7 *Neomys fodiens*, water shrew	Sf	Sh	
Primates			
8 *Homo* sp., man	Eh,Bg,Ta		
Lagomorpha			
9 *Lepus* cf. *europaeus*, common hare	Sf		
Rodentia			
10 *Spermophilus citellus*, ground squirrel	Eh		
11 *Castor fiber*, beaver	Bg,Ta,Sf	Sh	
12 *Glis glis*, fat dormouse	Eh,Bg		
13 *Cricetus cricetus*, common hamster	Eh,Bg,Ta,Sf		
14 *Clethrionomys glareolus*, bank vole	Eh, Ta,Sf	Sh,SM	BH
15 *Arvicola terrestris cantiana*, extinct water vole	Eh, Ta,Sf	Bi,Sh,SM	JM,TN, BH
16 *Pitymys* cf. *subterraneus*, pine vole	Eh Ta		
17 *Microtus oeconomus*, northern vole	Sf	SM	BH
18 *Microtus agrestis*, short-tailed vole		Bi,Sh,SM	JM,TN, BH
19 *Microtus arvalis*, common vole	Eh Ta		
20 *Microtus arvalis/agrestis*, a vole	Sf		
21 *Apodemus sylvaticus*, wood mouse	Eh,Bg,Ta,Sf	Sh,SM	BH
Carnivora			
22 *Martes martes*, pine marten	Eh Sf		
23 *Meles meles*, badger	Eh Sf	Bi	JM BH
24 *Lutra lutra*, otter	Sf		
25 *Panthera leo*, lion	Eh,Bg,Ta	Bi	JM,TN,VC,BH
26 *Crocuta crocuta*, spotted hyaena	Bg,Ta	Bi SM	JM,TN,VC,BH
27 *Ursus spelaeus*, cave bear	Eh, Ta		
28 *Ursus arctos*, brown bear	Eh,Bg,Ta	Bi	JM,TN,VC
29 *Canis lupus*, wolf	Eh,Bg Sf	Bi	JM,TN
30 *Vulpes vulpes*, red fox	Bg	Bi	JM,TN BH
31 *Felis silvestris*, wildcat	Bg		JM
Proboscidea			
32 *Palaeoloxodon antiquus*, straight-tusked elephant	Bg,Ta	Bi SM	JM, VC,BH
33 *Mammuthus primigenius*, mammoth	Eh ? Ta		BH
Perissodactyla			
34 *Stephanorhinus hemitoechus*, narrow-nosed rhinoceros	Eh,Bg,Ta	Bi ?SM	JM,TN,VC,BH
35 *Stephanorhinus kirchbergensis*, extinct rhinoceros	Bg,Ta		
36 *Coelodonta antiquitatis*, woolly rhinoceros	Eh		
37 *Equus ferus*, horse‡	Eh,Bg,Ta		
38 *Equus hydruntinus*, extinct ass	Bg,Ta		

Table 3. *Continued*

	Germany open sites	Britain fluvial deposits	Britain caves
Artiodactyla			
39 *Sus scrofa*, wild boar	Bg,Ta	Sh	JM
40 *Hippopotamus amphibius*, hippopotamus§		Bi SM	JM,TN,VC
41 *Cervus elaphus*, red deer	Eh,Bg,Ta,Sf	Bi,Sh,SM	JM,TN,VC,BH
42 *Megaloceros giganteus*, giant deer	Eh Ta	Bi	JM,TN,VC
43 *Capreolus capreolus*, roe*	Eh,Bg,Ta,Sf		
44 *Dama dama*, fallow deer	Bg,Ta	Bi SM	JM,TN ?BH
45 *Alces* sp., elk	Eh Ta		
46 *Bison priscus*, extinct bison	Eh,Bg,Ta	Bi	BH
47 *Bos primigenius*, aurochs‖		Bi SM	BH
48 *Bos/Bison* sp., aurochs and/or bison			JM,TN,VC

Bg, Burgtonna (Kahlke 1979; Heinrich & Jánossy 1978; Heinrich 1981); BH, Bacon Hole Layers 3–5 (Stringer 1977; Stringer *et al.* 1986); Bi, Barrington (Gibbard & Stuart 1975; Stuart 1982); Eh, Ehringsdorf Upper Travertine (Kahlke 1975; Steiner 1979; Heinrich 1981); JM, Joint Mitnor Cave (Sutcliffe 1960); Sf, Schönfeld (Heinrich 1991); Sh, Shropham Ipswichian detritus mud (Stuart in preparation); SM, Swanton Morley (Coxon *et al.* 1980; Stuart 1982); Ta, Taubach (Kahlke 1977; Heinrich 1981; Heinrich & Jánossy 1977); TN, Tornewton Cave Hyaena Stratum (Sutcliffe & Zeuner 1962); VC, Victoria Cave (Gascoyne *et al.* 1981).

*Other records from deposits of probable Ipswichian age at Shropham, Norfolk, include: *Hippopotamus amphibius* and *Capreolus capreolus* (Stuart in preparation)
†*Crocidura russula/leucodon* is recorded from fluvial deposits at Itteringham, Norfolk in association with a fauna attributed to the Ipswichian (Stuart in preparation)
‡Horse *Equus ferus* occurs in Layer 1 (base of sequence) at Bacon Hole (Stringer 1977)
§*Hippopotamus amphibius* is recorded from deposits attributed to the Last Interglacial from gravel workings in the Oberrheinebene of western Germany (Koenigswald 1988). Other records include: *Palaeoloxodon antiquus, Stephanorhinus kirchbergensis, Sus scrofa, Dama dama, Capreolus capreolus*, extinct buffalo *Bubalus murrensis* (Koenigswald 1986) and *Bos primigenius*
‖*Bos primigenius* is recorded from the base of the Upper Loess at the volcanic crater sites of Plaidter-Hummerich and Tönchesberg, western Germany, of Last Interglacial and/or early Weichselian age (Turner 1991). Other records include: *Crocuta crocuta, Mammuthus primigenius, Equus* sp., *Equus hydruntinus, Coelodonta antiquitatis, Stephanorhinus* cf. *hemitoechus, Dama dama, Capreolus capreolus. Bos primigenius* is also recorded from the Oberrheinebene sites (Koenigswald 1988).

Ipswichian records of such 'exotic' (i.e. non-Holocene) taxa as *Rana arvalis* and *Natrix maura/tessellata* (Holman 1993) perhaps suggests relatively later severance from the Continent than in the Holocene.

The following absentees in the British faunas especially merit consideration as witnesses of the insular status of Britain: humans *Homo* sp., horse *Equus ferus*, extinct ass *Equus hydruntinus*, extinct rhinoceros *Stephanorhinus kirchbergensis*, and pine voles *Pitymys* spp. White-toothed shrews *Crocidura* spp. are placed by Currant (1989) in his Faunal group 2, which is exclusively pre-Last Interglacial. If this is correct, these shrews would also be significant absentees from the British Last Interglacial. However, *Crocidura russula/leucodon* occurs in interglacial detritus muds at Itteringham, Norfolk. Preliminary study of water vole *Arvicola terrestris cantiana* cheek teeth from these deposits suggests that they belong to the Ipswichian/Last Interglacial.

At first sight all of these absences may be attributed to isolation during at least part of the Ipswichian. However, looking more closely, the lack of horse *Equus ferus* in the Ipswichian is enigmatic. Faunal remains from a carefully excavated cave at Bacon Hole, Gower (Stringer *et al.* 1986) (see notes to Table 3) indicate that horse was present late in the previous cold stage and/or early in the Ipswichian. Why then did horse not survive into the middle of the interglacial? There is ample evidence, from both pollen and mammal faunas, of open grassy vegetation in the middle of the Ipswichian (Gibbard & Stuart 1975; Stuart 1976, 1982), so it does not appear to have been excluded by lack of suitable habitat.

Table 4. *Provisional list of Last Cold Stage faunas of Britain and Ireland*
(Sources: Sutcliffe & Kowalski 1976; Stuart 1982; 1987; Stuart & Wijngaarden-Bakker 1985; McCabe *et al.* 1987; Currant 1991; Housley 1991)

Britain	Ireland
Primates	
1 *Homo sapiens*, man	
Lagomorpha	
2 *Lepus timidus*, arctic hare	*Lepus timidus*, arctic hare
3 *Ochotona pusilla*, pika*	
Rodentia	
4 *Spermophilus* sp., ground squirrel	
5 *Microtus agrestis*, field vole	
6 *Microtus gregalis*, tundra vole	
7 *Microtus oeconomus*, northern vole	
8 *Arvicola terrestris*, water vole	
9 *Lemmus lemmus*, Norway lemming	*Lemmus lemmus*, Norway lemming
10 *Dicrostonyx torquatus*, arctic lemming	*Dicrostonyx torquatus*, arctic lemming
Carnivora	
11 *Ursus arctos*, brown bear	*Ursus arctos*, brown bear
12 *Canis lupus*, wolf	*Canis lupus*, wolf
13 *Vulpes vulpes*, red fox	?
14 *Alopex lagopus*, arctic fox	?
15 *Gulo gulo*, wolverine	
16 *Panthera leo*, lion	
17 *Crocuta crocuta*, spotted hyaena	*Crocuta crocuta*, spotted hyaena
Proboscidea	
18 *Mammuthus primigenius*, woolly mammoth	*Mammuthus primigenius*, woolly mammoth
Perissodactyla	
19 *Equus ferus*, horse	*Equus ferus*, horse
20 *Coelodonta antiquitatis*, woolly rhinoceros	
Artiodactyla	
21 *Rangifer tarandus*, reindeer	*Rangifer tarandus*, reindeer
22 *Cervus elaphus*, red deer	
23 *Alces alces*, elk (moose)	
24 *Megaloceros giganteus*, giant deer	*Megaloceros giganteus*, giant deer
25 *Saiga tatarica*, saiga antelope	
26 *Ovibos moschatus*, musk-ox	*Ovibos moschatus*, musk-ox‡
27 *Bison priscus*, extinct bison	
28 *Bos primigenius*, aurochs †	

Note: More work is needed to establish the provenance of a number of cave records of taxa which may date from the Last Cold Stage.
* Lateglacial record only
† Lateglacial Interstadial record only
‡ Aghnadarragh (A. M. Lister pers. comm. 1993)

The lack of human remains or even artefacts (Wymer 1988), when both occur at several German sites (Table 3) is also difficult to explain. If pond tortoise was able to reach Britain, then surely humans also could have arrived early in the interglacial before separation from the Continent. On present evidence it is difficult to avoid the intriguing conclusion that horses, and probably also humans, became extinct in Britain before the mid-Ipswichian. The presence of a sea barrier would then have prevented recolonization. The cause or causes of such extinctions are highly speculative without further data.

An interesting final point is if and when, within the interglacial, connection with the Continent was re-established (see Lister (1995) for a Last Interglacial sea-level curve). At present the problem is the recognition of faunas which date from the later part of the Ipswichian. At Bacon Hole, Gower, woolly mammoth

Mammuthus primigenius was identified from Layer 5 (Stringer 1977; Stringer *et al.* 1986), which appears to date from the middle to late Ipswichian, in association with a number of typical interglacial elements (Table 3). Sutcliffe *et al.* (1987) suggest that this reappearance of mammoth implies the re-establishment of a land-bridge. However, it is also possible that mammoths recolonized from refugia in say Scotland, or perhaps swam across from the Continent.

Devensian/Last Cold Stage

The marked fall in sea-level of the Last Cold Stage and resultant broad reconnection to the Continent is dramatically demonstrated by the abundant finds of teeth and bones of mammoth *Mammuthus primigenius*, woolly rhinoceros *Coelodonta antiquitatis*, reindeer *Rangifer tarandus* and other typical 'cold' fauna trawled from the bed of what is now the southern North Sea (Mol & van Essen 1992)

British Devensian faunas (Currant 1991; Stuart 1982, 1991; Table 4) are inadequately known, especially the insectivores, rodents and smaller carnivores. However, bearing these deficiencies in mind, the British faunas appear very similar to those of the adjacent Continent, consistent with the existence of an extensive land-bridge and unimpeded immigration. The archaeological record indicates that people were present in Britain thoughout much of the Devensian. Following the arrival of Upper Palaeolithic 'anatomically modern' humans at about 35 000 radiocarbon years BP, the archaeological record suggests more or less continued human presence, except at the time of the main glaciation c. 20 000 to 15 000 radiocarbon years BP.

There is no definite record of vertebrates for the period of the main glaciation, and it is very probable that most mammals were unable to survive in Britain during this time. The reappearance of a diverse fauna in the Lateglacial, starting at c. 13 000 BP, is consistent with a continuing land connection. The British Lateglacial faunas are again similar to those of the adjacent Continent (e.g. Turner 1991; Aaris-Sørensen 1988). During this period several elements of the 'megafauna' became extinct both in Britain and elsewhere (Stuart 1991).

Holocene

The following brief account of Holocene faunas is confined to species present in mainland Britain. For detailed discussion of the modern distribution of mammals on offshore islands of the British Isles, the reader is referred to the excellent review by Yalden (1982).

The history of the British terrestrial vertebrate fauna during the Holocene is generally rather poorly known, especially for the smaller species. However, the fossil assemblages from Star Carr, East Yorkshire, and Thatcham, Berkshire (Yalden 1982; Stuart 1982), demonstrate that most elements of the British fauna (Table 5) either survived from the previous cold stage, or were rapidly established following the rise in temperature and resultant vegetational changes at the onset of the Holocene. The latter group included temperate species such as hedgehog *Erinaceus europaeus*, badger *Meles meles*, wild cat *Felis silvestris*, wild boar *Sus scrofa*, roe *Capreolus capreolus* and aurochs *Bos primigenius*. It is interesting that these species lived in birch and pine woodland, in 'advance' of the arrival of temperate deciduous forest – their expected biome.

Many of the large mammals ('megafauna'), present in previous interglacials, e.g. straight-tusked clephant *Palaeoloxodon antiquus*, extinct rhinoceros *Stephanorhinus hemitoechus* and giant deer *Megaloceros giganteus*, had become extinct everywhere during the Last Cold Stage (Stuart 1991). Other species such as lion *Panthera leo* and fallow deer *Dama dama*, which had avoided total extinction, had Holocene ranges much reduced from those of previous interglacials and did not extend beyond southern Europe.

Nearly all the large mammal species that did survive into the Holocene of northwestern Europe (see Aaris-Sørensen (1988) for a detailed account of the Danish records) were present in Britain. An exception, the European bison *Bison bonasus* recorded from the earliest Holocene (Preboreal) of Denmark, evidently did not range so far to the northwest, because at this time there was a land connection between Britain and the Continent.

Elk *Alces alces*, horse *Equus ferus*, and lynx *Lynx lynx* probably became extinct in Britain early in the Holocene. Many others have been exterminated by human activity within the last thousand years or so.

Holocene insularity is reflected mainly in the small vertebrate faunas. A number of small mammals occur on the adjacent Continent as far as the Channel coast, but are absent from Britain today and from the Holocene fossil record, e.g. white-toothed shrew *Crocidura russula*, pine vole *Pitymys subterraneus* and common vole *Microtus arvalis*. All of the species in question do not range far north into

Table 5. *Holocene amphibians, reptiles and mammals from Britain and Ireland*
(Based on Stuart 1979; 1982; Yalden 1982; Stuart Wijngaarden-Bakker 1985; Preece et al. 1986)

Britain	Ireland
AMPHIBIA	
1 *Triturus vulgaris*, smooth newt	?*Triturus vulgaris**
2 *Triturus helveticus*, palmate newt	
3 *Triturus cristatus*, great crested newt	
4 *Bufo bufo*, common toad	
5 *Bufo calamita*, natterjack toad	(*B. calamita:* introduced)
6 *Rana temporaria*, common frog	(*R. temporaria:* introduced)
REPTILIA	
1 *Emys orbicularis*, European pond tortoise.	
2 *Anguis fragilis*, slow worm	(*A. fragilis:* introduced)
3 *Lacerta vivipara*, viviparous lizard	?*Lacerta vivipara**
4 *Lacerta agilis*, sand lizard	
5 *Natrix natrix*, grass snake	
6 *Vipera berus*, adder	
7 *Coronella austriaca*, smooth snake	
MAMMALIA (excluding bats)	
Insectivora	
1 *Erinaceus europaeus*, hedgehog	(*E. europaeus:* introduced)
2 *Sorex araneus*, common shrew	
3 *Sorex minutus*, pygmy shrew	?*Sorex minutus**
4 *Neomys fodiens*, water shrew	
5 *Talpa europaea*, common mole	
Primates	
6 *Homo sapiens*, man	*Homo sapiens* (**M**)
Lagomorpha	
7 *Lepus timidus*, mountain hare	*Lepus timidus* (**M**)
8 *Lepus capensis*, brown hare	(*L. capensis:* introduced)
Rodentia	
9 *Sciurus vulgaris*, red squirrel	?*Sciurus vulgaris**
10 *Muscardinus avellanarius*, common dormouse	
11 *Castor fiber*, beaver†	
12 *Clethrionomys glareolus*, bank vole	(*C. glareolus:* introduced)
13 *Arvicola terrestris*, water vole	
14 *Microtus agrestis*, short-tailed vole	
15 *Apodemus sylvaticus*, wood mouse‡	*Apodemus sylvaticus* (**M**)
16 *Apodemus flavicollis*, yellow-necked mouse	
17 *Micromys minutus*, harvest mouse	
Carnivora	
18 *Canis lupus*, wolf†	*Canis lupus* (**M**)
19 *Vulpes vulpes*, red fox	*Vulpes vulpes* (**N**)
20 *Ursus arctos*, brown bear†	*Ursus arctos* (**M**)
21 *Mustela nivalis*, weasel	
22 *Mustela erminea*, stoat	*Mustela erminea* (**EC**)
23 *Mustela putorius*, polecat	
24 *Martes martes*, pine marten	*Martes martes* (**EC**)
25 *Meles meles*, badger	*Meles meles* (**N**)
26 *Lutra lutra*, common otter	*Lutra lutra* (**B/I**)
27 *Felis silvestris*, wild cat	*Felis silvestris*† (**N**)
28 *Felis lynx*, lynx†§	
Artiodactyla	
29 *Sus scrofa*, wild boar†	*Sus scrofa* (**M**)
30 *Cervus elaphus*, red deer	*Cervus elaphus* (**M**)

Table 5. *Continued*

Britain	Ireland
31 *Alces alces*, elk (moose)†‖	
32 *Capreolus capreolus*, roe deer	
33 *Bos primigenius*, aurochs†	

* Uncertain if native in Ireland
† Extinct in Holocene
‡ Recorded from early Holocene (Mesolithic) (Preece *et al.* 1986)
§ Probably present in Britain in the early Holocene (Jenkinson 1983)
‖ Perhaps present in Ireland in the early Holocene (Monaghan 1989)

Earliest record in Ireland (Stuart & Wijngaarden-Bakker 1985): (**M**), Mesolithic; (**N**), Neolithic; (**B/I**), Bronze/Iron Age; (**EC**), Early Christian

Horse *Equus ferus*, reindeer *Rangifer tarandus* and northern vole *Microtus oeconomus*, present in the Late Devensian, survived marginally into the earliest Holocene

Scandinavia, and it is reasonable to assume that these were 'latecomers' that did not penetrate sufficiently far north from their Last Cold Stage refugia to reach Britain prior to the formation of the Strait of Dover.

The Holocene amphibian and reptile fauna stands apart because of its reduced diversity in comparison with previous interglacials (Table 2). The effect is again due to isolation from the Continent before the 'latecomers' could arrive in the area, e.g. spadefoot toad *Pelobates fuscus*, tree frog *Hyla arborea* and the water frogs *Rana esculenta/ridibunda* and *Rana lessonae*. Water frogs have been successfully introduced to southern England (Frazer 1983) so that their failure to colonize Britain naturally cannot be attributed to adverse climatic conditions. In contrast, the natterjack *Bufo calamita* was able to colonize Britain at the beginning of the Holocene, when the combination of moderately high summer temperatures and lack of dense forest cover were favourable to this amphibian (Holman & Stuart 1991). Subsequently, as the development of forests 'caught up' with the climate, natterjacks became restricted to locally favourable habitats, i.e. coastal dunes and inland heaths.

The pre-Holocene records of taxa with southern distributions (Arnold & Burton 1978) – *Emys orbicularis*, *Elaphe longissima* and *Natrix maura/tessellata* (Table 2) – reflect phases of higher summer temperatures as well as connection to the Continent. With the notable exception of European pond tortoise *Emys orbicularis* (Stuart 1979, 1982), such 'exotics' are absent from the British Holocene. Although so far there is no fossil evidence prior to sub-zone VIIa (Flandrian FlII), presumably *Emys orbicularis* colonized Britain early in the Holocene before the severance of the land connection. Significantly it is widely recorded from the Holocene of Denmark (Boreal and Atlantic, c. 9000 to 5000 radiocarbon years BP) and southern Sweden, well north of its present range (Aaris-Sørensen 1988). Declining summer temperatures probably caused the extinction of pond tortoise in Britain, as in Denmark, at about the end of the Atlantic period.

Ireland

Midlandian/Last Cold Stage

The Irish Quaternary faunas have been discussed in some detail elsewhere (Stuart 1986; Wijngaarden-Bakker 1986; Stuart & van Wijngaarden-Bakker 1985), and will only be reviewed briefly here. However, there has been an important new discovery since these publications, at Aghnadarragh, County Antrim (McCabe *et al.* 1987), where Early Midlandian fluvial deposits have yielded molars of woolly mammoth *Mammuthus primigenius* and a skull of musk-ox *Ovibos moschatus* (A. M. Lister, pers. comm.). Only sparse finds of musk-ox have been found in Britain and elsewhere in Europe.

At present we have no information on the Irish Pleistocene vertebrate fauna earlier than the Last Cold Stage (Midlandian). This situation is no doubt due mainly to the fact that Ireland was extensively glaciated late in the Midlandian Stage, so that much of the older deposits were destroyed or buried (Stuart 1986; Stuart & van Wijngaarden-Bakker 1985). Moreover, the Midlandian faunas are imperfectly known, due both to the paucity of sites and to the inadequate standards of excavation and recording at many of the sites which have been investigated in the past.

However, even after making generous allowance for the imperfection of the fossil record, it is clear that the Irish Midlandian faunas were strikingly impoverished in comparison with those of the British Devensian (Table 5). In particular, the Irish faunas lack all species of vole, woolly rhinoceros *Coelodonta antiquitatis* and extinct bison *Bison priscus*, none of which are likely to have been overlooked in the Irish record. Other, rarer species that are also missing include *Spermophilus*, *Ochotona pusilla*, *Panthera leo*, *Gulo gulo*, *Cervus elaphus* and *Saiga tatarica*. Moreover, there is no evidence for humans in Ireland until the early Holocene (Woodman 1986). This situation is perhaps surprising, given that people crossed moderate water gaps, by boat or raft, to enter Australia before c. 40 000 years BP.

Devoy (1985, 1986) has extensively reviewed the geological evidence for a land-bridge, concluding that the only plausible route would have been the most northerly, from Islay to Donegal. He considers that this route, if it operated at all, would have been a 'low, soggy, possibly shifting and partially discontinuous linkage' only open in the Lateglacial some time between 11 400 and 10 200 radiocarbon years BP. The absence of voles in Ireland, especially the three species of *Microtus* widely recorded from Britain (Table 4), argues strongly against any continuous connection during the Lateglacial, or later.

Presumably all of the Irish Midlandian fauna arrived from Britain by various means across a water gap. The main glaciation of the Late Midlandian, which covered most of Ireland, almost certainly would have exterminated all or most of the mammal fauna. The Irish Lateglacial fauna – so far only *Megaloceros giganteus* and reindeer *Rangifer tarandus* are definitely known from this phase – must represent re-immigration from Britain, again with no land connection. It is interesting to note here that deer, together with elephants and hippopotamus, are better swimmers than most mammals (Lister 1995), which enabled them, for example, to colonize a number of Mediterranean islands in the Pleistocene.

Holocene

The Irish Holocene fauna (Wijngaarden-Bakker 1986; Stuart & Wijngaarden-Bakker 1985; Yalden 1982, 1986) is also severely impoverished in comparison with Britain. The remarkable absence of snakes was mentioned as early as the third century AD, significantly earlier than St Patrick. The absence of frogs (and snakes) was pointed out as early as the eighth century. As in the Midlandian, voles are entirely absent, as are nearly all insectivores, beaver, weasel, polecat, roe, aurochs and others (Table 5). Humans appear to have been absent until nearly 9000 radiocarbon years BP (Woodman 1986).

A major difficulty is in determining which species are genuinely native and which are human introductions. Unless we have early Holocene records in good stratigraphic context for a given species, we cannot be certain of its native status. Indeed, to be completely certain we need records that pre-date the arrival of people. From this time accidental or deliberate transport of alien species by boat becomes a possibility, although large and dangerous animals (unless perhaps young) would have been unlikely passengers in mesolithic boats (Yalden 1986). The more recent introductions are listed by Stuart & Wijngaarden-Bakker (1985).

At present, *Lepus timidus*, *Canis lupus*, *Ursus arctos*, *Sus scrofa* and *Cervus elaphus* are known from the Mesolithic (Table 5). In addition, *Apodemus sylvaticus* has been found in good stratigraphic context, younger than about 7600 radiocarbon years BP, in association with mesolithic artefacts at Newlands Cross, County Dublin (Preece *et al.* 1986). None of these records pre-dates human arrival, which is not surprising as they are nearly all from archaeological sites.

In order to explain the puzzling presence in Ireland today of pygmy shrew *Sorex minutus* but not the common shrew *Sorex araneus* (which is the more abundant in Britain), Yalden (1982, 1986) postulated a 'low-lying, waterlogged land bridge', most probably early in the Holocene. Such a land-bridge might have acted as a filter allowing the pygmy shrew to cross, but excluding the more fossorial common shrew, and also moles and perhaps other burrowers.

However, our current knowledge of the Irish Holocene fauna appears compatible with geological evidence for the isolation of Ireland from Britain throughout the Holocene (Devoy 1985, 1986). It is extremely difficult to imagine how animals such as frogs, snakes and all species of vole could have been excluded if there had been a land-bridge. Such animals as wild boar *Sus scrofa*, red deer *Cervus elaphus*, wolf *Canis lupus* and brown bear *Ursus arctos* probably swam across from Britain. Precisely which species made the crossing and which did not was probably in part a matter of chance, as well as of swimming ability. Red deer have been documented crossing water gaps of up to 7 km (Lister 1995). Wood mouse *Apodemus sylvaticus* and some other small mammals may have arrived on floating vegetation, although early human introduction is also a possibility. It is interesting to note that wood mouse occurs today on many Scottish islands, where other rodents are absent (Yalden 1982).

We need to consider carefully whether some species, generally accepted as Irish natives, could be human introductions. For example the pigmy shrew, discussed above, has no Irish fossil record whatsoever (Stuart & Wijngaarden-Bakker 1985) and perhaps was introduced fairly recently. Other species with no fossil record, and thus of questionable native status, include smooth newt *Triturus vulgaris*, viviparous lizard *Lacerta vivipara* and red squirrel *Sciurus vulgaris*. These are all species whose occurrence would be difficult to account for had they arrived naturally.

A final general point to be made about Ireland is that all immigrant species have a 'one-way ticket'. Climatic and resultant vegetational changes leave the Irish mammal populations with no escape routes, leading to local extinctions. Although we have no data before the Last Cold Stage, presumably such extinctions must have happened repeatedly during the Pleistocene. As mentioned previously, the main Lateglacial glaciation of Ireland and associated climatic changes must have exterminated most or all of the earlier mammal fauna.

Climatic changes in the Lateglacial and beginning of the Holocene also resulted in local extinctions. Giant deer *Megaloceros giganteus* died out in Ireland apparently in response to climatic deterioration c. 10 500 radiocarbon years BP (Stuart 1991). Conversely, reindeer probably disappeared at about the beginning of the Holocene with ameliorating climate and the spread of forests. Lemmings *Dicrostonyx torquatus* and *Lemmus lemmus* may also have disappeared at this time. That these local extinctions were not due to competition can be demonstrated by the absence of likely competitors. Reindeer were not replaced by red deer, nor lemmings by voles.

These unequivocal demonstrations of (local) extinctions, in the absence of human interference, provide valuable information towards understanding the very complex pattern of 'megafaunal' extinctions during the Last Cold Stage (Stuart 1991).

I am grateful to Dr A. M. Lister, Dr R. C. Preece and Dr A. J. Sutcliffe for information on relevant literature and for comments on the manuscript, and to T. Lord for helpful discussion.

References

AARIS-SØRENSEN, K. 1988. *Danmarks Forhistoriske Dyreverden, fra Istid til Vikingetid*. Gyldendal, Copenhagen.

ADAM, K. D. 1954. Die mittelpleistozänen Faunen von Steinheim an der Murr (Württemburg). *Quaternaria*, 1, 131–144.

ANDREWS, P. 1990. *Owls, Caves and Fossils*. Natural History Museum Publications, London.

ARNOLD, E. N. & BURTON, J. A. 1978. *A Field Guide to the Reptiles and Amphibians of Britain and Europe*. Collins, London.

ASHTON, N. M., BOWEN, D. Q., HOLMAN, J. A. ET AL. 1994. Excavations at the Lower Palaeolithic site at East Farm, Barnham, Suffolk 1989–92. *Journal of the Geological Society, London*, 151, 599–605.

BISHOP, M. J. 1982. The mammal fauna of the early Middle Pleistocene cavern infill site of Westbury-sub-Mendip, Somerset. *Special Papers in Palaeontology*, 28, 1–108.

COXON, P., HALL, A. R., LISTER, A. M. & STUART, A. J. 1980. New evidence on the vertebrate fauna, stratigraphy and palaeobotany of the interglacial deposits at Swanton Morley, Norfolk. *Geological Magazine*, 117 (6), 525–546.

CURRANT, A. P. 1989. The Quaternary origins of the modern British mammal fauna. *Biological Journal of the Linnean Society*, 38, 23–30.

—— 1991. A Late Glacial Interstadial mammal fauna from Gough's Cave, Cheddar, Somerset, England. *In:* BARTON, N., ROBERTS, A. & ROE, D. (eds). *The Late Glacial in northwest Europe: human adaptation and environmental change at the end of the Pleistocene*. Council for British Archaeology, Research Report no 77, 48–50.

DEVOY, R. J. 1985. The problem of a late Quaternary landbridge between Britain and Ireland. *Quaternary Science Reviews*, 4, 43–58.

—— 1986. Possible landbridges between Ireland and Britain: a geological appraisal. *Occasional Publications of the Irish Biogeographical Society*, 1, 15–26.

FORTELIUS, M., MAZZA, P. & SALA, B. 1993. *Stephanorhinus* (Mammalia: Rhinoceratidae) of the western European Pleistocene, with a revision of *S. etruscus*. *Palaeontographia Italica*, 80, 63–155.

FRAZER, D. 1983. *Reptiles and Amphibians in Britain*. Collins, London.

GASCOYNE, M., CURRANT A. P. & LORD, T. C. 1981. Ipswichian fauna of Victoria Cave and the marine palaeoclimatic record. *Nature, London*, 294, 652–654.

GIBBARD, P. L. & STUART, A. J. 1975. Flora and vertebrate fauna of the Barrington Beds. *Geological Magazine*, 112, 493–501.

HEINRICH, W. D. 1981. Systematische Zusammenstellung der in den thüringischen Interglazialtravertinen von Burgtonna, Taubach und Weimar-Ehringsdorf nachgewiesenen Kleinsäugarten. *Quartärpaläontologie*, 4, 127–30.

—— 1991. Paläoökologische und biostratigraphische Kennzeichnung der pleistozänen Säugetierfaunen von Schönfeld, Kreis Calau in der Niederlausitz. *In:* STRIEGLER, R. & STRIEGLER, U. (eds), *Natur und Landschaft in der Niederlausitz,*

190–199.

—— & JANOSSY, D. 1977. Insektivoren und Rodentier aus dem Travertin von Taubach bei Weimar. *Quartärpaläontologie*, **2** 401–411.

—— & —— 1978. Insektivoren und Rodentier aus dem Travertin von Burgtonna in Thüringen. *Quartärpaläontologie*, **3**, 167–170.

HOLMAN, J. A. 1993. British Quaternary herpetofaunas: a history of adaptations to Pleistocene disruptions. *Herpetological Journal*, **3**, 1–7.

—— 1994. A new record of the aescuplapian [sic] snake, *Elaphe longissima* (Laurenti), from the Pleistocene of Britain. *British Herpetological Society Bulletin*, **50**, 37–39.

—— & STUART, A. J. 1991. Amphibians of the Whitemoor Channel early Flandrian site near Bosley, east Cheshire, with remarks on the fossil distribution of *Bufo calamita* in Britain. *Herpetological Journal*, **1**, 568–573.

——, —— & CLAYDEN, J. D. 1990. A Middle Pleistocene herpetofauna from Cudmore Grove, Essex, England, and its paleogeographic and paleoclimatic implications. *Journal of Vertebrate Paleontology*, **10** (1), 86–94.

HOUSLEY, R. A. 1991. AMS dates from the Late Glacial and early Postglacial in North-west Europe: a review. *In:* BARTON, N., ROBERTS, A. & ROE, D. (eds). *The Late Glacial in northwest Europe: human adaptation and environmental change at the end of the Pleistocene.* Council for British Archaeology, Research Report no 77, 25–39.

JENKINSON, R. D. S. 1983. The recent history of northern lynx (*Lynx lynx* Linné) in the British Isles. *Quaternary Newsletter*, **41**, 1–7.

JONES, R. L. & KEEN, D. H. 1993. *Pleistocene Environments in the British Isles.* Chapman & Hall, London.

KAHLKE, H. D. (ed.) 1965. Das Pleistozän von Voigtstedt. *Paläontologische Abhandlungen A II*, **2/3**, 227–692.

—— (ed.) 1974. Das Pleistozän von Weimar Ehringsdorf. Part 2. *Paläontologische Abhandlungen*, **23**, 1–594.

—— (ed.) 1977. Das Pleistozän von Taubach bei Weimar. *Quartärpaläontologie*, **2**, 1–509.

—— (ed.) 1979. Das Pleistozän von Burgtonna in Thüringen. *Quartärpaläontologie*, **3**, 1–359.

—— & MANIA, D. 1994. Komplexe Interglazialfundstellen Thüringens (Exkursion B2). *Altenburger Naturwissenschaffliche Forschungen*, **7**, 357–377.

KOENIGSWALD, W. V. 1986. Beziehungen des pleistozänen Wasserbüffels (*Bubalus murrensis*) aus Europa zu den asiatische Wasserbüffeln. *Zeitschrift für Säugetierkunden*, **51**, 312–323.

—— 1988. Paläoklimatische Aussage letztinterglazialer Säugetiere aus der nördlichen Oberrheinebene. *In:* KOENIGSWALD, W. v. (ed.) *1988 Zur Paläoklimatolgie des letzen Interglazials im Nordteil der Oberrheinebene.* Gustav Fischer, Stuttgart, 205–314.

KOLFSCHOTEN, T. VAN 1985. The Middle Pleistocene (Saalian) and Late Pleistocene (Weichselian) mammal faunas from Maastricht-Belvédère, Southern Limburg, the Netherlands. *Mededelingen Rijks Geologische Dienst*, **39** (1), 45–74.

—— 1990. The evolution of the mammal fauna in the Netherlands and the Middle Rhine area (Western Germany) during the late Middle Pleistocene. *Mededelingen Rijks Geologische Dienst*, **43** (3), 1–69.

KURTÉN, B. 1986. *Crocuta* (Hyaenidae) from the Pleistocene of Voigtstedt, Thuringia (G.D.R.). *Quartärpaläontologie*, **6**, 99–100.

LISTER, A. M. 1993. The stratigraphical significance of deer species from the Cromer Forest-bed Formation. *Journal of Quaternary Science*, **8** (2) 95–108.

—— 1995. Sea-levels and the origin of island endemics: the dwarf red deer of Jersey. *This volume.*

—— & BRANDON, A. 1991. A pre-Ipswichian cold stage mammalian fauna from the Balderton Sand and Gravel, Lincolnshire, England. *Journal of Quaternary Science*, **6** (2), 139–157.

MCCABE, A. M., COOPE, G. R., GENNARD, D. E. & DOUGHTY, P. 1987. Freshwater organic deposits and stratified sediments between Early and Late Midlandian (Devensian) till sheets, at Aghnadarragh, County Antrim, Northern Ireland. *Journal of Quaternary Science*, **2**, 11–33.

MOL, D. & ESSEN, H. VAN 1992. *De Mammoet; Sporen uit de Ijstijd.* BZZTôh, Den Haag.

MONAGHAN, N. T. 1989. The elk *Alces alces* in Irish Quaternary deposits. *Irish Naturalists Journal*, **23** (3), 97–101

PREECE, R. C., COXON, P. & ROBINSON, J. E. 1986. New biostratigraphic evidence of the Post-glacial colonization of Ireland and for Mesolithic forest disturbance. *Journal of Biogeography*, **13**, 487–509.

ROBERTS, M. B. 1986. Excavations of the Lower Palaeolithic site at Amey's Eartham Pit, Boxgrove, West Sussex; a preliminary report. *Proceedings of the Prehistoric Society*, **52**, 215–245.

—— 1990. *In:* TURNER, C. (ed.) *SEQS: The Cromer Symposium, Norwich 1990.* Field Excursion Guidebook, Cambridge.

——, STRINGER, C. B. & PARFITT, S. A. 1994. A hominid tibia from the Middle Pleistocene sediments at Boxgrove, U.K. *Nature*, **369**, 311–313.

STRINGER, C. B. 1977. Evidence of climatic change and human occupation during the Last Interglacial at Bacon Hole Cave, Gower. *Gower*, **28**, 32–37.

——, CURRANT, A. P., SCHWARCZ, H. & COLCUTT, S. N. 1986. Age of Pleistocene faunas from Bacon Hole, Wales. *Nature*, **320**, 59–62.

STUART, A. J. 1976. The history of the mammal fauna during the Ipswichian/Last Interglacial in England. *Philosophical Transactions of the Royal Society of London B*, **276**, 221–250.

—— 1979. Pleistocene occurrences of the European pond tortoise (*Emys orbicularis* L.) in Britain. *Boreas*, **8**, 359–371.

—— 1980. The vertebrate fauna from the interglacial deposits at Sugworth, near Oxford. *Philosophical Transactions of the Royal Society of*

London B, **289**, 87–97.

—— 1981. A comparison of the Middle Pleistocene vertebrate faunas of Voigtstedt (Thuringia, G.D.R.) and West Runton (Norfolk, England). *Quartärpaläontologie*, **4**, 155–163.

—— 1982. *Pleistocene Vertebrates in the British Isles*. Longman, London.

—— 1986. Pleistocene Mammals in Ireland. (pre-10,000 years BP) *Occasional Publications of the Irish Biogeographical Society*, **1**, 28–33.

—— 1991. Mammalian extinctions in the Late Pleistocene of northern Eurasia and North America. *Biological Reviews*, **66**, 453–562.

—— 1992. The Pleistocene vertebrate faunas of West Runton, Norfolk, England. *Cranium*, **9** (2), 77–84.

—— 1993. An elephant skeleton from the West Runton Freshwater Bed (early Middle Pleistocene: Cromerian Temperate Stage). *Bulletin of the Geological Society of Norfolk*, **41**, 75–90.

—— 1995. Vertebrate faunas from the early Middle Pleistocene of East Anglia. *In:* TURNER, C. (ed.). *The Early Middle Pleistocene of Europe*, Balkema, Rotterdam, in press.

—— & WIJNGAARDEN-BAKKER, L. H. VAN 1985. Quaternary Vertebrates. *In:* EDWARDS, K. & WARREN, W. (eds) *The Quaternary History of Ireland*, Academic Press, London.

——, WOLFF, R. G., LISTER, A. M., SINGER, R. & EGGINTON, J. M. 1993. Fossil Vertebrates. *In:* SINGER, R., GLADFELTER, B. G. & WYMER, J. J. (eds) *The Lower Paleolithic Site at Hoxne, England*. University of Chicago Press, 163–206.

SUTCLIFFE, A. J. 1960. Joint Mitnor Cave, Buckfastleigh. *Transactions and Proceedings of the Torquay Natural History Society*, **13**, 1–26.

—— 1964. The mammalian fauna. *In:* OVEY, C. D. (ed.) *The Swanscombe Skull. A Survey of Work on a Pleistocene Site*. Royal Anthropological Society of Great Britain and Ireland, Occasional Paper, **20**.

—— & KOWALSKI, K. 1976. Pleistocene rodents of the British Isles. *Bulletin of the British Museum Natural History (Geology)*, **27**, 33–147.

—— & ZEUNER, F. E. 1962. Excavations in the Torbryan Caves, Devonshire. 1, Tornewton Cave. *Proceedings of the Devon Archaeological Exploration Society*, **5**, 127–145.

——, CURRANT, A. P. & STRINGER, C. B. 1987. Evidence of sea-level change from coastal caves with raised beach deposits, terrestrial faunas and dated stalagmites. *Progress in Oceanography*, **18**, 243–271.

THIEME, H., MANIA, D., URBAN, B. & KOLFSCHOTEN, T. VAN 1993. Schöningen (Nordharzvorland) eine altpaläolithische Fundstelle aus dem mittleren Eiszeitalter. *Archäologische Korrespondenzblatt*, **23**, 147–163.

TURNER, A. 1995. Evidence for Pleistocene contact between the British Isles and the European Continent based on distributions of large carnivores. *This volume*.

TURNER, E. 1991. Pleistocene stratigraphy and vertebrate faunas from the Neuwied Basin region of Western Germany. *Cranium*, **8** (1), 21–34.

WIJNGAARDEN-BAKKER, L. 1986. The colonization of islands. The mammalian evidence from Irish archaeological sites. *Occasional Publications of the Irish Biogeographical Society*, **1**, 38–41.

WOODMAN, P. C. 1986. Man's first appearance in Ireland and his importance in the colonization process. *Occasional Publications of the Irish Biogeographical Society*, **1**, 34–37.

WYMER, J. J. 1988. Palaeolithic archaeology and the British Quaternary sequence. *Quaternary Science Reviews*, **7**, 79–88.

YALDEN, D. W. 1982. When did the mammal fauna of the British Isles arrive? *Mammal Review*, **12** (1), 157.

—— 1986. How could mammals become Irish? *Occasional Publications of the Irish Biogeographical Society*, **1**, 49–52.

Insularity of the British Isles 250 000–30 000 years ago: the mammalian, including human, evidence

ANTONY J. SUTCLIFFE

Natural History Museum, Cromwell Road, London SW7 5BD, UK

Abstract: During the Middle–Late Quaternary the British Isles have alternately been part of the Continent of Europe and separated from it by the sea. Comparison of mammalian faunas from sites on both sides of the present English Channel can contribute to the reconstruction of Britain's island history. The most suitable chronological framework for this study is discussed. Attention is drawn to problems of applying the 1973 scheme of the Geological Society of London, which is used here only in conjunction with the more straightforward chronology of the deep-sea oxygen isotope record. Special attention is directed to events during the period of time *c.* 250 000 to 30 000 years BP. The chalk land-bridge that had previously joined the British Isles to the Continent of Europe had by this time been breached, and any further changes in the land connection are likely to have been the result of eustatic changes of sea-level, associated with glacial–interglacial fluctuations of climate.

Whether the British Isles became isolated from the Continent during the high sea-level phase that occurred during isotope stage 7 is uncertain from the mammalian evidence. The occurrence of human artefacts and remains of pond tortoise *Emys orbicularis*, which is unlikely to have tolerated salt-water, in deposits representing the earliest part of this stage, suggests a land connection then or not very long previously. Human presence continues into isotope stage 6, characterized also by the arrival of steppe rodents from eastern Europe; evidence of a further episode of low sea-level.

During the Last Interglacial, oxygen isotope stage 5e, sea-level rose once more, isolating the British Isles from the Continent. Remarkably, humans, horses and the mollusc *Corbicula fluminalis*, so common previously, fail to reappear, a consequence, perhaps, of being cut off by rapidly rising sea-level at the end of stage 6. A very characteristic fauna, including hippopotamus, straight-tusked elephant, narrow-nosed rhinoceros, fallow deer and abundant carnivores, did nevertheless manage to become established throughout England and Wales at this time, some as far north as Durham.

Falling sea-level associated with the Devensian (isotope stages 5d–2) re-established the land connection. Hippopotamus had by this time disappeared and been replaced with a fauna including woolly mammoth, woolly rhinoceros and reindeer. Humans and horses returned once more. It was not until sea-level rose again during the Holocene that the British Isles became finally isolated from the Continent, as they are today.

As Turner (1995) points out, the larger mammal species of the British Pleistocene (and this would apply to smaller species and humans also) mostly 'originated elsewhere and dispersed into the region, where their presence in fossil deposits represents no more than the extreme north-western tip of a larger distribution. The pattern of their appearance may therefore hold some clues to the changing nature of the link between the British Isles and the rest of the European continent'.

The successful interpretation of such data is nevertheless not without its problems. Most notably it is dependent on an accurate chronological framework and can be developed only as far as this has been established. Erroneous interpretations of the order of events within the British Isles, or of the age relationships of deposits there and on the Continent of Europe, can invalidate such studies. Exact chronological control is therefore necessary.

In this study, special attention will be directed to Britain's island history during the period of time approximately 250 000–30 000 years BP, with only brief reference to earlier events.

A chronological scheme for this study

Until a decade ago, the chronological scheme most widely employed in studies of the British Quaternary was that of the Geological Society's sub-committee on the Quaternary Era (Mitchell *et al.* 1973; see Table 1). In the formulation of this scheme, sub-division into zones within the temperate stages was based on pollen assemblages, which at that time were considered the best tool available for zonation of predominantly terrestrial deposits. Sub-division of the cold stages was based principally on lithostratigraphy. Its provisional nature was nevertheless emphasized, and Mitchell *et al.* believed that, since stratigraphical correlation and chronology is always aiming to achieve synchroneity when tracing formations across the world, it was right to apply the radiometric clock as soon as it became available. For the time being, however, only the radiocarbon method, with its time range effectively limited to the Holocene and Late Devensian, was applicable to the types of deposit available for study in the British Isles. The classification was framed so that new stages could be inserted as evidence accumulated (West 1981).

Severe difficulties in the application of this scheme were nevertheless soon to arise. Principal among these were: (a) suggestions that the deposits of the Wolstonian stratotype near Coventry (Shotton 1953) were in fact of pre-Hoxnian age (Sumbler 1983; Rose 1987), potentially leaving the period of time between the Ipswichian and Hoxnian without any stratotypes and unnamed; and (b) suggestions that more than one interglacial was represented among the many sites listed as Ipswichian (stratotype Bobbitshole, near Ipswich, Suffolk (Sparks 1957, West 1957)), the additional unnamed interglacial having a flora of Ipswichian character but actually being of pre-Ipswichian and post-Hoxnian age (Sutcliffe 1975, 1976). This apparent confusion of two successive interglacial episodes was to have a far-reaching influence on subsequent chronological studies, since it was to obscure the events of a major cold period which occurred between them.

West (1981), in a discussion of an analagous problem affecting the Eemian (last) Interglacial on the Continent, pointed out that a misconception had arisen that there had been more than one Eemian temperate phase. 'There can, however, be only one, and that is the one represented at the type site ... Others may have been wrongly correlated with the type site, but that is a matter which has to be proved'. It follows that we can accept as Ipswichian only sites which can be shown to correlate with the type site of Bobbitshole. If additional interglacials are proven, then new type sites and new names must be established.

The Ipswichian problem is further compounded by the fact that the sequence of deposits at the type locality does not span the entire interglacial. Whereas there do exist excellent data relevant to its early and middle parts, what happened during its later part creates a problem. Deposits occur at the type locality only up to sub-stage Ip IIb, so that the rest of the interglacial had to be reconstructed from other sites, and pollen diagrams (for example in West (1968)) were composite. The most important site to have been assigned on palynological evidence to the later part of the Ipswichian – Histon Road, Cambridge (Walker 1953; Sparks & West 1959) – is situated 70 km from Bobbitshole and its chronological position in relation to the deposits on that site is equivocal. The possibility that the two series of deposits do not date from the same temperate stage (Histon Road possibly being earlier than Bobbitshole), cannot, from the palynological evidence, be discounted.

Two decades later, even though much new field data have become available, still no new stratotypes have been proposed to represent any remaining part of post-Hoxnian time, although names appear in the literature from time to time from writers anxious to have something to call an additional interglacial. 'Ilfordian' (Bowen 1977), 'Minchin Hole Interglacial' (Green & Walker 1991), 'Middle Terrace of the Thames Interglacial' (Sutcliffe & Kowalski 1976) and simply 'an interglacial stage within the Wolstonian' (Boylan 1977) are among terms that have been employed. Bowen *et al.* (1989) assigned the site of Stanton Harcourt, Oxfordshire, to this chronological position, but likewise without new stratotype designation. In order to avoid ambiguity, some writers (e.g. Jones & Keen 1993; Bridgland 1994) now employ terms such as 'Ipswichian (*sensu* Mitchell *et al.* 1973)' or 'Ipswichian (*sensu* Trafalgar Square')' – an important hippopotamus locality – as a means of expressing greater chronological precision.

Since it seems unlikely that the additional new stratotypes that would be necessary to make the Geological Society scheme workable can be designated and assigned absolute dates within a reasonable time, some other chronological basis is necessary for the present study.

The great advances in absolute dating methods that have since taken place, especially the uranium series method (Ivanovich & Harmon 1982), applicable to travertines (including cave travertines) and corals back to more than 300 000 years BP, supplemented by relative methods such as amino acid racemization studies (Bowen *et al.* 1989), now provide calibration for many Middle–Late Quaternary sites, whilst the absolute ages of most of the

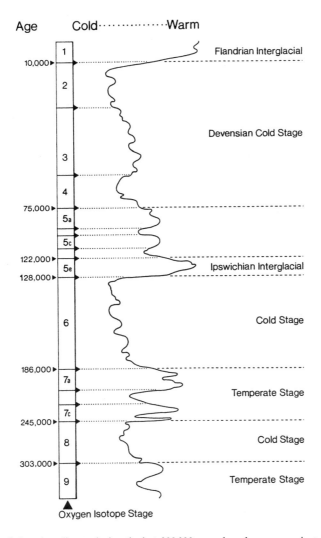

Fig. 1. The history of changing climate during the last 300 000 years based on oxygen isotopic data, reproduced with permission from Green & Walker (1991).

Geological Society stratotypes are still subject to discussion.

It had already been observed by Shackleton & Turner (1967) that 'Nobody today would seriously attempt to review methods available for dating the glacial events of the Pleistocene without considering the evidence which has been obtained from the (oxygen isotope) study of deep sea cores'. In this scheme, odd numbers were assigned to warm phases and even numbers to the cold phases between them. By 1969 this work was at an advanced stage and Shackleton (1969) had been able to divide Emiliani's oxygen isotope stage 5 of 1961 into five sub-stages (e–a), with an exceptionally high peak of warmth at 5e. With the subsequent extension and age calibration of this scheme, later to be described by Imbrie & Imbrie (1979) as the 'Rosetta Stone' of late Pleistocene climate, to provide a climatic record for the whole of the Quaternary (Fig. 1, showing latest part only), it is apparent that many more fluctuations of climate actually occurred than can be accommodated in the Geological Society's scheme. In this context the apparent duplication of events assigned to the Ipswichian becomes less controversial. The recent investigations of the Summertown–Radley terrace of the River Thames, including the highly fossiliferous sites of Stanton Harcourt, Oxfordshire (Briggs et al. 1985; Briggs 1988; Bridgland 1994; K. Scott, pers. comm.) and of Marsworth, Buckinghamshire (Green et al. 1984), with both interglacials demonstrably separated by an episode of periglacial conditions, provide field confirmation of previous over-simplification.

For the purpose of the present paper, the author, in common with others (e.g. Bowen 1978; Wymer 1985; Jones & Keen 1993; Stringer & Gamble 1993; Bridgland 1994), accepts the oxygen isotope chronology as the most satisfactory chronological framework currently available. The Geological Society's scheme (1973) nevertheless cannot be abandoned in its favour, since it has been so widely applied in the literature of the 1970s and 1980s, and its legacy must continue to be accommodated, as far as is possible, with the prospect of complete calibration and correlation of existing and new stratotypes being a task for some time in the future. It is a widely held view that the Holocene is oxygen isotope stage 1, the Devensian 2, 3, 4 and sub-stages 5a–d, with the Ipswichian stratotype of Bobbitshole probably representing sub-stage 5e. This last deduction has nevertheless not passed unchallenged, Currant (1989) presenting evidence that an earlier interglacial may in fact be represented. It is argued here that deposits of both sub-stages 5e and 7 are represented among the sites listed as Ipswichian by the Geological Society. It would be out of place here to enter a detailed discussion of the reasons for their separation, but a more complete account will be published elsewhere. Before Stage 5, correlations between the Geological Society's stratotypes and the oxygen isotope chronology are mostly unproven. The suggestions of Bowen (1989) and of Singer et al. (1993) that the Hoxnian stratotype may be of stage 9 age, that of Bowen (1989) that the Swanscombe human site dates from stage 11, and that of Roberts et al. (1994) that the Boxgrove human site may be of stage 13 age, provide a first approximation of what might be expected in the future.

The acceptance of stage 7 as an interglacial episode apparently not represented in the Geological Society's scheme goes only part way towards resolving problems during this part of the Quaternary. From the oxygen isotope evidence, stage 7 is itself seen to be composite, with two warm sub-stages (7c and 7a, each with two peaks) divided by a cooler sub-stage (7b), in all spanning 60 000 years, from c. 245 000–186 000 BP (Fig. 1). With increasing evidence of a Quaternary composed of many climatic fluctuations of short duration, 'stage 7' must not be allowed to become a 'dumping ground' for sites of more than one climatic episode, as has occurred with the Ipswichian. The present paper can only be a very simple approximation of the fine-tuning that future decades will produce.

Sea-level evidence

Evidence provided by mammals of the former insularity of the British Isles needs to be considered in conjunction with whatever sea-level evidence is also available. An important factor was the final breakdown of the chalk land bridge connecting Britain and the Continent at the site of the present Strait of Dover. Smith (1985) suggests that this was probably a catastrophic event at about the Anglian/Hoxnian boundary. Both Smith (1985) and Gibbard (1988, 1995) describe a large ice-dammed lake occupying the southern part of the North Sea during Anglian times, which could not have been retained had a barrier not existed. After the breaching of the barrier, high sea-levels would have been continuous between the English Channel and the North Sea, and eustatic fluctuations would have been the most likely cause of changing land connections. Eustatic highs of the interglacials can be expected to have presented greater barriers to mammalian movements than the low sea-levels of the cold stages; this would have been so to a greater extent when preceded by a rapid rise of sea-level than when this occurred more slowly.

Sea-levels marginally higher than at the present day can be distinguished in southern England and south Wales during both oxygen isotope stage 7 and sub-stage 5e (Catt & Penny 1966; Sutcliffe 1981; Currant et al. 1984; Bowen et al. 1985; Davies & Keen 1985; Stringer et al. 1986; Mottershead et al. 1987; Sutcliffe et al. 1987; Proctor & Smart 1991, Keen 1995).

Human and other mammalian evidence

When attempting comparisons of the mammalian faunas of different sites, it becomes immediately apparent that severe taphonomic difficulties can arise. Some of the most important faunas to be discussed here, for example that from Ilford, Essex, were collected more than a century ago, before it had become fashionable to sieve for small mammals, so that only large species are represented. It is only during the present century that collecting of small mammals, the importance of which was highlighted by Hinton (1926), has been widely undertaken, and some of the most productive sites (for example Cudmore Grove, Essex) have produced very few large mammals. A further taphonomic problem is encountered when making comparison between cave faunas (often with a high proportion of carnivores, owl pellet bones and bats), and assemblages from open sites, where these are but poorly represented. The facts that the clawless otter *Cyrnaonyx* from Tornewton Cave, Devon (Willemsen 1992) is unique in the British record, and that the felid *Panthera* aff. *pardus* has been recorded from only a few sites, emphasizes the incompleteness of the palaeontological record.

Before oxygen isotope stage 7

As pointed out by Stuart (1995), the close similarity of mammalian faunas of the Cromerian stratotype at West Runton, Norfolk, and of Voigtstedt, Germany, during the Cromerian Interglacial, is evidence of a land connection at that time. With a chalk land-bridge still intact, such continuity might seem hardly surprising. Nevertheless, humans had apparently not yet arrived in the British Isles, where the earliest evidence for their presence is found at Boxgrove, Sussex (Roberts et al. 1994). That they may not even have reached western central Europe before this time is suggested by Roebroeks & Van Kolfschoten (1994). Sutcliffe (1964) drew attention to the similarity between the mammalian fauna from the human skull site of Swanscombe, England, and that at Steinheim an der Murr, Germany (Adam 1954), suggesting later continuity. Holman et al. (1990) found the richness of the herpetofauna of the Hoxnian site of Cudmore Grove, Essex, so great as to suggest that free movement from the Continent was possible at that time. The faunal remains at this site nevertheless rest on estuarine deposits representing a high sea-level (H. M. Roe, pers. comm.), making a direct land connection surprising.

Oxygen isotope stage 7, c. 245 000–186 000 years BP

As previously observed, oxygen isotope 7, with three sub-stages (c–a), represents a substantial period of time, so that any attempt to define a typical 'stage 7' mammalian fauna would run the risk of the over-simplification that confused studies of the Ipswichian. A study of the degree of homogeneity of the mammalian faunas from sites currently considered as of probable stage 7 age may provide a lead towards determining whether any separation is possible.

Open-air sites, with mammalian remains, which it is suggested here are of stage 7 age, include: the Lower Channel of Marsworth, Buckinghamshire (Green et al. 1984); Aveley (West 1969; Bridgland 1994), Ilford (Sutcliffe 1964; West et al. 1964), Lexden (Shotton et al. 1962) and West Thurrock (Carreck 1976) in Essex; Northfleet, Kent (Burchell 1957; Bridgland 1994); Stanton Harcourt, Oxfordshire (Bridgland 1994; K. Scott, pers. comm.); Brundon, (Moir & Hopwood 1939), Stoke Tunnel and Maidenhall (Wymer 1985), Harkstead and Stutton (Sparks & West 1964; Wymer 1985) in Suffolk; the Brighton raised beach (Turner 1980; Davies & Keen 1985; Keen 1994, 1995) and Selsey (West & Sparks 1960) in Sussex; and Bielsbeck, Yorkshire (Boylan 1977).

Suggested stage 7 mammalian cave sites include the 'Otter Stratum' of Tornewton Cave (Sutcliffe & Kowalski 1976) in Devon, and Pontnewydd Cave, North Wales (Currant 1984; Green 1984).

Whereas the mammalian faunas of the open-air and cave sites listed above are difficult to compare for taphonomic reasons, the similarity among those of the former group (within the limits of the differing number of species represented) is remarkable. While it would be premature to suggest that all these sites represent the same temperate episode, for the time being they cannot be distinguished on the mammalian evidence alone.

Characteristic mammals are a mammoth of the *Mammuthus trogontherii – primigenius* lineage, with small teeth, and horse *Equus ferus* (both very common); straight-tusked elephant *Palaeoloxodon antiquus*; narrow-nosed rhinoceros *Stephanorhinus hemitoechus*; giant ox *Bos primigenius*; *Bison* sp.; red deer *Cervus elaphus*; giant deer *Megaloceros giganteus*; roe deer *Capreolus capreolus*; a very large lion *Panthera leo*; brown bear *Ursus arctos*; and wolf *Canis lupus*. Human artefacts, often showing Levallois technique, are recorded from six of these sites (West Thurrock, Northfleet, Brundon, Stoke Tunnel, Stutton and Selsey), and remains of pond tortoise *Emys orbicularis* from three (Harkstead, Stutton and Selsey). At sites where Mollusca are preserved, *Corbicula fluminalis* (of relevance to discussion of oxygen isotope stage 5 faunas, to follow) is almost invariably present.

It is good fortune that several of these sites have been subject to palynological study, providing scope for assessing potential change of mammalian species within the interglacial, a method previously applied by Stuart (1976, 1982) in his studies of the Ipswichian. Selsey sub-stages I–IIb (West & Sparks (1960), Ipswichian of these authors) has *Palaeoloxodon antiquus*; *Stephanorhinus hemitoechus*, a bovid, beaver *Castor fiber* and pond tortoise *Emys orbicularis*. Hippopotamus, listed by Sutcliffe in an appendix to the above paper, is discounted here on the grounds of an erroneous determination of a single imperfect specimen (Sutcliffe & Parfitt in preparation). Aveley (sub-stages II–III (West 1969), Ipswichian of this author), has *P. antiquus*, *Equus*, a bovid, and white-toothed shrew *Crocidura*, and, in overlying deposits, small-toothed mammoth and red deer. Stutton (sub-stage III (Sparks & West 1964), Ipswichian of these authors) has *P. antiquus*, *Mammuthus* lineage, rhinoceros sp., horse, bovid, red deer, giant deer, lion, bear and *Emys orbicularis*. Although comparison between these faunas is limited by the small number of species represented, their apparent uniformity throughout the interglacial is surprising. Some possible changes are detectable within the above-mentioned sequences, but no consistent pattern emerges. At Aveley, *P. antiquus* in sub-stage IIb (two complete skeletons) is replaced by *Mammuthus* (at least three individuals) in sub-stage III. At Stutton (where the 'IpIII' pollen sample was collected almost at the base of the deposit), on the other hand, remains of *Mammuthus* were found below those of *P. antiquus* (H. E. P. Spencer, in an unpublished letter, property of the Natural History Museum). Whether the apparent alternation of the two species of prosboscidean during the course of the interglacial actually occurred (possibly in response to minor fluctuations of climate) or is a manifestation of insufficient sample size, with both species present together, is a problem for future study. The proboscidian fauna of the sub-stage III deposits of the Stanton Harcourt Channel (K. Scott, pers. comm.) is predominantly mammoth but also includes *P. antiquus*. The occurrence of *Emys*, the eggs of which can hatch only with summer temperatures slightly higher than experienced in the British Isles today (Stuart 1979), at both the beginning and end of the interglacial, confirms warm climate conditions, at least during the summer.

Two of the above-mentioned sites, Brighton and Selsey, are of special importance in the study of stage 7 events, since their mammalian faunas can be directly related to associated sea-level data. At Brighton, remains of horse (specimens in the Brighton Museum) have been found in a raised beach, indicating a sea-level a few metres higher than at the present day. At Selsey (although direct relationship cannot be observed because of present-day sea defences and shingle), extensive raised beach deposits apparently overlie the now intertidal peat in which the fauna listed above was found. From this evidence it would seem that, by the beginning of the interglacial, humans and pond tortoise, which is not known from marine associations (E. N. Arnold, pers. comm.), had already arrived in England or had survived from an earlier immigration. A substantial rise of sea-level then followed. Whether or not the British Isles became isolated from the Continent at this stage remains uncertain from the mammalian evidence.

Although comparison of stage 7 mammalian faunas from the open sites discussed above and from caves is difficult for taphonomic reasons, the cave evidence is of special importance since absolute ages can often be obtained by uranium series dating of interstratified layers of flowstone. This method was applied with especial success at Pontnewydd Cave in North Wales, a richly mammaliferous site with Neanderthal skeletal remains and artefacts, interpreted by Green (1984) as a site of occupation beginning before 250 000 years BP (oxygen isotope sub-stage 7c) and continuing until about 230 000 years BP (sub-stage 7b). Studies of the associated mammalian remains by Currant (1984) suggest an open woodland environment for the horizon known as the 'Intermediate Complex', changing to an open steppe environment, with Norway lemming *Lemmus lemmus*, northern vole *Microtus oeconomus* and pika *Ochotona* sp., at the time of deposition of the following 'Lower Breccia'.

The 'Otter Stratum' of Tornewton Cave, Devon (Sutcliffe & Zeuner 1962; Sutcliffe & Kowalski 1976) is a further important cave deposit, with potential stage 7 affinities, shown to pre-date a horizon with remains of hippopotamus, assumed to be of sub-stage 5e age. Its remarkable mammalian fauna includes the white-toothed shrew *Crocidura* sp. (Rzebik 1968) (the first record of this species in the British Isles, although it has been found at other sites subsequently), wood mouse *Apodemus sylvaticus*, common hamster *Cricetus cricetus*, bank vole *Clethrionomys glareolus*, water vole *Arvicola* sp., short-tailed field vole *Microtus agrestis*, northern vole *M. oeconomus*, wolf *Canis lupus*, brown bear *Ursus arctos*, and clawless otter *Cyrnaonyx* sp. (the only known British record). This is a predominantly interglacial assemblage.

From the evidence outlined above some preliminary thoughts about the possible subdivision of the mammalian fauna of stage 7 are now feasible. Currant's studies of the Pontnewydd faunas is suggestive of a change from open woodland to open steppe during the middle of this stage. It may be significant that among all the sites listed above there are only two records of Merck's rhinoceros *Stephanorhinus kirchbergensis* (a common species during earlier interglacials, for example at Clacton and Swanscombe); from Ilford (Sutcliffe 1964) and a single specimen from Pontnewydd Cave. The possibility that this species persisted only during the early part of stage 7 requires investigation. The bear from Ilford is the brown bear *Ursus arctos* (Sutcliffe 1964) whereas that from Pontnewydd has spelaeoid (cave bear) characteristics (Currant 1984). The relative rarity of *Crocidura* in the British Quaternary might be seen as an aid towards dating the Otter Stratum of Tornewton Cave. The occurrence of this species at sites shown on palynological evidence to date from more than one time – Aveley (Stuart 1976) with Ipswichian affinities (West 1969), and Cudmore Grove (Currant in Bridgland *et al.* 1988) with Hoxnian affinities (H. M. Roe, pers. comm.) – nevertheless precludes any precise chronological placing of the Otter Stratum on the evidence of *Crocidura*, for the time being. The fallow deer *Dama* sp. is unrecorded from any of the sites discussed above, with the exception of Histon Road (A. M. Lister, pers. comm.), the precise chronological position of which is disputed. Absence of fallow deer may prove to be a criterion diagnostic of stage 7. The great abundance of remains of this animal in the Burtle Beds, Somerset (marine and estuarine deposits representing a sea-level probably 9–12 m higher than at the present day (Bulleid & Jackson 1937, 1941; Kidson *et al.* 1978)) make a stage 7 age unlikely, a greater age being more probable than a younger one.

Oxygen isotope stage 6, c. 186 000–128 000 years BP

Although the oxygen isotope evidence (Fig. 1) suggests a long period of cold conditions following stage 7, the field evidence for actual glaciation is still obscure. The Ridge Acre Boulder Clay at Quinton, near Birmingham (Horton 1989) may be a candidate. The evidence for periglacial conditions is nevertheless more conclusive. Part of the thick solifluction deposit which overlies the Brighton raised beach is probably of this age, as are the periglacial horizons that overlie the interglacial deposits at Stanton Harcourt (the Stanton Harcourt Gravel) and that above the Lower Channel at Marsworth, both of which have been shown to be earlier than deposits with remains of hippopotamus attributed to oxygen isotope sub-stage 5e. Ice-wedge casts at Balderton (Brandon & Sumbler 1991) provide further evidence of this periglacial episode.

Although there are some rich mammalian sites that may represent this period of time, few have good chronological control and their inclusion in stage 6 can be regarded only as provisional. The following sites apparently belong here.

(i) The Stanton Harcourt Gravel with associated ice wedges, overlying the interglacial Stanton Harcourt Channel deposits. Horse and mammoth persist; woolly rhinoceros *Coelodonta antiquitatis* is a new arrival.

(ii) The Balderton Sand and Gravel, Lincolnshire (Lister & Brandon 1991) with mammoth, woolly rhinoceros, reindeer *Rangifer tarandus*, musk-ox *Ovibos moschatus* and other species.

(iii) The Warwickshire Avon Terrace 4 (Whitehead 1989) with reindeer and musk-ox.

(iv) The Taplow Gravel of the River Thames (Gibbard 1986) with mammoth, woolly rhinoceros and musk-ox.

(v) Crayford, Kent. Kennard (1944) described the following sequence of deposits, which he considered to be later than those of Ilford: 3, 'Upper Brickearth', apparently not fluviatile but the result of sludging; 2, 'Lower Brickearth' and 'Corbicula Bed', laid down in a sluggish stream and then more strongly running water, respectively; 1, 'Lower Gravel', exposed notably at Slades Green, deposited by a fast-flowing river. Its top was a land surface, occupied by humans, who left many artefacts of Levallois technique which occur also in the 'Lower Brickearth'. A broken jaw of a woolly rhinoceros was found lying on a heap of flint flakes at this horizon.

Mammalian species from this locality include extinct ground squirrel or suslik *Spermophilus primigenius*, a species different from that of the Devensian (Gromov in Sutcliffe & Kowalski 1976), collared and Norwegian lemmings *Dicrostonyx torquatus* and *Lemmus lemmus*, northern vole *Microtus oeconomus*, wolf *Canis lupus*, brown bear *Ursus arctos*, a very large lion *Panthera leo*, mammoth *Mammuthus primigenius*, horse *Equus ferus*, narrow-nosed rhinoceros *Stephanorhinus hemitoechus*, woolly rhinoceros *Coelodonta antiquitatis*, giant deer *Megaloceros giganteus*, red deer *Cervus elaphus*, giant ox *Bos* sp., bison *Bison* sp., bison and musk-ox *Ovibos moschatus*. The absence of reindeer *Rangifer tarandus* was specially noted by Kennard. Also present was a rich molluscan fauna.

The climatic interpretation of this site is beset with problems, the most notable being the occurrence of remains of lemmings, usually associated with cold conditions, in association with supposedly interglacial *Corbicula* (which Kennard regarded as the last appearance of this genus in the British Pleistocene) in the '*Corbicula* Bed', under circumstances that cannot be explained by secondary derivation of this mollusc or by burrowing into a pre-existing land surface by the lemmings. From the evidence of the land molluscs, Kennard inferred an open grassland environment without woodland or marshes, with summers that may have been as warm as those of the south of France and winters probably milder than those of Kent today. In contrast, it has been pointed out by Currant (1986) that there are few better 'cold' mammalian faunas (mammoth, woolly rhinoceros, musk-ox, suslik and lemming) known from the British Isles. We do not necessarily have to infer an environment the same as that occurring anywhere in the world at the present day. A steppe-like environment might explain these contrasting associations, with *Corbicula* surviving into oxygen isotope stage 6.

(vi) Tornewton Cave, Devon, 'Glutton Stratum'. Although from the same cave as the 'Otter Stratum', the precise relationship of these two deposits is unfortunately unproved, as they are situated in different chambers. A 'penultimate glaciation' age (i.e. later than the 'Otter Stratum') is inferred (Sutcliffe & Kowalski 1976). Brown bears and other carnivores make up most of the contained fauna, but there is also an unusual assemblage of rodents, including the steppe lemming *Lagurus lagurus* (the only British record of this species), common hamster *Cricetus ericetus* and extinct hamster *Allocricetus bursae*, typically species of a steppe environment. This deposit is disturbed and mixed so that caution is necessary in faunal interpretation. On the basis of available evidence, however, the occurrence of a major steppe episode at this time still seems acceptable.

(vii) Hutton Cave, Somerset, with teeth of a mammoth showing archaic characters and of extinct hamster *Allocricetus bursae*, the only other British record of which is from the 'Glutton Stratum' of Tornewton Cave.

(viii) Bacon Hole Cave, South Wales. The earliest horizon of this coastal cave, the 'Coarse Sands', may be a further stage 6 deposit: there is evidence of wind deposition and a fauna with horse *Equus ferus*, and a large form of northern vole *Microtus oeconomus*, suggesting relatively cold conditions, and land snails indicating bare rock scree and little vegetation (Stringer *et al.* 1986).

The occurrence at a number of sites, regarded as of 'penultimate glaciation' age, in the Neuweid Basin, Germany, of remains of woolly rhinoceros, mammoth, reindeer and *Lagurus*, suggestive of a cold steppe environment (Roebroeks *et al.* 1992), is closely matched by the faunal assemblages of the British sites discussed above. From the data then available, Kowalski (in Sutcliffe & Kowalski 1976) already inferred an invasion of eastern European rodents from the Continent at this time – evidence of a land connection and low sea-level. During the coldest part of this period, all rodents connected with the forest environment disappeared from the British Isles.

Oxygen isotope stage 5, c. 128 000–75 000 years BP

As previously mentioned, Shackleton (1969), on the basis of the oxygen isotope evidence, subdivided stage 5 into three warm peaks (e, c and a, 5e being especially warm), separated by two cold phases (d and b) (Fig. 1).

In contrast to earlier oxygen isotope stages, the events and faunas of stage 5, especially of sub-stage 5e, are better understood. It has long been recognized that in the Late Pleistocene there was an interglacial episode which was warmer than the present day, during which hippopotamus spread over England and Wales. That this is equivalent to deep-sea oxygen sub-stage 5e is well established from the work of Gasgoyne *et al.* (1981) who, by the uranium series method, determined an age of *c.* 120 000 years for the 'Lower Cave Earth' of Victoria Cave, Yorkshire, a deposit with remains of hippopotamus and narrow-nosed rhinoceros.

At other sites the mammalian fauna associated with these two species (henceforth to be called the 'hippopotamus fauna') is of very consistent composition and, unless it should be shown at some time that it was repeated (for which there is currently no evidence), it does seem to provide a most excellent temporal marker, to which other events can be related. Characteristic faunal elements include hippopotamus *Hippopotamus amphibius*, straight-tusked elephant *Palaeoloxodon antiquus*, narrow-nosed rhinoceros *Stephanorhinus hemitoechus*, bison

Bison priscus, giant ox *Bos primigenius*, pig *Sus scofa*, red deer *Cervus elaphus*, fallow deer *Dama dama*, giant deer *Megaloceros giganteus*, lion *Panthera leo*, spotted hyaena *Crocuta crocuta*, wolf *Canis lupus*, fox *Vulpes vulpes*, wildcat *Felis silvestris*, badger *Meles meles*, brown bear *Ursus arctos* and hare *Lepus* sp. The rodent fauna of this time was species-poor – in the south of England are found only water vole *Arvicola* sp. and field vole *Microtus agrestis*, with possibly wood mouse *Apodemus sylvaticus* and bank vole *Clethrio-nomys glareolus*, suggesting isolation from the Continent of Europe.

From the evidence of the great number of sites where this distinctive Late Pleistocene faunal assemblage (with hippopotamus) has been found in the British Isles, this episode appears to have been a major event in the climatic history of the islands. Sites have been recognized over an area extending from Kent in southeast England to Plymouth in the southwest, in south and north Wales, and as far north as Yorkshire and Durham. Important examples include Eastbourne, Sussex; Swalecliff, Kent; Peckham, Trafalgar Square and Brentford (Zeuner 1945) in London; East Mersea (Bridgland *et al.* 1988) and Walton-on-the-Naze, Essex; Waterhall Farm, Hertfordshire; Barrington, Cambridgeshire; Durdham Down and Milton Hill Caves, Somerset; Eastern Torrs Quarry Cave, Joint Mitnor Cave (Sutcliffe 1960), the 'Hyaena Stratum' of Tornewton Cave and Honiton, Devon; Swanton Morley, Norfolk; Fulbeck, Lincolnshire; Victoria Cave, Kirkdale Cave (Buckland 1823), Raygill Fissure and the Sewerby raised beach, Yorkshire; Ravenscliff Cave, South Wales and Cefn Cave, North Wales (both Falconer in Murchison 1868).

Although it has become widely accepted (e.g. West 1988) that the Ipswichian stratotype of Bobbitshole represents the same climatic event as the 'hippopotamus episode' oxygen isotope sub-stage 5e, the mammalian fauna of that site is sparse and hippopotamus is not among the mammals represented. Such contemporaneity remains unproven from the mammalian evidence, although the pollen, molluscan (Sparks 1957) and coleopteran (Coope 1974) associations are in agreement. Currant's (1989) suggestion, based on mammalian evidence, that the Bobbitshole deposits might be of earlier than 5e age, nevertheless cannot be ruled out.

Several of the above-mentioned sites have pollen data: Ip IIb for Trafalgar Square (Franks 1960) and Barrington (Gibbard & Stuart 1975) and Ip III for Swanton Morley (Stuart 1976). The hippopotamus apparently persisted from at least Ip IIb to III of the interglacial.

Evidence of relatively high sea-level with potential isolation of Britain from the Continent at this time is provided by the occurrence of remains of hippopotamus, straight-tusked elephant and narrow-nosed rhinoceros in the Sewerby Raised Beach, Yorkshire (Catt & Penny 1966; Boylan 1967).

The 'hippopotamus fauna' has not been recognized in Ireland, suggesting that there was no land-bridge at that time or, if there were, it could only have been from a part of Scotland north of the spread of hippopotamus.

A series of Late Pleistocene sites with hippopotamus in Germany, attributed by Von Koenigswald (1982) to the Last Interglacial, may mark part of the migration route whereby this animal arrived in Britain.

If we now compare the sub-stage 5e mammalian fauna described above with the most recent fauna preceding it that is recognizable (i.e. Crayford and the Tornewton Cave 'Glutton Stratum'), the difference between them is seen to be quite astonishing. With the exception of the principal carnivores (lion, bear and wolf, which carry on), the change is almost total. Especially remarkable disappearances were those of humans (Currant 1986; Wymer 1988), horse and the mollusc *Corbicula fluminalis*, all last seen at Crayford. Something quite fundamental appears to have occurred, with climatic change and the high sea-level of sub-stage 5e likely components.

Evidence concerning what happened during the later part of stage 5 is best provided from studies of the two Welsh coastal caves, Minchin Hole and Bacon Hole (Sutcliffe & Bowen 1973; Sutcliffe 1981; Currant *et al.* 1984; Sutcliffe & Currant 1984; Stringer *et al.* 1986; Sutcliffe *et al.* 1987). Uranium series dating of flowstone deposits from these caves by Schwarcz (in Currant *et al.* 1984; Stringer *et al.* 1986) and P. Smart (pers. comm.) indicate the survival of some of the typical 'hippopotamus episode' mammals, including *Palaeoloxodon, Stephanorhinus hemitoechus* and *Dama dama*, and continuation of a relatively high sea-level (later to fall again) for some time after sub-stage 5e and hence into the early part of the Devensian. The occurrence of remains of mammoth and northern vole at a horizon within this otherwise interglacial sequence at Bacon Hole is interpreted as evidence of a temporary minor deterioration of climate.

Remains of wolverine and reindeer from deposits with a uranium series date of $83\,000 \pm 6000$ years BP at Stump Cross Cave, Yorkshire (Sutcliffe *et al.* 1985) indicate a significant climatic deterioration within the younger part of isotope stage 5.

Oxygen isotope 4 onwards, from c. 75 000 years BP

With the further deterioration of climate and falling sea-level of the Devensian, woolly mammoth, woolly rhinoceros, reindeer and other cold-climate species spread across the British Isles. The abundance of mammoths' teeth recovered from the floor of the North Sea by trawlers is evidence of former dry land where this is now situated. Humans and horses returned to the British mainland. The faunal isolation apparent during isotope stage 5e had by this time been broken. It was not to be established again until the Holocene.

Discussion and conclusions

From the evidence of the mammals there emerges a story of repeated connection of the British mainland to the Continent of Europe, and isolation from it, with the latter probably a relatively rare event (Table 1). Until the breaching of the chalk land-bridge at about the Anglian/Hoxnian boundary, the two land areas were joined and the similarity of the mammalian faunas from the Cromerian stratotype and of contemporary faunas in Germany is unsurprising. The faunas of Swanscombe, Kent, and Steinheim an der Murr, Germany, are similar, and the richness of the herpetofauna of the Hoxnian or near-Hoxnian site of Cudmore Grove, Essex, is also suggestive of land connection.

Oxygen isotope stage 7 is seen to have been a time when sea-level rose appreciably above that of the present day, with potential isolation of the British Isles, but whether this actually occurred cannot be demonstrated from the mammalian evidence. That a land connection had been present not very long previously is suggested by the presence of pond tortoise *Emys orbicularis* (which is unlikely to have tolerated salt-water) and the occurrence of human artefacts in sub-stage II of this interglacial at Selsey.

The arrival of eastern European steppe species during oxygen isotope stage 6 (Crayford and Tornewton Cave) suggests a further low sea-level and land connection. What happened after this to cause the disappearance of humans (who had previously been present in the British Isles, on and off, back to about 500 000 years BP, from the time of Boxgrove), horses and *Corbicula* can only be a source of speculation. If they had been driven south by the severe cold apparent from the oxygen isotope curve (Fig. 1), why are glacial deposits of this age not widely recognized across the British Isles? Does the evidence for steppe conditions provided by the mammals provide a lead to this problem, associated low precipitation preventing the build-up of major ice bodies, just as occurred repeatedly throughout the Pleistocene in central Alaska? Any glacial deposits that were laid down are then likely to have been obscured by those of the Devensian. Whatever the explanation, something very drastic occurred at this time to cause humans to abandon the British Isles.

Stage 5e marks the arrival in the British Isles of a suite of interglacial mammal species, including hippopotamus and fallow deer, but now remarkably without humans, horses or the mollusc *Corbicula*. Why, when conditions might seem to have been so ideal for them, were they unable to join the other animals as the climate ameliorated? Both the sea-level and mammalian evidence indicate that the British Isles were isolated from the Continent at the height of the interglacial. If the rapidity with which the climate is shown from the oxygen isotope evidence to have become warmer at the termination of stage 6 were to have been matched by a corresponding rapidity of sea-level rise, then any mammals that were tardy in attempting the crossing to the British Isles, or could not swim, risked being left behind. With the Strait of Dover being gradually widened, was this what happened to humans and the other absent animals? Roebroeks *et al.* (1992) and Wymer (discussion in Roebroeks *et al.*) think this more likely than human inability to adapt to the forest conditions of the interglacial, another possible deterrent. *Homo sapiens* had not yet arrived in Europe and there is no evidence that the Neanderthals, already there, had learned to make boats. *Corbicula* could probably not have crossed salt-water. Some mammals, notably elephants, hippopotamus and deer, are known to have been better colonizers of islands than others, and for those that did succeed in reaching Britain, interglacial conditions warmer than today offered a rich living.

With the exposure of the floor of the southern part of the North Sea that accompanied the oncoming of the Devensian, movement of humans and other animals between the British Isles and the Continent again became unrestricted. Not all newcomers, however, need necessarily have come from the Continent. It has been pointed out by Stuart (1995) that the temporary appearance of mammoth during a minor episode of climatic deterioration apparent within the upper part of the interglacial sequence at Bacon Hole (Stringer *et al.* 1986), could have resulted from recolonization of mammoths from

Table 1. *Britain's island status during the Middle–Upper Pleistocene, based on sea level and fossil mammalian evidence.*

Geological Society scheme (Mitchell et al. 1973)	Oxygen isotope chronology	Some important sites	Mammalian faunas	Climate	Human presence	Land connection between British Isles and Continent
Holocene	1 10 000 years BP				+	Isolation (later part)
Devensian	5d–2 120 000 years BP	'Cave Earth' of Kent's Cavern	Woolly mammoth Woolly rhinoceros, Reindeer fauna	Cold	+	Connection
Ipswichian	5e 130 000 years BP	Barrington Joint Mitnor Cave Trafalgar Square Victoria Cave	Hippopotamus fauna	Interglacial	Absent	Isolation
	6 186 000 years BP	Crayford Tornewton Cave 'Glutton Stratum'	Woolly rhinoceros, Musk-ox fauna	Steppe	+	Connection
	7 245 000 years BP	Marsworth 'Lower Channel' Pontnewydd Cave Stanton Harcourt	Horse, mammoth fauna	Interglacial	+	?
	8					
Hoxnian?	9	Hoxne ?Cudmore Grove		Interglacial	+	Connection (not entire stage)
	10					
	11	?Swanscombe		Interglacial	+	Connection (not entire stage)
Anglian?	12			Massive glaciation	+	Connection
	13	?Boxgrove		Interglacial	+	
Cromerian						

No finality is claimed for the correlations suggested, which are presented for the purpose of encouraging discussion. The table should be read from the bottom upwards.

refugia in, say, Scotland, without any land-bridge to the Continent being necessary.

The precise date of the return of humans to the British Isles after stage 5 is for the time being unproven. By this time, Neanderthals and *Homo sapiens* are considered to have become separate lineages, the latter not arriving in Europe until about 40 000 years ago (Stringer & Gamble 1993). The earliest British dated skeletal remains (30 900 ± 900 years BP for a maxilla from Kent's Cavern; 26 350 ± 550 years BP for a humerus from Paviland Cave, South Wales (Stringer 1990), are both of *Homo sapiens*. Human presence earlier during the Devensian is also suggested by the occurrence of an industry including hand-axes and Levallois flakes, that archaeologists usually refer to as an 'insular British Mousterian of Acheulian tradition', at sites such as Bramford Road, Suffolk (Wymer 1985). Did the Neanderthals, after a long absence, find their way back to England during the last few millennia of their survival, to be the manufacturers of this industry, or was this the product of newly arrived modern man?

Many problems still attend the understanding of Britain's island status during the Quaternary.

The author thanks Patrick Boylan, David Bridgland, David Bowen, the late John Carreck, Russell Coope, Andy Currant, Chris Green, David Keen, Wighart von Koenigswald, Thijs van Kolfschoten, Adrian Lister, Tom Lord, Fred Owen, Simon Parfitt, Richard Preece, Chris Proctor, Helen Roe, Jim Rose, Kate Scott, Danielle Schreve, Pete Smart, Chris Stringer, Tony Stuart, Alan Turner, Elizabeth Walker, Graham Ward and John Wymer for information and helpful discussion.

References

ADAM, K. D. 1954. Die mittelpleistozänen Faunen von Steinheim an der Murr (Wurttemberg). *Quaternaria*, **1**, 131–144.

BOWEN, D. Q. 1977. Hot and cold climates in prehistoric Britain. *Geographical Magazine*, **49**, 685–698.

—— 1978. *Quaternary Geology*, Pergamon, Oxford.

—— 1989. The Last Interglacial–Glacial cycle in the British Isles. *Quaternary International*. **3/4**, 41–47.

——, HUGHES, S., SYKES, G. A. & MILLER, G. H. 1989. Land–sea correlations in the Pleistocene based on isoleucine epimerization in non-marine molluscs. *Nature*, **340**, 49–51.

——, SYKES, G. A., REEVES, A., MILLER, G. A., ANDREWS, G. T., BREW, J. S. & HARE, P. E. 1985. Amino acid geochronology of raised beaches in south west Britain. *Quaternary Science Reviews*, **4**, 279–318.

BOYLAN, P. J. 1967. The Pleistocene Mammalia of the Sewerby-Hessle Buried Cliff, East Yorkshire. *Proceedings of the Yorkshire Geological Society*, **36**, 115–125.

—— 1977. *The Ice Age in Yorkshire and Humberside*. Yorkshire Museum.

BRANDON, A. & SUMBLER, M. G. 1991. The Balderton Sand and Gravel: pre-Ipswichian cold stage fluvial deposits near Lincoln. *Journal of Quaternary Science*, **6**, 117–138.

BRIDGLAND, D. R. 1994. *Quaternary of the Thames*. Chapman & Hall, London.

——, ALLEN, P., CURRANT, A. P. ET AL. 1988. Report of Geologists' Association field meeting in north-east Essex, May 22nd–24th, 1987. *Proceedings of the Geologists' Association*, **99**, 315–333.

BRIGGS, D. J. 1988. The environmental background to human occupation in the upper Thames valley during the Quaternary period. *In:* MACRAE, R. J. & MOLONEY, N. (eds). *Non-flint Tools in the Palaeolithic Occupation of Britain*. British Archaeological Report, British Series, **189**, 167–185.

——, COOPE, G. R. & GILBERTSON, D. D. 1985. *The Chronology and Environment of Early Man in the Upper Thames Valley: a New Model*. British Archaeological Report, British Series, **137**.

BUCKLAND, W. 1823. *Reliquiae Diluvianae*. John Murray, London.

BULLEID, A. & JACKSON, W. 1937. The Burtle Beds of Somerset. *Proceedings of the Somersetshire Archaeological and Natural History Society*, **83**, 171–195.

—— & —— 1941. Further notes on the Burtle Sand-beds of Somerset. *Proceedings of the Somersetshire Archaeological and Natural History Society*, **87**, 111–116.

BURCHELL, J. P. T. 1957. A temperate bed of the last interglacial period at Northfleet, Kent. *Geological Magazine*, **94**, 212–214.

CARRECK, J. N. 1976. Pleistocene mammalian and molluscan remains from 'Taplow' Terrace deposits at West Thurrock, near Grays, Essex. *Proceedings of the Geologists' Association*, **87**, 83–92.

CATT, J. A. & PENNY, L. F. 1966. The Pleistocene deposits of Holderness, East Yorkshire. *Proceedings of the Yorkshire Geological Society*, **35**, 375–420.

COOPE, G. R. 1974. Interglacial coleoptera from Bobbitshole, Ipswich, Suffolk. *Journal of the Geological Society of London*, **130**, 333–340.

CURRANT, A. P. 1984. The mammalian remains. *In:* GREEN, H. S. (ed.) *A Lower Palaeolithic Hominid Site in Wales*. National Museum of Wales, Cardiff, 171–180.

—— 1986. Man and Quaternary interglacial faunas of Britain. *In:* COLLCUTT, S. N. (ed.) *The Palaeolithic of Britain and its Nearest Neighbours*. University of Sheffield, 50–52.

—— 1989. The Quaternary origins of the modern British mammal fauna. *Biological Journal of the Linnean Society*, **38**, 23–30.

——, STRINGER, C. P. & COLLCUTT, S. W. 1984. Bacon Hole Cave. *In:* BOWEN, D. Q. & HENRY, A. (eds) *Quaternary Research Association Annual Field Meeting Guide. Wales: Gower*. Cambridge, 38–45.

DAVIES, K. H. & KEEN, D. H. 1985. The age of the Pleistocene marine deposits at Portland, Dorset. *Proceedings of the Geologists' Association*, **96**, 217–225.

FRANKS, J. W. 1960. Interglacial deposits at Trafalgar Square, London. *New Phytologist*, **59**, 145–152.

GASGOYNE, M., CURRANT, A. P. & LORD, T. 1981. Ipswichian fauna of Victoria Cave and the marine palaeoclimate record. *Nature*, **294**, 652–654.

GIBBARD, P. L. 1986. *The Pleistocene History of the Middle Thames Valley*, Cambridge University Press, Cambridge.

—— 1988. The history of the great northwest European rivers during the past three million years. *Philosophical Transactions of the Royal Society of London*, **B318**, 559–602.

—— 1995. The formation of the Strait of Dover. *This volume*.

—— & STUART, A. J. 1975. Flora and vertebrate fauna of the Barrington Beds. *Geological Magazine*, **112**, 493–501.

GREEN, C. P., COOPE, G. R., CURRANT, A. P. ET AL. 1984. Evidence of two temperate episodes in the late Pleistocene deposits at Marsworth, U.K. *Nature*, **309**, 778–781.

GREEN, H. S. 1984. *Pontnewydd Cave. A Lower Palaeolithic Hominid Site in Wales*. National Museum of Wales, Cardiff.

GREEN, S. & WALKER, E. 1991. *Ice Age Hunters, Neanderthals and Early Modern Hunters in Wales*. National Museum of Wales, Cardiff.

HINTON, M. A. C. 1926. *Monograph of the voles and lemmings (Microtinae), living and extinct*. Vol. 1, British Museum (Natural History), London.

HOLMAN, A. J., STUART, A. J. & CLAYDEN, J. D. 1990. A Middle Pleistocene herpetofauna from Cudmore Grove, Essex and its paleogeographic and paleoclimatic implications. *Journal of Vertebrate Paleontology*, **10**, 86–94.

HORTON, A. 1989. Quinton. *In:* KEEN, D. H. (ed.) *The Pleistocene of the West Midlands: Field Guide*. Quaternary Research Association, Cambridge.

IMBRIE, J. & IMBRIE, K. P. 1979. *Ice Ages, Solving the Mystery*. Macmillan.
IVANOVICH, M. & HARMON, R. S. (eds) 1982. *Uranium Series Disequilibrium: Applications to Environmental Problems*. Oxford University Press, 69–76.
JONES, R. L. & KEEN, D. H. 1993. *Pleistocene Environments in the British Isles*. Chapman & Hall, London.
KEEN, D. H. 1994. The Brighton Raised Beach, Black Rock. *Excursion scientifique en Angleterre*. Association Français pour l'Étude du Quaternaire, 12 au 15 Mai, 1994.
—— 1995. Raised beaches and sea-levels in the English Channel in the Middle and Late Pleistocene: problems of interpretation and implications for the isolation of the British Isles. *This volume*.
KENNARD, A. S. 1944. The Crayford brickearths. *Proceedings of the Geologists' Association*, **55**, 121–69.
KIDSON, C., GILBERTSON, D. D., HAYNES, J. R., HEYWORTH, A., HUGHES, C. E. & WHATLEY, R. C. 1978. Interglacial marine deposits of the Somerset Levels, South West England. *Boreas*, **4**, 215–228.
KOENIGSWALD, W. v. 1982. Jungpleistozäne *Hippopotamus* – Funde der Oberrheinebene und ihre biogeographische Bedeutung. *Neues Jahrbuch für Geologie und Paläontologie Abhandlungen*, **163**, 331–348.
LISTER, A. M. & BRANDON, A. 1991. A pre-Ipswichian cold stage mammal fauna from the Balderton Sand and Gravel, Lincolnshire, England. *Journal of Quaternary Science*, **6**, 139–157.
MITCHELL, G. F., PENNY, L. F., SHOTTON, F. W. & WEST, R. G. 1973. *A Correlation of Quaternary Deposits in the British Isles*. Geological Society of London, Special Report, **4**.
MOIR, J. R. & HOPWOOD, A. T. 1939. Excavations at Brundon, Suffolk (1935–37). *Proceedings of the Prehistoric Society*, n.s. **5**, 1–32.
MOTTERSHEAD, D. H., GILBERTSON, D. D. & KEEN, D. H. 1987. The raised beaches and shore platforms of Tor Bay: a re-evaluation. *Proceedings of the Geologists' Association*, **98**, 241–256.
MURCHISON, C. (ed.) 1868. *Palaeontological Memoirs and Notes of the Late Hugh Falconer, A.M., M.D.* (2 vols). Hardwicke, London.
PROCTOR, C. J. & SMART, P. L. 1991. A dated cave sediment record of Pleistocene transgressions on Berry Head, Southwest England. *Journal of Quaternary Science*, **6**, 233–244.
ROBERTS, M. B., STRINGER, C. B. & PARFITT, S. A. 1994. A hominid tibia from Middle Pleistocene sediments at Boxgrove, U.K. *Nature*, **369**, 311–313.
ROEBROEKS, W., CONRAD, N. J. & VAN KOLFSCHOTEN, T. 1993. Dense forests, cold steppes and the Palaeolithic settlement of Northern Europe. *Current Anthropology*, **33**, 551–586.
—— & VAN KOLFSCHOTEN, T. 1994. The earliest occupation of Europe: a short chronology. *Antiquity*, **68**, 489–503.
ROSE, J. 1987. The status of the Wolstonian in the British Quaternary. *Quaternary Newsletter*, **54**, 15–20.
RZEBIK, B. 1968. *Crocidura* Wagler and other Insectivora (Mammalia) from the Quaternary deposits of Tornewton Cave in England. *Acta Zoologica Cracoviensia*, **13**, 251–263.
SHACKLETON, N. J. 1969. The Last Interglacial in the marine and terrestrial records. *Proceedings of the Royal Society of London*, **B174**, 135–154.
—— & TURNER, C. 1967. Correlation between marine and terrestrial Pleistocene successions. *Nature*, **216**, 1079–1082.
SHOTTON, F. W. 1953. Pleistocene deposits of the area between Coventry, Rugby and Leamington and their bearing on the topographic development of the Midlands. *Philosophical Transactions of the Royal Society of London*, **B237**, 209–260.
——, SUTCLIFFE, A. J. & WEST, R. G. 1962. The fauna and flora of the Brick Pit at Lexden, Essex. *Essex Naturalist*, **31**, 15–22.
SINGER, R., GLADFELTER, B. G. & WYMER, J. 1993. *The Lower Paleolithic Site at Hoxne, England*. University of Chicago Press.
SMITH, A. J. 1985. A catastrophic origin for the palaeovalley system of the eastern English Channel. *Marine Geology*, **64**, 65–75.
SPARKS, B. W. 1957. The non-marine mollusca of the interglacial deposits at Bobbitshole, Ipswich. *Philosophical Transactions of the Royal Society of London*, **B241**, 33–44.
—— & WEST, R. G. 1959. The palaeoecology of the interglacial deposits at Histon Road, Cambridge. *Eiszeitalter und Gegenwart*, **10**, 123–43.
—— & —— 1964. The Interglacial deposits at Stutton, Suffolk. *Proceedings of the Geologists' Association*, **74**, 419–432.
STRINGER, C. B. 1990. British Isles. *In:* ORBAN, R. (ed.) *Hominid remains: an Update*. Université Libre de Bruxelles, **3**, 1–40.
—— & GAMBLE, C. 1993. *In Search of the Neanderthals*. Thames & Hudson, London.
——, CURRANT, A. P., SCHWARCZ, H. P. & COLLCUTT, S. N. 1986. Age of Pleistocene faunas from Bacon Hole, Wales. *Nature*, **320**, 59–62.
STUART, A. J. 1976. The history of the mammal fauna during the Ipswichian/Last interglacial in England. *Philosophical Transactions of the Royal Society of London*, **B267**, 221–250.
—— 1979. Pleistocene occurrences of the European pond tortoise (*Emys orbicularis* L.) in Britain. *Boreas*, **8**, 359–371.
—— 1982. *Pleistocene Vertebrates in the British Isles*. Longman, London.
—— 1995. Insularity and Quaternary vertebrate faunas in Britain and Ireland. *This volume*.
SUMBLER, M. G. 1983. A new look at the type Wolstonian glacial deposits of Central England. *Proceedings of the Geologists' Association*, **94**, 23–31.
SUTCLIFFE, A. J. 1960. Joint Mitnor Cave, Buckfastleigh. *Transactions and Proceedings of the Torquay Natural History Society*, **13** (for 1958–59), 1–26.
—— 1964. The mammalian fauna. *In:* OVEY, C. D. (ed.) *The Swanscombe Skull*. Royal Anthropo-

logical Institute of Great Britain and Ireland, London, 85–111.

—— 1975. A hazard in the interpretation of glacial–interglacial sequences. *Quaternary Newsletter*, **17**, 1–3.

—— 1976. The British glacial–interglacial sequence. *Quaternary Newsletter*, **18**, 1–7

—— 1981. Progress report on excavations in Minchin Hole, Gower. *Quaternary Newsletter*, **33**, 1–17.

—— & BOWEN, D. Q. 1973. Preliminary report on excavations in Minchin Hole, April–May 1993. *Newsletter of the William Pengelly Cave Studies Trust*, **21**, 12–25.

—— & CURRANT, A. P. 1984. Minchin Hole Cave. *In:* BOWEN, D. Q. & HENRY, A. (eds) *Quaternary Research Association Annual Field Meeting Guide. Wales: Gower.* Cambridge, 33–37.

—— & KOWALSKI, K. 1976. Pleistocene rodents of the British Isles. *Bulletin of the British Museum of Natural History (Geology)*, **27**(2), 33–147.

—— & ZEUNER, F. E. 1962. Excavations in the Torbryan Caves, Devonshire. 1. Tornewton Cave. *Proceedings of the Devon Archaeological Exploration Society*, **5**, 127–145.

——, CURRANT, A. P. & STRINGER, C. B. 1987. Evidence of sea-level change from coastal caves with raised beach deposits, terrestrial faunas and dated stalagmites. *Progress in Oceanography*, **18**, 243–271.

——, LORD, T. C., HARMON, R. S., IVANOVICH, M., RAE, A. & HESS, J. W. 1985. Wolverine in northern England about 83,000 yr B.P.: Faunal evidence for climatic change during Isotope Stage 5. *Quaternary Research*, **24**, 73–86.

TURNER, A. 1995. Evidence for Pleistocene contact between the British Isles and the European continent based on distributions of larger carnivores. *This volume.*

TURNER, C. 1980. Surface and sedimentary processes. Palaeoclimatology Case Study. Glaciation and the Ice Age. Open University Press, Milton Keynes.

WALKER, D. 1953. The interglacial deposits at Histon Road, Cambridge. *Quarterly Journal of the Geological Society of London*, **108**, 273–282.

WEST, R. G. 1957. Interglacial deposits at Bobbitshole, Ipswich. *Philosophical Transactions of the Royal Society of London*, **241**, 1–31.

—— 1968. *Pleistocene Geology and Biology.* Longman, London.

—— 1969. Pollen analyses from interglacial deposits at Aveley and Grays, Essex. *Proceedings of the Geologists' Association*, **80**, 217–218.

—— 1981. Palaeobotany and Pleistocene stratigraphy in Britain. *New Phytologist*, **87**, 127–137.

—— 1988. The record of the cold stages. *Philosophical Transactions of the Royal Society of London*, **B318**, 505–522.

—— & SPARKS, B. W. 1960. The coastal interglacial deposits of the English Channel. *Philosophical Transactions of the Royal Society of London*, **B243**, 95–133.

——, LAMBERT, C. A. & SPARKS, B. W. 1964. Interglacial deposits at Ilford, Essex. *Philosophical Transactions of the Royal Society of London*, **B247**, 185–212.

WHITEHEAD, P. F. 1989. The development and sequence of deposition of the Avon valley river-terraces. *In:* KEEN, D. H. (ed.) *The Pleistocene of the Midlands: Field Guide*, Quaternary Research Association, Cambridge, 37–41.

WILLEMSEN, G. F. 1992. A revision of the Pliocene and Quaternary Lutrinae from Europe. *Scripta Geologica*, **101**, 1–115.

WYMER, J. 1985. *Palaeolithic Sites of East Anglia.* GeoBooks, Norwich.

—— 1988. Palaeolithic archaeology and the British Quaternary sequence. *Quaternary Science Reviews*, **7**, 79–98.

ZEUNER, F. E. 1945. *The Pleistocene Period, its Climate, Chronology and Faunal Successions.* Ray Society, London.

Evidence for Pleistocene contact between the British Isles and the European Continent based on distributions of larger carnivores

ALAN TURNER

Department of Human Anatomy and Cell Biology, University of Liverpool, PO Box 147, Liverpool L69 3BX, UK

Abstract: The generalized ecological tolerances of larger terrestrial carnivores, reflected in their wide distributions in fossil and living faunas, make them good potential indicators of the changing pattern of links between Britain and the continent of Europe over time. Extensive contact is indicated until the earlier part of the Middle Pleistocene, that is prior to the Anglian–Elsterian glaciation. The pattern after that is more complex, with evidence for some form of marine barrier by the time of the Hoxnian and for complete isolation during the Last Interglacial.

With few exceptions, the larger mammal species of the British Pleistocene originated elsewhere and dispersed into the region, where their presence in fossil deposits represents no more than the extreme northwestern tip of a larger distribution. The pattern of their appearance may therefore hold some clues to the changing nature of the link between the British Isles and the rest of the European continent.

In this paper I shall concentrate on the timing of appearances of some of the larger members of the Carnivora during the Middle Pleistocene. These ecological generalists are among the most widely distributed of the Pleistocene mammals (Turner 1990, 1992), and are ideally suited to act as indicators of contact. The base of the Pleistocene is here taken as the base of the Eburonian Stage in The Netherlands at 1.6 Ma BP, the likely local equivalent of the Italian stratotype boundary (Gibbard *et al.* 1991, p. 47).

Patterns of appearance

Although a diverse fauna of larger mammals may be broadly assigned to the later Pliocene and perhaps the earlier part of the Lower Pleistocene in Britain, the haphazard recovery of many of the specimens makes allocation to particular stages rather difficult (Gibbard *et al.* 1991). In any event, the inescapable conclusion from a recent assessment of the British and Dutch evidence is that a major hiatus in the sequence of fossiliferous deposition within the Cromer Forest Bed Formation seems to exist between the beginning of the Pleistocene (as defined above) and the Cromerian (s.s.) (Gibbard *et al.* 1991). Lister (1993) suggests that some mammal-bearing gravels in the CF-bF may have been deposited within this general depositional 'hiatus'. Nonetheless, the discussion of Pleistocene contacts for the present purpose is necessarily largely confined to Middle and Late Pleistocene events. This restriction may be less of an impediment to understanding the timing of mammalian dispersions than would at first appear, since Gibbard (1988, 1995) has recently summarized the evidence for the formation of the Strait of Dover, and therefore for lengthy periods of isolation induced by subsequent changes in sea-level, first taking place during the Elsterian–Anglian glacial stage. In this scheme, the breach would have resulted from erosion by the outlet of an ice-dammed lake in the southern North Sea into which rivers such as the Thames and Rhine then flowed. It is apparent from Gibbard's discussion that the subsequent history of the Strait, and thus the extent of contact between Britain and Europe, is likely to have been complex, although any efforts to specify the absolute sequence of events involve a number of subsidiary assumptions about the number of interglacials after the Elsterian–Anglian stage. I therefore propose to give a broad outline of the pattern of appearances by the larger carnivores (including gaps in appearance) in order to relate this pattern to the most likely sequence of temperate stages and then to examine some of the implications of the two taken together.

Evidence of contact

Earliest evidence for contact based on larger Carnivora (see Table 1) comes from a small,

Table 1. *Appearances of larger carnivores in the British Isles.*
The temperate Group assemblages (1–5) of Currant (1989) are employed together with their likely correlation to the standard interglacial sequence of Mitchell *et al.* (1973). Also shown are three of the standard cold stages and a generalized pre-Cromerian stage.

	Pre-Cro	5 (Cro)	4	Ang	3 (Hox)	'Wol'*	2	'Wol'*	Dev	1 (Ips)
Felidae										
Acinonyx pardinensis	+									
Homotherium latidens	+	+	+							
Panthera leo	?	+	+		+	+	+	+	+	+
Panthera pardus							+		+	+
Panthera gombaszoegensis	+	+	+							
Hyaenidae										
Pachycrocuta perrieri	+									
Pachycrocuta brevirostris	?	+								
Crocuta crocuta		+	+		?		+	?	+	+
Canidae										
Canis mosbachensis	+	+	+		+					
Canis lupus						+	+	+	+	+
Canis (Xenocyon) lycaonoides			+							
Ursidae										
Ursus deningeri		+	+							
Ursus spelaeus					+		?			
Ursus arctos						+	+	+	+	+

*Allocations to the 'Wolstonian' cold stage are based on material from the levels below the Hyaena Stratum at Tornewton (Sutcliffe & Zeuner 1962; Stuart 1982). See text for discussion.

broken premolar from the cliff deposits at Easton Bavents in Suffolk (Spencer 1959) ranging from Antian to early Baventian in age. Although the specimen, M20227 in the collections of the Natural History Museum, has been previously identified as *Felis pardoides*, the high crown and sharp cresting show that it is clearly a P/3 of a large cheetah, and should be referred to the European species *Acinonyx pardinensis*. This is the first evidence for the species in Britain, although it is known throughout the Villafranchian faunal span in many other parts of Europe (Turner 1992).

The earliest European appearance of the spotted hyaena *Crocuta crocuta*, may be in the transitional faunas of the end-Villafranchian faunal event in Italy (Azzaroli 1983), where it is recorded from the karst fissure at Selva Vecchia in Verona (D. Torre, *in litt.*), and just above the Brunhes/Matuyama boundary in the TD3 deposits at Atapuerca in Spain (Aguirre *et al.* 1990). In Britain, the earliest records are from the West Runton Freshwater Bed deposits of the Cromerian type locality as well as from 'Cromerian' deposits at Palling. It is also recorded from the rootlet (= Freshwater Bed) deposits at Corton Cliff by Newton (1883), and is present in small numbers in the latest levels of the Calcareous member at Westbury-sub-Mendip (Turner in press *a*).

The Westbury assemblage may also record the earliest occurrence of the lion *Panthera leo* in Britain, although Stuart (1982, p. 115) mentions its presence at Pakefield in Suffolk in what may be Cromerian deposits. There is also a large, somewhat rolled felid distal humerus, found on Cromer beach and now in the Colman Collection at Norwich Castle Museum, that appears to be of lion. The 'undetermined large cat' recorded from the West Runton Freshwater Bed (Stuart 1990), represents two other large felid taxa, as discussed further below. The earliest European appearance of the lion would seem to be at Vallonnet on the southern French coast, where the mammalian assemblage was found in a deposit for which a date of formation during the Jaramillo event seems increasingly likely (de Lumley *et al.* 1988).

The leopard *Panthera pardus* is also first clearly seen in Europe at Vallonnet and, like the lion, may have dispersed from Africa at around that time (Turner 1992). Although it appears throughout the Middle and Upper Pleistocene of Europe it is never a common or abundant member of assemblages, and it appears to have been entirely absent from Cromerian deposits and is not recorded at Westbury (Turner in press *a*). Apart from a few isolated specimens from Mendip localities of Upper Pleistocene age (Sanford 1867), the only other (and therefore the oldest) British material comes from the Intermediate complex and Lower Breccia of Pontnewydd Cave, with a date of c. 225 000 years BP (Currant 1984; Green 1984).

The larger *Panthera gombaszoegensis*, intermediate in size between lion and leopard and often referred to as the Eurasian jaguar, is first recorded in the earliest part of the Lower Pleistocene at Olivola and in what seems to have been the uppermost levels at Tegelen (Von Koenigswald 1960; Azzaroli *et al.* 1988). It is known in some numbers from Westbury (Bishop 1982), and is probably the smaller of two large felids from the type Cromerian deposits at West Runton (Turner in press *a*). A third possible occurrence is at Swanscombe, where lion is also clearly present (Turner in press *a*), which may be the jaguar's latest European record.

Dogs of the genus *Canis* provide potential evidence of two distinct events, the first appearance of the genus itself and the later-occurring earliest record of larger animals. Masini & Torre (1990) identify the former, which they term the 'wolf event', in the Olivola faunal association as the first appearance of a small, jackal-sized, wolf-like animal variously referred to *Canis etruscus* or *C. mosbachensis*. The second event takes place at some point towards the end of the Middle Pleistocene when there is a marked increase in the size of specimens identified as the true extant wolf of the Holarctic region, *C. lupus*.

Material from Forest Bed deposits at Sidestrand may represent the earliest appearance of the smaller *Canis mosbachensis* in Britain, since this may be of broadly similar age to Olivola (Stuart 1982; Gibbard *et al.* 1991). The species is also present in the West Runton Freshwater Bed and is matched by specimens of similar size from Westbury and Swanscombe (Turner in press *a*) and also from Boxgrove (Roberts 1990). In contrast, typically large representatives of *C. lupus* are seen in numerous deposits of Devensian and Last Interglacial age at sites such as Kent's Cavern, Joint Mitnor and the Tornewton Hyaena Stratum. Smaller numbers of specimens from the underlying Bear/Glutton Stratum at Tornewton, together with the undated but arguably pre-Last Interglacial specimen from Crayford, also appear to be large, although one specimen from the Bear Stratum at Tornewton does fall within the range of Westbury measurements (Turner in press *a*). Currant (1984) has pointed out that the fragmentary material of *Canis* from the Intermediate complex at Pontnewydd, dated with some consistency to c. 225 000 years BP (as summarized by Green (1984)) falls within the larger size range.

The European hunting dog *Canis (Xenocyon) lycaonoides* is known in Britain only from the deposits at Westbury (Bishop 1982; Turner in press *a*), where it makes an important addition to the sparse European record of the species at sites such as Mosbach 2 and Gombaszoeg. It is also now recorded from the Lower Pleistocene deposits at the German site of Untermassfeld (M. Sotnikova, in prep.) and similar remains are known from Asian deposits (Sher 1986; Vangengejm *et al.* 1990). Masini & Torre (1990) suggest that the progenitor of *C. (X.) lycaonoides* may be the large *C. (X.) falconeri*, first recorded in late Villafranchian faunas of the Italian Tasso unit.

The gigantic hyaena *Pachycrocuta brevirostris* first appears in Europe in earliest Pleistocene deposits at Olivola (Azzaroli 1983), but is recorded from a number of very late Villafranchian/Galerian faunal sites in continental Europe, including Vallonnet (de Lumley *et al.* 1988), Gombaszoeg (Kretzoi 1938), Süssenborn (Kurtén 1969), Stránska Skála (Kurtén 1972) and Untermassfeld (Turner in press *b*). In Britain, it is represented in the Cromer Forest Bed deposits at Bacton, Mundesley and Sidestrand, as well as in the lowermost (Siliceous Member) deposits at Westbury (Bishop 1982: Turner in press *a*) generally regarded as pre-Cromerian in age on the basis of the small bovid sample (A. Gentry, pers. comm.), but is not recorded from the Freshwater Bed deposits of the Cromerian (s.s.). In view of the date of the earliest European appearance at Olivola and the apparent hiatus in the Cromer Forest Bed Formation, the British specimens from East Anglia must therefore belong either to the Cromerian or to the very earliest Pleistocene, although the Westbury specimen may fall within the intervening period.

The sabre-toothed cat *Homotherium latidens*, one of the longest-lived of the Plio-Pleistocene carnivores and known widely from the earliest Villafranchian onwards, occurs in various British deposits of the Cromer Forest Bed Formation and at the Derbyshire cave locality of

Doveholes, associated with the Pliocene gomphothere *Anancus arvernensis* (Spencer & Melville 1974). It is likely to be the larger of the two previously unidentified large felids present in the West Runton Freshwater Bed deposits (Turner in press *a*). It is also present at Westbury, where it is found in some of the very latest deposits at the site in Unit 19 (Turner in press *a*).

Ursus deningeri, a likely forerunner of the later cave bear *U. spelaeus*, is recorded at Vallonnet. In Britain it is known from the West Runton Freshwater Bed but the largest sample, of approximately Cromerian age, is probably that from Bacton (Bishop, 1982). It occurs in abundance at Westbury (Bishop 1982), where its denning activity is very much in evidence (Andrews & Turner 1992), and is also recorded at Boxgrove (Roberts 1990). *U. spelaeus* is recorded at Swanscombe (Sutcliffe 1964), one of the few instances of this species in Britain. The living brown bear *U. arctos* is known from all later deposits and is certainly present in Last Interglacial and Devensian-age deposits. Currant (1984) has referred to the bear from the Intermediate complex and Lower Breccia at Pontnewydd as having 'spelaeoid characters', but also acknowledged that such features in isolation may offer little help in establishing specific identity.

The presence of these species therefore points to extensive contact between Britain and the Continent during the later Pliocene and earliest Pleistocene and, in the case of the lion, the spotted hyaena and perhaps the smaller dogs and *Canis (Xenocyon)*, from the Jaramillo event to the formation of the Freshwater Bed and the deposits at Westbury inclusive. The large dogs referred to *Canis lupus* also point to a later contact some time between the formation of the Swanscombe deposit and the formation of the Intermediate complex deposit at Pontnewydd at *c.* 225 000 years BP, a contact perhaps supported by the presence of the leopard at Pontnewydd.

Evidence of barriers

While the timing of first appearances may point with some clarity to contacts between two regions, it is inherently more difficult to adduce evidence of barriers since the reasons for any absences from the fauna of a given region may be various. However, in the case of the larger carnivores, the very breadth of their ecological preferences that produces such wide geographic dispersions suggests a specific, physical impediment as the cause of any major gaps in the apparent range. While aspects of the discussion that follows are necessarily somewhat speculative, it seems that two distinct features of the larger carnivore fauna, augmented by the evidence of biometry and age structure in *Crocuta crocuta*, may suggest such an impediment, or such impediments, to occupation.

The first feature is the scarcity of *Crocuta crocuta* in Britain after its appearance at Westbury until it is recorded again in vast numbers from Last Interglacial deposits in sites such as Tornewton, Joint Mitnor and Kirkdale. In fact there are three records of it in the intervening period. The first is in the Lower Breccia deposit at Pontnewydd with a likely date of *c.* 225 000 years BP (Currant 1984; Green 1984). The second is from the deposits below the Hyaena Stratum at Tornewton Cave in Devon (Sutcliffe & Zeuner 1962). The precise stratigraphical relationships of these latter deposits, often referred to as 'Wolstonian' (Stuart 1982; Lister 1984) is presently under review (A. Currant, pers. comm.), but it is apparent that a colder-period fauna that may include *C. crocuta* is represented in the Glutton Stratum. The third instance is in the Gray's Thurrock assemblage from Essex, thought to come from a pre-Last Interglacial stage (Currant 1989). (A fourth potential instance from Lawford in Warwickshire, first described by Buckland (1823) and found in deposits ascribed by Shotton (1953) to the earliest stages of the Wolstonian glaciation, is most likely to have come from Devensian deposits within the Lawford pit (Sumbler 1983; Lister 1989)). These instances aside, its absence holds good until the Last Interglacial, whatever scheme is adopted for the chronology of the British Pleistocene (see discussion below); it is not present at Hoxne nor at sites most plausibly correlated with the deposits there, although it is known to have been present in many parts of continental Europe.

The second feature is the complete absence of *Pachycrocuta perrieri*, the third large hyaena of the European Plio-Pleistocene, from Middle Pleistocene deposits in the British Isles. This species has a peculiar pattern of appearance in Europe as a whole, since it first disappeared at the end of the Pliocene after a lengthy and apparently successful occupation of most parts of the Continent (Turner 1992) which included a number of appearances in the Red Crag deposits of Suffolk (Lydekker 1886; Newton 1891). Stuart (1982) suggests that Red Crag material may be placed within the Pre-Ludhamian of the British sequence, equating with the late Pliocene Praetiglian of The Netherlands (Gibbard *et al.* 1991). The species then reappeared in the earlier part of the European Middle Pleistocene at sites

Table 2. Lengths of lower second (P/2) and third (P/3) premolars from British Last Interglacial and Devensian localities compared with a sample of Last Glaciation age from Goyet in Belgium.

	P/2	P/3
	N Mean ± SEM (mm)	N Mean ± SEM (mm)
Britain		
Last Interglacial		
Barrington	11 17.40 ± 0.35	16 21.66 ± 0.24
Joint Mitnor	12 16.97 ± 0.23	29 21.46 ± 0.09
Tornewton H.S.	23 17.38 ± 0.18	45 21.59 ± 0.09
Kirkdale	29 17.33 ± 0.20	46 21.46 ± 0.11
Devensian		
Wookey Hole	49 16.76 ± 0.13	85 21.86 ± 0.12
Uphill	46 16.37 ± 0.15	68 21.73 ± 0.10
Kent's Cavern	187 16.49 ± 0.07	264 22.17 ± 0.06
Coygan	62 16.64 ± 0.10	118 22.05 ± 0.08
Sandford Hill	50 16.41 ± 0.11	57 21.92 ± 0.10
Pinhole	36 16.37 ± 0.11	41 21.61 ± 0.15
Belgium		
Goyet *	57 16.44 ± 0.13	58 22.25 ± 0.12

SEM, standard error of mean
* calculated from measurements supplied by B. Kurtén. Other measurements made by author.

such as Petralona in Greece, Mauer and Mosbach 2 in Germany, and L'Escale and Lunel Viel in southern France, where it has previously been identified as a large striped hyaena *Hyaena prisca* (Turner 1990). However, and perhaps significantly, it does not seem to have recolonized Britain during the Middle Pleistocene, nor for that matter did it reappear in Spain or Italy.

While one might argue for some kind of habitat barrier, such as mountains, as the reason for the absence of *P. perrieri* from Italy and Spain, it is difficult to believe that the various absences of both hyaenid taxa from Britain during at least part of the latter portion of the Middle Pleistocene, at the time of formation of the Freshwater Bed of the Cromerian, the Westbury calcareous deposits and during the Hoxnian interglacial, were the result of anything other than real, physical impediments.

Some support for the presence of a barrier may possibly be seen in differences in the morphometric characteristics and age structures of the populations of *Crocuta crocuta* from Last Interglacial and Devensian deposits. Several years ago, Kurtén (1963) pointed out that specimens from the Tornewton Cave 'Hyaena Stratum' differed from those of Devensian age in the absolutely and relatively greater size of the lower second premolar, a relationship confirmed by my own measurements on a somewhat larger sample (Table 2). In comparison, the Devensian hyaenas bear much more resemblance to those from the Belgian Last Glaciation site of Goyet (Table 2), and were most plausibly part of a larger NW European population during the lowered sea-level of this stage. Unfortunately, there are as yet no suitable samples of *Crocuta* teeth of Last Interglacial age known from continental Europe for comparison, but the British Last Interglacial morphology is unique in comparison with all other European samples known to me so that it seems most reasonably interpreted as a reflection of an isolated population.

If it turns out that the British Last Interglacial hyaenas were indeed isolated, then the discontinuity with the Last Glaciation populations is perhaps explained by the age structure of the death assemblages from the two periods in Britain. These have been placed in four adult age groups judged by tooth wear, based on work by Kruuk (1972) on modern populations in eastern Africa. The results for sites with samples large enough for the purpose are given in Table 3, where the bases for the age groupings are outlined. It may be seen that the pattern of age at death is for increased numbers in the lower age groups at the denning sites of Tornewton,

Table 3. *Percentages of hyaena teeth in adult age classes.*

	N	\multicolumn{4}{c}{Age class}
		II	III	IV	V
Last Interglacial					
Joint Mitnor	103	39	28	11	22
Tornewton H.S.	226	45	25	16	15
Kirkdale	198	46	20	17	17
Torcourt	147	48	23	15	14
Devensian					
Wookey Hole	271	33	25	21	22
Uphill	211	35	23	18	24
Kent's Cavern	909	36	24	21	18
Coygan	353	37	24	17	22
Sandford Hill	224	15	28	31	26
Pinhole	164	35	17	17	32

Total numbers are summed lower teeth (P/2–M/1), left and right, in each of the four adult age classes. Age classes, based on those used by Kruuk (1972), employ tooth wear stages, from initial slight wear of young adults (II) to heavy wear and senility (V).

Kirkdale and Torcourt caves, a pattern slightly altered by the assemblage at Joint Mitnor where the hyaenas evidently fell into a trap (Sutcliffe 1960). The Devensian sites, in contrast, show increased proportions of deaths in the older age groups, exaggerated in the case of Sandford Hill, another probable trap locality. (The heightened proportion of age-class V animals at both Joint Mitnor and Sandford Hill is probably explained by the attraction of the carcases of earlier victims of the trap for older and less able animals.) It is conceivable that the comparatively heightened mortality of the younger adults in the Last Interglacial population may have contributed to a local extinction of the species during the physical isolation of the interglacial, since even animals within age-class III had only just reached reproductive age, based on comparisons with modern representatives and the inter-birth interval tends to be long (Kruuk 1972).

Numbers of temperate stages

Quite clearly, the standard sequence of British Pleistocene stages proposed by Mitchell *et al.* (1973) now requires updating, a conclusion supported by mammalian as well as other lines of evidence. The oxygen isotope evidence (Shackleton 1987) is of particular relevance to any debate over stages, since it provides an independent and global framework, but for present purposes I shall confine my attention to the question of how many terrestrially based temperate stages should be placed between the Cromerian, as defined at West Runton, and the Last Interglacial in order to discuss the patterns of larger carnivore occurrences.

Currant (1989) has suggested that three such stages are warranted, with the Ipswichian (the Last Interglacial in the standard sequence) as the second and the Cromerian as the fifth of his five-stage scheme based on the division of mammalian assemblages into five groups. In this arrangement, the Calcareous member at Westbury, together with Boxgrove 4c and Ostend, would equate with Group 4, while Group 3 would include assemblages from sites such as Swanscombe, Hoxne and Clacton. Interestingly, Currant (1984, p. 179) suggested that the Pontnewydd Intermediate Complex assemblage, although possibly belonging to the Hoxnian stage of the classic British sequence, could equally well lie between the Hoxnian and the Last Interglacial. The smaller size of the Hoxnian-age dog at Swanscombe (Turner in press *a*) in comparison with the material from Pontnewydd (Currant 1984) would strengthen that suggestion, and result in placement of the assemblage within his Group 2. If the bear from those levels at Pontnewydd is indeed *Ursus spelaeus*, then this would indicate a post-Hoxnian presence in Britain.

Andrews (1990) has also concluded that a post-Cromerian, pre-Hoxnian age is indicated for the Calcareous Member deposits at Westbury, and has argued (1990, p. 177) for an 'interglacial complex, with at least two peak interglacials'. The first of the two would be the temperate phase of Currant's Group 4 based on the Unit 11 'Pink Breccia', although Andrews was also of the opinion that the second pre-dated the Hoxnian of the standard scheme. Other recent attempts at synthesis also seem to agree on placement of the Westbury, Ostend and Boxgrove faunas together in a phase after the Cromerian and before the Anglian–Elsterian glaciation and the subsequent Hoxnian (Lister *et al.* 1990; Gibbard *et al.* 1991).

Discussion

In Table 1, I have elected to place the appearances of the various larger carnivores in five temperate groups following Currant (1989), together with a generalized earlier group (pre-Cromerian) to mark specimens from Pliocene and perhaps earliest Lower Pleistocene deposits. I have also interpolated the Devensian and Anglian cold stages, the former anchored by its position as the last glacial stage and the latter anchored by the direct observation that it is conformably overlain by the Hoxnian at Hoxne (with which Currant correlated his Group 3) and the general consensus that it occurs after the deposition of assemblages found at Westbury, Ostend and Boxgrove (Lister *et al.* 1990; Gibbard *et al.* 1991) (placed by Currant in his Group 4). In this scheme I have retained the Ipswichian as the last temperate stage before the Devensian, although Currant has argued that the fauna from the type locality might better equate with his Group 2.

One further complication in Table 1 concerns the position of the Wolstonian cold stage, defined in the scheme of Mitchell *et al.* (1973) as that between the Hoxnian and the Ipswichian. If a further warm stage is now to be recognized, based not only on the terrestrial evidence but also on the oxygen isotope data as Stage 7 (Shackleton 1987), then the Wolstonian could be said to occur either side of that warm stage. The whole issue of the chronological and relative placement of the Wolstonian (Sumbler 1983; Lister 1989) is beyond treatment here, and I have simply acknowledged the problem by showing the taxa known to have occurred between the Hoxnian and the Ipswichian in two alternative places indicated by 'Wol'.

If the Anglian–Elsterian glaciation after the temperate assemblage of Group 4 saw the origin of the Strait of Dover, then the absence of *Crocuta crocuta* from the Hoxnian might well be explicable as a failure to cross the area of the Strait in time, combined with the local extinction of the population present earlier at sites such as Westbury. If my interpretation is correct, then the age structure of the Last Interglacial spotted hyaenas in comparison with those from Devensian deposits is consistent with the possibility that interglacial populations always found survival in the more wooded conditions somewhat more difficult, a clear implication from behavioural studies of the species today (Kruuk 1972).

It is worth stressing that the isolation of the Last Interglacial spotted hyaenas implied by their dental biometrics, whether or not the rate of mortality indicated by the number of deaths among young adults led to their local extinction, is reinforced by one very marked feature of the ungulate fauna: the complete absence of horse from the hippopotamus-rich assemblages of the Last Interglacial (Sutcliffe 1960, 1995; Turner 1981; Currant 1989; Stuart 1995). The closed, wooded conditions that may have heightened the mortality rate among the hyaenas would also have found little favour with the horses, although conditions of cool grassland during the previous stage would have been ideal for them, as indicated by their presence in the Balderton Sand and Gravel (Lister & Brandon 1991). Most of the Last Interglacial assemblages, and certainly those in caves, were actually accumulated by the hyaenas (Turner 1981), for whom horses are likely to have been an attractive prey. It is true, as Lister & Brandon (1991, p. 141) remark, that pre-Last Cold Stage horses were generally large, especially as compared with the zebras encountered by modern spotted hyaenas. This may have reduced the likelihood of their being hunted by hyaenas, especially in wooded conditions where the hunting success that comes with large group size in open terrain cannot be attained. However, it is not plausible that horses, if present, were not incorporated in hyaena-accumulated bone assemblages. The Last Interglacial absence of horse is therefore real, and points in itself to a barrier to the Continent following the disappearance of the local population at the end of the previous cold stage.

One further feature of Table 1 that may be worth emphasizing is the continued presence of the lion from the time of its earliest appearance in either the Cromerian or temperate Group 4. Such continuity might be taken to imply the long-term existence of a resident population, albeit perhaps in conjunction with some additional immigrations, and yet that may be questioned. The extreme conditions at the height of the last glaciation (West 1988) must make it very unlikely that any larger mammals could have survived, while the conditions even in the less extreme phases are likely to have led to seasonal movements of prey and therefore of predators. If the pattern were a repeat of that occurring during the previous cold stages, then the apparently continual presence of any large mammal species as an isolated, ancestral-descendant population is likely to be an illusion. Whether that is also likely to have been true of the Lower Pleistocene is less clear.

I am grateful to P. Andrews, A. P. Currant and A. J. Stuart for access to material and discussion, and especially to A. P. Currant for comments on an earlier draft. I also thank A. M. Lister for a constructive and helpful review. This research is supported by The Leverhulme Trust.

References

AGUIRRE, E., ARSUAGA, J. L., BERMUDEZ DE CASTRO, J. M. ET AL. 1990. The Atapuerca sites and the Ibeas hominids. *Human Evolution*, **5**, 55–73.

ANDREWS, P. 1990. *Owls, Caves and Fossils*. Natural History Museum Publications, London.

—— & TURNER, A. 1992. Life and death of the Westbury bears. *Annales Zoologici Fennici*, **28**, 139–149.

AZZAROLI, A. 1983. Quaternary mammals and the 'end-Villafranchian' dispersal event – a turning point in the history of Eurasia. *Palaeogeography, Palaeoclimatology, Palaeoecology*, **44**, 117–139.

——, DE GIULI, C., FICCARELLI, G. & TORRE, D. 1988. Late Pliocene to early Mid-Pleistocene mammals in Eurasia: faunal succession and dispersal events. *Palaeogeography, Palaeoclimatology, Palaeoecology*, **66**, 77–100.

BISHOP, M. J. 1982. The mammal fauna of the early Middle Pleistocene cavern infill site of Westbury-sub-Mendip, Somerset. *Special Papers in Palaeontology*, **28**, 1–108.

BUCKLAND, W. 1823. *Reliquiae Diluvianae*. John Murray, London.

CURRANT, A. 1984. The mammalian remains. *In:* GREEN, H. S. (ed.) *Pontnewydd Cave*. National Museum of Wales, Cardiff, 171–180.

—— 1989. The Quaternary origins of the modern British mammal fauna. *Biological Journal of the Linnean Society*, **38**, 23–30.

DE LUMLEY, H., KAHLKE, H.-D., MOIGNE, A.-M. & MOULLE, P.-E. 1988. Les faunes de grands mammifères de la grotte du Vallonet Roquebrune-Cap Martin, Alpes Maritimes. *L'Anthropologie*, **92**, 465–496.

GIBBARD, P. L. 1988. The history of the great northwest European rivers during the past three million years. *Philosophical Transactions of the Royal Society of London*, **B318**, 559–602.

—— 1995. The formation of the Strait of Dover. *This volume*.

——, WEST, R. G., ZAGWIJN, W. H. ET AL. 1991. Early and Early Middle Pleistocene correlations in the southern North Sea basin. *Quaternary Science Reviews*, **10**, 23–52.

GREEN, H. S. 1984. Summary and discussion. *In:* GREEN, H. S. (ed.), *Pontnewydd Cave*. National Museum of Wales, Cardiff, 199–218.

KRETZOI, M. 1938. Die Raubtiere von Gombaszoeg nebst einer Ubersicht der Gesamtfauna. *Annales Museum Nationale Hungarici*, **31**, 88–157.

KRUUK, H. 1972. *The Spotted Hyaena*. Chicago University Press, Chicago.

KURTÉN, B. 1963. The cave hyaena, an essay in statistical analysis. *In:* BROTHWELL, D. & HIGGS, E. S. (eds) *Science in Archaeology*. Thames & Hudson, London, 224–234.

—— 1969. Die Carnivoren-Reste aus den Kiesen von Süssenborn bei Weimar. *Paläontologische Abhandlungen*, A **3**, 735–756.

—— 1972. Fossil Hyaenidae from the excavations at Stránska Skála. *Studies Museum Moraviae (Anthropos)*, **20**, 113–120.

LISTER, A. M. 1984. Evolutionary and ecological origins of British deer. *Proceedings of the Royal Society of Edinburgh*, **82B**, 205–229.

—— 1989. Mammalian faunas and the Wolstonian debate. *In:* KEEN, D. H (ed.) *The Pleistocene of the West Midlands*. Field Guide. Quaternary Research Association, 5–12.

—— 1993. The stratigraphical significance of deer species in the Cromer Forest-bed Formation. *Journal of Quaternary Science*, **8**, 95–108

—— & BRANDON, A. 1991. A pre-Ipswichian cold stage mammalian fauna from the Baldeston Sand and Gravel, Lincolnshire, England. *Journal of Quaternary Science*, **6**, 139–157.

——, McGLADE, J. M. & STUART, A. J. 1990. The early Middle Pleistocene vertebrate fauna from Little Oakley, Essex. *Philosophical Transactions of the Royal Society of London B*, **328**, 359–385.

LYDEKKER, R. 1886. Note on some vertebrata from the Red Crag. *Quarterly Journal of the Geological Society of London*, **42**, 364–368.

MASINI, F. & TORRE, D. 1990. Large mammal dispersal events at the beginning of the late Villafranchian. *In:* LINDSAY, E., FAHLBUSCH, V. & MEIN, P. (eds) *European Neogene Mammal Chronology*. NATO ASI series A: Life Sciences, **180**, 31–38.

MITCHELL, G. F., PENNY, L. F., SHOTTON, F. W. & WEST, R. G. 1973. *A Correlation of Quaternary Deposits in the British Isles*. Geological Society, London, Special Report, **4**, 1–99.

NEWTON, E. T. 1883. On the occurrence of the cave hyaena in the 'Forest Bed' at Corton Cliff, Suffolk. *Geological Magazine*, **10**, 433–435.

—— 1891. *The Vertebrata of the Pliocene Deposits of Britain*. Memoirs of the Geological Survey of the United Kingdom. Stationery Office, London.

ROBERTS, M. B. 1990. Amey's Eartham Pit, Boxgrove. *In:* TURNER, C. (ed.) *Cromer Symposium Field Excursion Guidebook*, 62–81.

SANFORD, W. A. 1867. On the Pleistocene mammalia of Somerset. Catalogue of bones in the museum of the Archaeological and Natural History Society at Taunton. No. 1 – *Felis*. *Proceedings of the Somersetshire Archaeological and Natural History Society*, **14**.

SHACKLETON, N. J. 1987. Oxygen isotopes, ice volume and sea level. *Quaternary Science Reviews*, **6**, 183–190.

SHER, A. V. 1986. On the history of mammal fauna of Berengida. *Quartärpaläontologie*, **6**, 185–193.

SHOTTON, F. W. 1953. The Pleistocene deposits of the area between Coventry, Rugby and Leamington and their bearing upon the topographic development of the Midlands. *Philosophical Transactions of the Royal Society of London B*, **237**, 209–260.

SPENCER, H. E. P. 1959. The prehistoric Mammalia of Suffolk. *Transactions of the Suffolk Naturalists' Society*, **11** (3), 237–240.

—— & MELVILLE, R. V. 1974. The Pleistocene mammalian fauna of Dove Holes, Derbyshire. *Bulletin of the Geological Survey of Great Britain*, **48**, 43–53.

STUART, A. J. 1982. *Pleistocene Vertebrates in the British Isles*. Longman, London.

—— 1988. *Preglacial Pleistocene vertebrate faunas of East Anglia. In:* GIBBARD, P. L. & ZALASIEWICZ, J. A. (eds) *Pliocene–Middle Pleistocene of East Anglia*, Quaternary Research Association, Cambridge, 57–64.

—— 1990. Vertebrate fauna. *In:* TURNER, C. (ed.) *Cromer Symposium Field Excursion Guidebook*, 27–33.

—— 1995. Insularity and Quaternary vertebrate faunas in Britain and Ireland. *This volume*.

SUMBLER, M. G. 1983. A new look at the type Wolstonian glacial deposits of central England. *Proceedings of the Geologists' Association*, **94**, 23–31.

SUTCLIFFE, A. J. 1960. Joint Mitnor cave, Buckfastleigh. *Transactions and Proceedings of the Torquay Natural History Society*, **13**, 1–26.

—— 1964. The mammalian fauna. *In:* OVEY, C. D. (ed.) *The Swanscombe Skull*, Royal Anthropological Institute, Occasional Paper, **20**, 85–111.

—— 1995. Insularity of the British Isles 250 000–30 000 years ago: the mammalian, including human, evidence. *This volume*.

—— & ZEUNER, F. E. 1962. Excavations in the Torbryan caves, Derbyshire. 1. Tornewton Cave. *Proceedings of the Devon Archaeological and Exploration Society*, **5**, 127–145.

TURNER, A. 1981. Ipswichian mammal faunas, cave deposits and hyaena activity. *Quaternary Newsletter*, **33**, 17–23.

—— 1990. The evolution of the guild of larger terrestrial carnivores during the Plio-Pleistocene in Africa. *Geobios*, **23**, 349–368.

—— 1992. Villafranchian–Galerian larger carnivores of Europe: dispersions and extinctions. *Courier Forschungsinstitut Senckenberg*, **153**, 153–150.

—— 1995a. Larger Carnivores (Mammalia, Carnivora) from Westbury-sub-Mendip, Somerset. *In:* ANDREWS, P., COOK, J., CURRANT, A. P. & STRINGER, C. B. (eds) *The Pleistocene Cave at Westbury-sub-Mendip*, in press.

—— 1995b. Remains of *Pachycrocuta brevirostris* (Mammalia, Hyaenidae) from the Lower Pleistocene site of Untermassfeld near Meiningen. *Quärtarpaläontologie*, in press.

VANGENGEJM, E. A., ERBAEVA, M. A. & SOTNIKOVA, M. V. 1990. Pleistocene mammals from Zashino, western Transbaikalia. *Quartärpaläontologie*, **8**, 257–264.

VON KOENIGSWALD, G. H. R. 1960. Fossil cats from Tegelen clay. *Publ. Natuurhistorisch Genootschap Limburg*, **12**, 19–27.

WEST, R. G. 1988. The record of the cold stages. *Philosophical Transactions of the Royal Society of London*, **B318**, 505–522.

Sea-levels and the evolution of island endemics: the dwarf red deer of Jersey

ADRIAN M. LISTER

Department of Biology, University College London, Gower Street, London WC1E 6BT, UK

Abstract: The dwarfing of red deer on Jersey in the Last Interglacial allows a detailed investigation of the 'island dwarfing' phenomenon. Uniquely, the island furnishes normal-sized deer in stratified deposits both preceding and succeeding the dwarfing episode. Timing the dwarfing process depends on a number of assumptions about land-sea history, but suggests a maximum duration of 6000 years. The changes in size and body form of the deer are discussed in terms of adaptation, development and evolutionary rate, and are compared with modern analogues.

The faunas of small islands have long been favourites for the study of evolutionary processes. Examples include the Galapagos archipelago, which has inspired famous studies of adaptation and diversification (Darwin 1859; Grant 1981), Mayr's (1963) model of allopatric speciation, which was based largely on the distribution of endemic birds in island archipelagos, and MacArthur & Wilson's (1967) theory of island biogeography, a landmark in community ecology.

Among the advantages of small islands for evolutionary research are: (i) relatively simple (species-poor) communities and ecosystems, which are easier to model; (ii) small geographical area, which can be thoroughly and intensively surveyed; (iii) immigration and emigration of terrestrial species is negligible or can be estimated; (iv) environmental perturbations are strongly felt by the fauna and flora, because of small population sizes and limited possibilities of moving into a different habitat. By the same token, these features make island systems peculiar. While we may learn much about evolution from their study, direct extrapolation to mainland conditions, even if analogous 'islands' of habitat can be perceived there, is hazardous. Above all, island biology exists for its own sake, providing an insight into the fascinating floras and faunas which islands provide.

Britain is an island, and is surrounded by its own offshore islands. Present-day populations of mammals on the offshore islands are in some cases sufficiently distinct genetically and/or morphologically that they are recognized as endemic sub-species. Examples include the common vole *Microtus arvalis* Pallas on Orkney and Guernsey, field vole *Microtus agrestis* L. on Islay, wood mouse *Apodemus sylvaticus* L. on Rhum, and house mouse *Mus musculus* L. on St Kilda (Berry 1977; Yalden 1982; Corbet & Harris 1991). Many other island populations have been given sub-specific or even specific status in the past, but in most cases these are no longer regarded as valid. On mainland Britain, the only currently recognized endemic sub-species are those of the brown hare *Lepus timidus* L. and red squirrel *Sciurus vulgaris* L. The prevalence of endemic forms on offshore islands compared to the mainland demonstrates the effects of isolation and small population size, but it is interesting that none of these sub-species has reached full speciation since the Holocene isolation of the British Isles.

In the Pleistocene fossil record, endemic mammalian taxa have been recognized on many islands around the world. One of the striking common threads of this phenomenon is the occurrence of large mammal species which have been reduced to very small body size. In the Mediterranean, for example, 16 islands (or former islands) have been found to contain fossils of distinctive (endemic) mammalian populations. There are elephants on ten islands, deer on nine, hippopotamus on four, and antelope-like bovids on two; most of these creatures are dwarfed (Sondaar 1986). Elsewhere, dwarfed proboscideans, derived from at least three distinct genera, are known from many islands of SE Asia (van den Bergh *et al.* 1995), while miniature mammoths have been found in

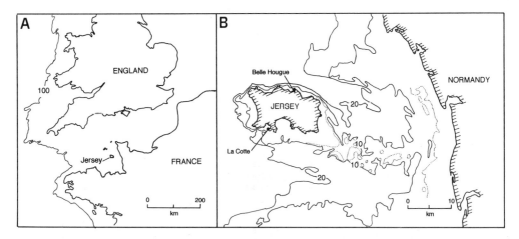

Fig. 1. (A) Location map of Jersey, showing 100 m submarine contour, the approximate position of the coastline for much of the Late Pleistocene cold stages. (B) Submarine contours at 20 m and 10 m around Jersey (Hydrographer of the Navy 1989). Depths are reduced to Chart Datum, which is approximately the level of the lowest astronomical tide. A narrow submerged isthmus is evident at 10 m. Contours at 5 m are shown as dotted lines.

islands off NE Siberia (Vartanyan et al. 1993) and California (Stock & Furlong 1928; Roth 1992, 1995). At least four Japanese islands hosted endemic populations of deer (Sondaar 1977).

These forms have much intrinsic interest and have provided fascinating insights into island adaptations and biogeography. However, our understanding of the tempo and mode of their evolution is severely limited in most cases because their stratigraphical context is poor and dating virtually non-existent. In this respect, the dwarf deer of the Belle Hougue Caves, Jersey, are of particular interest. Extensive Quaternary research on Jersey (Keen 1993), including especially the mammalian localities of Belle Hougue and La Cotte, has provided a dated sequence within which the dwarfing process can be observed, and its tempo and mode explored.

The precursors: Saalian deer of La Cotte

Jersey is an island with an area of c. 130 km^2, situated some 25 km from the coast of Normandy, northwest France, though politically affiliated to the United Kingdom. The sea between Jersey and France is shallow, and for much of the Quaternary Jersey has been part of the French mainland, becoming isolated only during episodes of high sea-level (Figs 1 & 2).

Red deer *Cervus elaphus* L. has been a common component of European mammalian faunas since the early Middle Pleistocene (Lister 1984, 1986). A species of relatively broad environmental tolerances, it occurred both in wooded interglacial conditions, and in more open, cool phases, though not in the most arctic conditions. Most bone samples from Britain and elsewhere in Europe indicate animals of relatively large size, similar to red deer living under optimal conditions today. These were larger than present-day western European animals, but similar to those currently inhabiting eastern Europe and the Caucasus (cf. Heptner et al. 1988; Whitehead 1993). The fossils indicate male shoulder heights of c. 125–130 cm, and body weights of c. 200–250 kg. Some variation in size occurred within mainland populations through the Pleistocene; for example in Britain, Hoxnian animals were relatively small while some populations of the Last Cold Stage were particularly large (Lister 1987a, 1993a).

The site of La Cotte de St Brelade, on the southern side of Jersey, has yielded rich mammalian faunas in association with palaeolithic industries (Callow & Cornford 1986). The most important faunal layers are broadly Saalian in age. Scott (1986) identified 28 remains of red deer, distributed among six successive faunal layers – D, C, B, A, 3 and 5. The lowest level, D, has been dated by thermoluminescence to 238 ± 35 ka BP (Ox-TL 222); the highest layer (5) is succeeded by a raised beach deposit referred to the Eemian. Layers D and C have therefore been referred to oxygen isotope stage

Fig. 2. View from the SW coast of Jersey at low tide. An extensive area of sea-floor is exposed (cf. Fig. 1B).

7, while layers B, A, 3 and 5, with faunal and lithostratigraphical evidence of cold conditions, are placed in Stage 6 (Callow & Cornford 1986).

At the time of the Saalian deposits at La Cotte, Jersey was broadly connected to France and, across the dry Channel, to Britain. The red deer remains from La Cotte are fragmentary, and only three bones and five antler bases are susceptible to measurement (Figs 3 & 4). All of these, however, are of large size, falling within the range of other Saalian samples from NW Europe, for example the extensive red deer material from La Fage, France (Delpech & Heintz 1976; Bouchud 1978; Fig. 3). Additional red deer fragments from La Cotte (20 in total), while not measurable, visually conform to the 'normal' large size.

The dwarfs: Eemian deer of Belle Hougue

With the rise in sea-level early in the Eemian, a marine transgression is recorded in Jersey, and the island became separated from the mainland. At La Cotte, raised beach deposits at 8 m above Ordnance Datum (OD) are referred to the Eemian, but they contain no mammalian remains (Callow & Cornford 1986). At Belle Hougue, however, on the north of the island, beach deposits at this height have yielded remains of dwarf deer.

The history of research at Belle Hougue has been described by Mourant (1984). The first of the caves to be explored, Belle Hougue I, was discovered by Father H. Morin in 1914. A few years later another Jesuit priest, Father P. Teilhard de Chardin, visited the cave and was the first to collect mammal bones from the raised beach. By 1939, when Dr F. Zeuner visited the cave in the company of Dr A. Mourant, much deposit remained, but no further bones were to be found, nor have any been discovered to this day (A. Mourant pers. comm.). This suggests that the deer bones came from a fairly localized area within the raised beach. The taphonomy of the deposit is uncertain; the position of the cave on a rocky coast suggests that bones or partial carcases may have fallen through a hole in the roof, rather than having been brought in through the seaward entrance (Zeuner 1946).

In 1965 a second cave, Belle Hougue II, was discovered nearby, containing similar beach deposits with further bones of dwarf deer (Mourant 1984).

The first detailed account of the dwarf deer was by Zeuner (1946), based only on material from Belle Hougue I. Zeuner (1946) deduced, from the height of the beach at $+4$ to $+8$ m OD and the contained molluscan fauna, that the

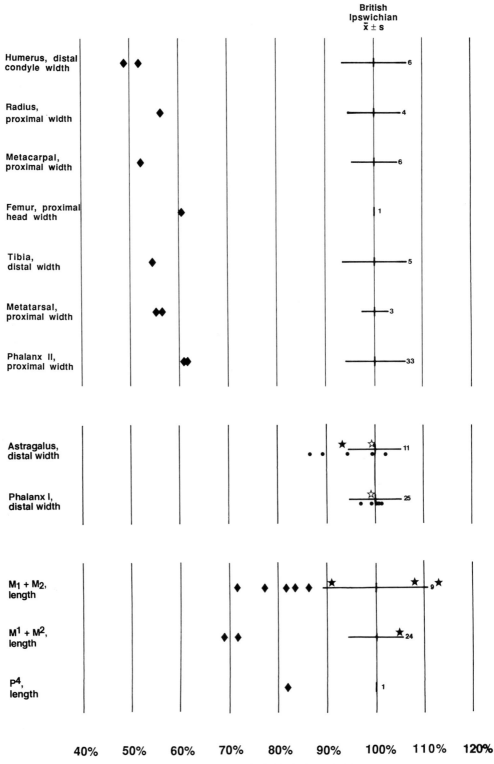

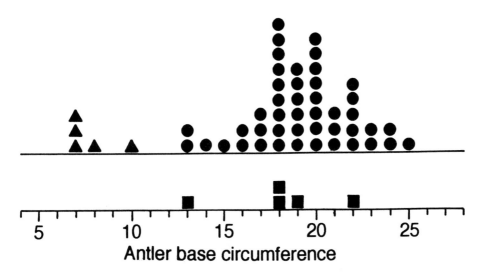

Fig. 4. Comparison of antler size in red deer. Antler base circumference measured below the rose, around the pedicle scar. Circles: composite sample from sites of Ipswichian (Eemian) age on the British mainland, spanning a range of growth stages from young (smaller) to old (larger) (cf. Lister 1990). Triangles: Eemian antlers from Belle Hougue Cave, Jersey. The Belle Hougue antlers are all mature, as shown by their tine development and short pedicles (cf. Lister 1990). Squares: Saalian antlers from La Cotte, Jersey. Data original; measurements in mm.

deposit was Last Interglacial in age. Raised shoreline deposits referred to the Eemian are at c. +8 m OD on both the Sussex and Normandy coasts (West & Sparks 1960; West 1972; Zagwijn 1983). These correspond in height to the Jersey beach, although their correlation to each other and to Jersey depends in part on assumptions about tectonic stability, non-repetition of maximum stands, and the relation of beach height to contemporaneous mean sea-level. Recently, several authors have suggested that beaches at similar height, or differing by only a few metres, were deposited in southern Britain on at least two separate occasions in the period c. 200–100 ka BP, probably corresponding to oxygen isotope stages 5e (Eemian) and 7 (Sutcliffe 1981; Davies & Keen 1985; Sutcliffe et al. 1987; Mottershead et al. 1987; Proctor & Smart 1991; Sutcliffe 1995).

Other evidence of Eemian age at Belle Hougue, however, was given by Keen et al. (1981), who presented a detailed account of the stratigraphy inside Belle Hougue I, and obtained a uranium-series date of $121 ^{+14}_{-12}$ ka on travertine cementing the raised beach. Although strictly a minimum age for the beach deposits, this date suggests correlation to oxygen isotope stage 5e. The occurrence in the beach of the mollusc *Astralium rugosum* (L.), today with a northern limit some 350 km to the south, indicating water temperatures 3–4°C warmer than today, corroborates the interglacial nature of the deposit (Keen et al. 1981). While an earlier interglacial cannot be totally ruled out, an Eemian age for the beach seems highly probable.

In the present study, 41 fossils of small deer from Belle Hougue were identified in the collections of the Hougue Bie Museum, Jersey.

Fig. 3. Comparison of skeletal and dental size in red deer. A composite sample from sites of Ipswichian (Eemian) age on the British mainland is used as a standard of comparison. The mainland Ipswichian mean for each measurement is plotted as 100%, and shown plus or minus one standard deviation, with sample size. Ten Eemian limb bones from Belle Hougue Cave, Jersey (diamonds) average 55.9% the diameter of their mainland contemporaries. Nine cheek teeth from Belle Hougue (diamonds) are relatively larger, averaging 77.4% the length of their mainland counterparts. First and second molars are often indistinguishable, so have been pooled to increase sample size. Two Saalian bones from La Cotte, Jersey (open stars) are very close in size to the mainland Ipswichian mean, as are a Saalian sample from La Fage, France (small solid circles). One limb bone and four teeth from the Weichselian of La Cotte (solid stars) show the reversion to large size, again within the range of mainland Ipswichian deer. All data original except La Fage, from Bouchud (1978).

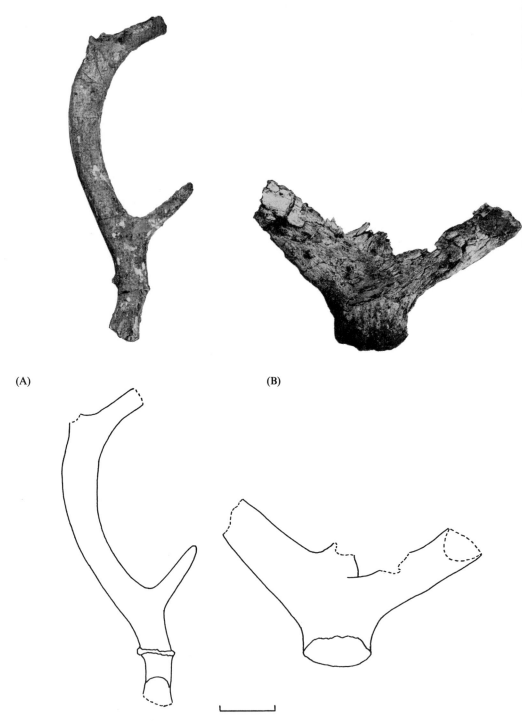

Fig. 5. Mature antler bases of red deer from Jersey. **(A)** Eemian, Belle Hougue Cave, no. 5; **(B)** Saalian, La Cotte, no. 525. The Belle Hougue antler shows a short (adult) pedicle at the base and a narrow antler circumference with high, acute first (brow) tine. The La Cotte antler is shed, and shows a large circumference, low and obtuse first (brow) tine, and the broken stump of a second (bez) tine above it. Scale bar 5 cm.

These were 12 antler parts, seven isolated teeth, two lower jaws with teeth, an atlas vertebra, two scapulae, four parts of humeri, one radius, one metacarpal, one pelvic fragment, three parts of femora, one tibia, one astragalus, three metatarsals, and two second phalanges. Based on antler bases, the minimum number of individuals is five. It is not now possible to separate bones which came from Belle Hougue I from those which came from Belle Hougue II. However, comparison of the above list with that provided by Zeuner (1946), who had access only to material from the first cave, indicates that the bulk of the currently available material corresponds to that seen by him, and is therefore from Belle Hougue I. The main exceptions appear to be the two lower jaws (of different individuals), the atlas and the astragalus, which are presumably from Belle Hougue II.

The identity of the deer was discussed at length by Zeuner (1946), based principally on the morphology of preserved antler bases (Fig. 5). Four specimens (nos 5, 6, 8 and 9) show the lower part of the beam and tines; of these, two can be recognized as adult by their stout pedicle proportions (cf. Lister 1990). The antlers bear a single basal tine, positioned several centimetres above the rose. Above this, the beam curves gently before producing a second tine. The upper parts are not preserved in any specimen.

As shown by Zeuner (1946), this morphology is completely different from three of the small- to medium-sized deer of the European Pleistocene – fallow *Dama dama* (L.), roe *Capreolus capreolus* (L.) and reindeer *Rangifer tarandus* (L.) (for antler descriptions see Lister (1987*b*)). It also differs from typical red deer *Cervus elaphus*, which normally has two basal tines (brow and bez) in adult individuals, usually positioned close to the rose. However, several populations of modern European red deer have antlers which approach the Jersey fossils. These are generally from small-bodied populations such as those in Spain, North Africa and Corsica. Here, the bez (second) tine is commonly absent, and the brow (first) tine is frequently placed several centimetres above the rose. Mattioli (1993) described a population of small red deer from the Po delta, northern Italy, in which both bez tine and distal crown were consistently absent, and in which the brow tine sometimes occupied a high position. It seems that the antler reduction which accompanies a decrease in body size tends in red deer to produce antlers of this form. Even individuals of other populations, such as those of Scotland, occasionally produce antlers of this type. Zeuner (1946) identified the Jersey fossils as a small-bodied form of red deer, which he named *Cervus elaphus jerseyensis*. The only other European species with comparable antlers are Pliocene to Early Pleistocene forms such as *Pseudodama* spp. (Azzaroli 1992), but the very preservation of the beach, as well as its molluscan fauna (Zeuner 1946), make such an age highly improbable.

Bones and teeth of the Jersey deer have been scored for characters established for the separation of red, fallow and reindeer (Lister 1981, 1990). In almost all significant characters which could be observed, the fossils corresponded to red deer: in the lower molars, buccal column and cingulum are very weak; in the scapula, the coracoid is straight rather than bent; in the anterior ligament pits of the distal humerus, the medial is larger than the lateral; in the proximal radius, the lateral tuberosity does not reach the proximal border; in the astragalus, the lateral ridge is stronger in its proximal than distal part. While it is conceivable that this combination of characters might have occurred in another, unknown species, the correspondence to red deer corroborates the identification based on antler form.

In Table 1, measurements of tooth length and limb-bone diameters for the Belle Hougue deer are given, in comparison with data from contemporaneous red deer in Britain. The latter sample is compiled from various sites dated by pollen, 'hippopotamus fauna', U-series or lithostratigraphy to the Last Interglacial: Histon Road and Barrington, Cambridgeshire; Easton Torrs, Tornewton and Joint Mitnor Caves, Devon; Bacon Hole, Gower; Trafalgar Square, London; Swanton Morley, Norfolk; Milton Hill, Somerset; Kirkdale and Victoria Caves, Yorkshire. For details of these localities see Sutcliffe (1960, 1985) and Stuart (1982). The Ipswichian red deer on the mainland were of comparable size to those of the preceding Saalian cold phase (Fig. 3), as were specimens from Late Pleistocene interglacial sites of disputed age, such as the 'brickearths' of Ilford, Essex (Lister 1981).

For estimating body weight of fossil artiodactyls, limb-bone diameters give the most accurate results (Damuth & MacFadden 1990). Table 1 and Fig. 3 indicate that the various limb bones show a consistent difference in diameter between the Belle Hougue and British Ipswichian samples. The Jersey material has been reduced to 56% of the mainland size (not 'by 56%' as stated in Lister (1989)). Scott (1987) found that among deer, the diameters of most long bones scale to body weight with approximately the 0.33 exponent expected from a simple cube-law relationship. This indicates a weight reduction to 0.56^3, or about one-sixth, in the Jersey popula-

Table 1. *Measurements of dwarf deer from Belle Hougue Cave, in comparison with contemporary mainland British Ipswichian (Eemian) samples.*

Element	Measurement	Belle Hougue specimen	British Ipswichian Mean	British Ipswichian s.d.	n	Belle Hougue as % of British Ipswichian mean
Teeth						
P^4	Length	10.0	12.20	–	1	82.0
M^1	Length	15.7	20.49	0.884	12	76.6
M^2	Length	15.1	22.09	1.026	12	68.4
M_1	Length	15.0	18.85	0.889	4	79.6
M_2	Length	17.1	22.50	1.584	5	76.0
						Mean ($n=5$) **76.5%**
Limb bones						
Scapula	Glenoid width	25.8	–	–	0	–
Scapula	Glenoid width	25.0	–	–	0	–
Humerus	Distal condyle depth	23.4	45.18	3.008	6	51.8
Humerus	Distal condyle depth	22.1	45.18	3.008	6	48.9
Radius	Proximal width	35.5	62.95	3.465	4	56.4
Metacarpal	Proximal width	22.5	43.17	2.104	6	52.1
Femur	Proximal head width	23.0	38.00	–	1	60.5
Tibia	Distal width	29.0	53.10	3.513	5	54.6
Metatarsal	Proximal width	22.0	39.83	1.097	3	55.2
Metatarsal	Proximal width	22.5	39.83	1.097	3	56.5
Phalanx II	Proximal width	13.3	21.84	1.322	33	60.9
Phalanx II	Proximal width	13.4	21.84	1.322	33	61.6
						Mean ($n=10$) **55.9%**

Each row represents a single Belle Hougue specimen. Mainland sample composite, pooled from various sites detailed in the text. In the extreme right column, Belle Hougue measurements are expressed as a percentage of the mean for the corresponding element in the Ipswichian sample, and then pooled to provide an estimate of the average reduction in tooth length and limb-bone diameter; see Lister (1993a) for a discussion of the technique. Data original; all measurements in mm.

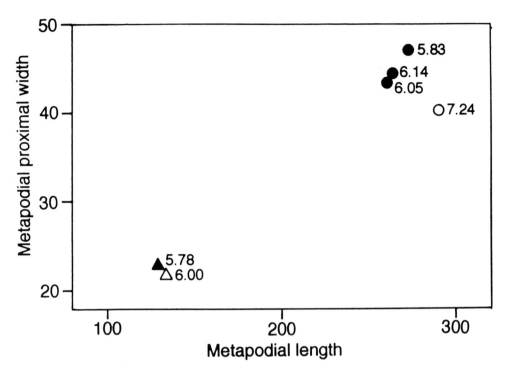

Fig. 6. Proportions of red deer metapodial bones. Solid symbols: metacarpals; open symbols: metatarsals. Triangles: Belle Hougue Eemian; circles: mainland British Ipswichian. The figure by each point is the ratio of length to width. Data original; all measurements in mm.

tion. Given the estimated 200 kg adult male body weight of the mainland animals, the Jersey animals averaged around 36 kg. In shoulder height, they would have stood around 70 cm, compared to 125 cm on the mainland. Their small size is emphasized by the small antler circumference, narrower in the Belle Hougue adults even than in juveniles on the mainland (Fig. 4).

On the other hand, Table 1 and Fig. 3 show that in tooth size, the Jersey deer were considerably less reduced than in limb bones. The average molar length is 76.5% that on the mainland. This indicates a higher ratio of tooth size to body size, a phenomenon commonly observed in small-bodied populations of mammals (Gould 1975; Fortelius 1985).

Another common feature of insular dwarfed mammals is the shortening of distal limb bone lengths, the metacarpals and metatarsals in particular becoming relatively short for their diameter (Sondaar 1977). Unfortunately, only one metacarpal and one metatarsal from Belle Hougue can be measured. Their ratio of length to width is slightly lower than in the few available mainland Ipswichian bones (Fig. 6), in the expected direction, but the sample is too small for statistical significance and the trend cannot have been pronounced (*contra* Zeuner 1946).

Return of the grown-ups: Devensian deer of La Cotte

In the Devensian Cold Stage, the fall in sea-level again connected Jersey to the mainland. Evidence of the Devensian mammalian fauna is provided by material from the upper layers at La Cotte, excavated in the early part of the century by Marett (1912, 1916). The fauna, stratigraphical position and flint industry clearly point to the Last Cold Stage, but any sub-division within this has been lost because the remains from different layers were merged together by Marett who regarded them as indistinguishable (A. E. Mourant, pers. comm. 1989). The resulting mammalian assemblage is nonetheless typical of the early to middle Devensian of NW Europe: horse, mammoth, woolly rhinoceros, bison,

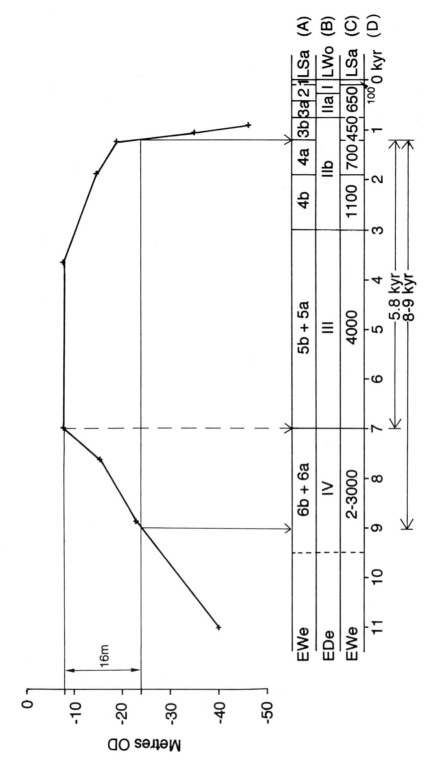

Fig. 7. Curve of Eemian sea-level change based on marine transgressions and regressions in The Netherlands in relation to pollen zonation. The curve is replotted from Zagwijn (1983) onto an absolute timescale provided by varved Eemian lake sediments in Germany (Müller 1974). The vertical axis shows the present height of the deposits in The Netherlands, the entire sequence having been downwarped since the Eemian. The horizontal axes show time running from right to left: (**A**) Dutch pollen zones (Zagwijn 1983); (**B**), equivalent British pollen zones (Phillips 1974); (**C**), duration of pollen zones in years, from Müller (1974); (**D**), timscale in ka since the beginning of the interglacial, based on (C). Solid guide lines indicate times at which the isthmus, 16 m below the Belle Hougue raised beach, would have been submerged and re-exposed. Dashed guide line indicates the latest possible date for the Belle Hougue beach, assuming it to have been deposited during the maximum stand. LSa/LWo = Late Saalian/Wolstonian, EWe/EDe = Early Weichselian/Devensian.

reindeer, red deer, hyaena, bear, lemming, ground squirrel, and a number of birds (Marett, (1912, 1916); and examination of material at the Hougue Bie Museum, Jersey). Marett (1916) stated that the industry was entirely of Mousterian type, implying a date before about 30 ka BP.

Red deer material identified in this study comprises eight teeth, two mandibles and five fragmentary limb bones. In size (Fig. 3), this material is of the 'normal' large type, falling within the range of the British Ipswichian sample, though towards the smaller end of the even larger deer from the Devensian of Kent's Cavern, Devon (Lister 1987a). Marett (1916) listed the presence of other deer ('? *Capreolus capreolus*', and '*Cervus* sp.') in addition to red deer and reindeer. However, detailed examination of the collection failed to reveal the presence of such species. Small artiodactyl bones were identifiable either as reindeer or, in one or two cases as sheep/goat which presumably came from later horizons, as suggested by Marett (1916) himself.

Whether the dwarf deer disappeared at the end of the Eemian, or survived for a while in the early Devensian is not known. However, at least by the time of the Devensian layers excavated by Marett, they seem to have disappeared, having been replaced by 'normal' large animals.

Timing the dwarfing process

The phenomenon of island dwarfing offers the possibility of pinning down the duration of a relatively rapid evolutionary transition, of relevance to debates about the rate of evolution in species origins (Lister 1993b). The Jersey deer, in particular, provide a better chance than most, because of the detailed background of research in dating and palaeoenvironments. The attempt to pin down the duration of the dating shows, however, how difficult this can be even in these favourable circumstances, and how many assumptions still need to be made.

First, while absolute dating has been important in providing the broad chronostratigraphical context (corroborating the Saalian and Eemian ages of the La Cotte and Belle Hougue deposits, respectively), it cannot at present take us further towards timing the dwarfing of the deer. For this, we would need a sequence of bone-bearing deposits encompassing the transition from large to small deer, and incorporating material suitable for dating. Such material is not available. Even if it were, the transition may have taken only a few thousand years or even less, beyond the resolution (i.e. within the error ranges) of available methods for dating an episode before 100 ka BP.

We must therefore take a different approach, and ask how long Jersey was isolated during the Last Interglacial. This figure will provide a maximum estimate of the time available for dwarfing, on the assumption that the process did not begin until the island deer were genetically isolated from the mainland, or at least that gene flow was greatly restricted.

Today, the sea between Jersey and the Normandy coast is quite shallow (Figs 1 & 2). A drop of only 8 m from present sea-level would be sufficient to connect Jersey to the mainland by a narrow isthmus at the lowest tides (Fig. 1). Thus, unless the present situation is the result of major uplift, Jersey was isolated in the Late Pleistocene only during episodes of very high sea-level, close to those of today. This result is very useful in tying down the period of isolation in the Last Interglacial.

The Last (Eemian) Interglacial has been equated with oxygen isotope stage 5e (Shackleton 1969), and has been deduced to last approximately 11 ka. This has been based on several lines of evidence. First, lake sediments covering the entire interglacial and varved in their early part have been extrapolated, on the assumption of constant sedimentation rate, to a total of c. 11 ka (Müller 1974; Fig. 7). Second, marine oxygen isotope ratios covering stage 5e span c. 126 to 115 ka BP, a duration of 11 ka, on the basis of interpolation between horizons with absolute dating, again assuming constant sedimentation rate (Shackleton & Pisias 1985; Martinson et al. 1987; Fig. 8). Recently, however, based on a stalagmite sequence at Devil's Hole, Nevada, Winograd et al. (1992) have suggested a duration of c. 20 ka for the Last Interglacial, but since the timing and duration of Late Pleistocene events at this site are out of keeping with those based on several other lines of evidence, the data may be the result of local factors (Broecker 1992). Similarly, based on Greenland ice core data, GRIP (1993) and Dansgaard et al. (1993) suggested a division of Stage 5e into five sub-stages, with a total duration of nearly 20 ka. However, due to evidence of flow deformation, the veracity of this lowest part of the core has since been called into question (Grootes et al. 1993). In the present contribution, therefore, an 11 ka estimate for the duration of the Eemian will be retained.

The shape of the eustatic sea-level curve during this interval has been modelled by several means. Zagwijn (1983) took a series of terrestrial sites in The Netherlands with evidence of marine transgression or regression, each dated to a

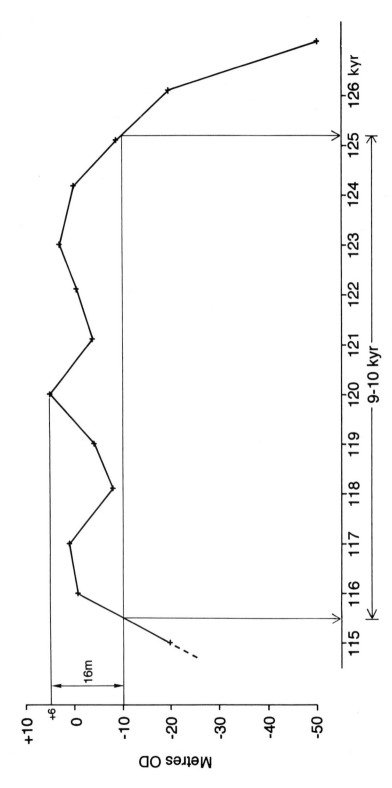

Fig. 8. Curve of Eemian sea-level change based on oxygen isotope ratios and interpolated dates, from Shackleton (1987). Time runs from right to left. Guide lines indicate times at which the isthmus, 16 m below the Belle Hougue raised beach, would have been submerged and re-exposed.

different sub-stage of the Eemian on the basis of pollen zonation. By plotting the height of the deposit relative to current OD, he obtained a sea-level curve through the interglacial. Although the entire Dutch sequence has been subsequently downwarped, the relative heights are valid, giving a shape to the curve. (Lister (1989) attempted to provide an 'absolute' height axis for this curve, by assuming that the peak corresponded to the maximum Eemian eustatic stand of +6 m OD (see below), so that the whole Dutch sequence had been downwarped by 14 m.) The graph (Fig. 7) does depend crucially on the correct assignation of all pollen sites to the appropriate sub-stage of a single (Eemian) climatic cycle. The same is true of coastal deposits at various sites in Britain and northern France which have been referred to the Ipswichian, and which between them suggest a sea-level curve similar to the Dutch one. At Selsey, Sussex, West & Sparks (1960) found a marine transgression at −1.8 m OD early in pollen sub-stage IIb. Similarly, estuarine deposits of sub-stage IIb age occur at Stone, Hampshire between −0.2 and −2.4 m OD (West & Sparks 1960). At Bembridge, Isle of Wight, there is evidence of a transgression during sub-stage II, and a mean sea-level of +4 m OD in late sub-stage IIb to early III (Preece et al. 1990). The maximum stand was deduced by West and Sparks (1960) and West (1972), on the basis of former shoreline deposits on the Sussex coast between Selsey and Brighton, to have been at +7.5 m OD. A regression is shown at Arromanches, Normandy, in late Eemian deposits (cf. sub-stage IV) at −2.5 m OD (West & Sparks, 1960).

Confidence in the attribution of these deposits to the Eemian is clearly crucial for reconstructing an Eemian sea-level curve. Sutcliffe (1995), for example, has suggested that Selsey may date to a previous (oxygen isotope stage 7) interglacial. For Bembridge, at least, where evidence suggests a mean sea-level of c. +4 m OD in late zone IIb-early zone III (i.e. just below the maximum), the Eemian attribution is supported by a thermoluminescence date of c. 115 ka BP on sand lenses within the raised beach.

Evidence from raised beaches at some sites suggests that the maximum stand was relatively similar in stages 5e and 7 (Davies & Keen 1985), but other studies in southern Britain indicate that sea-level reached a few metres higher in stage 7 than in stage 5e (Sutcliffe et al. 1987; Proctor & Smart 1991; Sutcliffe, 1995). The accidental inclusion of stage 7 sites in the Eemian curve therefore might have either little effect, or would bias it toward incorrectly high sea-levels by a few metres. This in turn would bias our calculated time of isolation of Jersey towards an overly long interval.

A quite independent sea-level curve was provided by Shackleton (1987, fig. 5) from calculations based on deep-sea oxygen isotope ratios (Fig. 8). This indicates a global eustatic maximum of +6 m, corroborating independent estimates based on coral terraces (Shackleton 1987; Chappell & Shackleton 1986).

The pollen-based and oxygen-based curves both show a steep rise in sea-level in the late Saalian and early Eemian, a plateau reached by about one-quarter of the way into the interglacial, approximate constancy (with fluctuations indicated in the oxygen data), and a decline within the last quarter of the interglacial (somewhat less steep in the pollen curve), continuing into the early Weichselian. This implies full isolation of Jersey for something between 50% and 100% of the length of the Eemian. For a more accurate estimate, land-sea relations on Jersey need to be examined more closely.

Figure 1 shows current submarine contours between Jersey and the Normandy coast. The depths used here are reduced to Chart Datum (CD), which is approximately the level of the lowest astronomical tide, and about 5 m below OD (Hydrographer of the Navy 1989). Chart Datum is an appropriate measure for present purposes, since it gives the sea-level at which, even at the lowest tide of the year, the land link to Jersey would have been broken. Point soundings indicate that an isthmus occurs at −8 m CD (Hydrographer of the Navy 1989; Lister 1989, fig.1) but, allowing a margin of error, we can take a lowest astronomical tide of higher than −10 m CD as providing full isolation of Jersey. At this point, *mean* sea-level would similarly have been at −10 m below its present level (roughly OD) at the point of isolation.

Two models are possible for translating this present-day figure into a duration of isolation in the Eemian.

(1) Assuming tectonic stability

The simplest model assumes that the Channel area has suffered little tectonic displacement since the Eemian. This assumption can be tested by comparing the height of Eemian sea-levels in this area, deduced from raised-beach deposits, with the Eemian maximum global eustatic sea-level of +6 m obtained from coral terrace and oxygen isotope data (Shackleton 1987). Eemian beaches in southern England are at +5.8 m in Torbay (Proctor & Smart 1991) and +7 to +11 m at Portland East (Davies & Keen 1985);

the Belle Hougue beach itself is at +4 to +8 m OD (Keen et al. 1981). These heights are similar to the Shackleton figure, although the beaches might have been formed above the mean sea-level of the time (see below). At Bembridge, Preece et al. (1990) deduced an actual Eemian sea-level of +4 m (see above), while at Minchin Hole, Sutcliffe et al. (1987) obtained a level somewhat below this, at about +2 m above today's level. There is no evidence here for major tectonic disturbances.

If regional tectonic stability is assumed, this implies that the same sea-level (−10 m OD) was required in the Eemian as today to reveal the isthmus, and the times of isolation and reconnection can be read off the −10 m level on the eustatic curves (Lister 1989). This level will be found at 16 m below the maximum height of the curves, calculated as the sum of the +6 m OD global eustatic maximum, and the −10 m OD depth of the isthmus (Lister 1989). Note that OD is being used here simply as a zero point for the measurements. Transferring this 16 m drop to the sea-level curves of Zagwijn (1983) and Shackleton (1987), we drop 16 m from the maximum point, irrespective of the absolute values of the vertical axis. Assuming Müller's (1974) and Shackleton's (1987) 11 ka timescale, the oxygen graph (Fig. 8) gives isolation and reconnection of Jersey about 9–10 ka apart, while on the pollen graph (Fig. 7) the figure is 8–9 ka (Lister, 1989). If stage 5e was longer than 11 ka as some workers have suggested (perhaps as long as 20 ka), the isolation figures would need to be stretched accordingly.

(2) No assumption of tectonic stability

The assumption of tectonic stability has been challenged by some recent work which suggests uplift of this area through the Pleistocene (e.g. Preece et al. 1990). If so, this would mean that Eemian sea-levels were in fact lower than implied by the height of the fossil beaches, so that our estimate of Jersey's isolation was again biased towards longer time intervals. Note that, because of the steepness of the sea-level curves around the times of separation and reconnection (Figs 7 & 8), a few metres of tectonic movement since the Eemian would make very little difference to the duration of the island's isolation calculated by the model in the previous section.

It is possible to circumvent the problem of tectonic instability, however, by taking a different approach to the Jersey beaches. This model requires the assumption, on purely depositional grounds, that the Eemian beaches represent the period of maximum stand in the interglacial (Lister 1993b, c). This seems likely from their survival as beach relicts, as well as the coincidence in height of preserved Eemian beaches at Belle Hougue and La Cotte, and their approximate correspondence to the highest Eemian deposits elsewhere in the region (see above). The presence of Mediterranean molluscs in the Belle Hougue beach (Keen et al. 1981) might also be taken as supportive of a peak interglacial age, although this depends on an assumption of congruence between the temperature and sea-level curves.

Under this scenario, the difference in sea-level between the maximum stand, and the exposure of a land-bridge is approximately the vertical displacement between the Jersey beaches and the submarine isthmus (Lister 1993b, c). Again using OD as a measuring stick, the total displacement is 6 m (beach to OD) plus 10 m (OD to isthmus). The resulting figure of 16 m depends on the beach height representing the mean sea-level of the time (see below), but is valid irrespective of any subsequent regional tectonics, provided Jersey and its sea-floor adjacent to France moved up or down together. The figure of 16 m is the same as that obtained under the assumption of tectonic stability, so the same duration of isolation, (8–10 ka) can be read off the sea-level curves by dropping 16 m from the maximum stand.

This calculation, however, depends on the beach approximately reflecting contemporary mean sea-level. There is evidence, e.g. at Bembridge (Preece et al. 1990), that storm beaches can extend to many metres above mean sea-level. An instructive comparison is with Minchin Hole, Gower (Sutcliffe et al. 1987; Sutcliffe, 1995). This is a sea-cave with a narrow entrance similar to Belle Hougue. Today, storm waves carry shingle into the cave to as high as +10 m OD. The Eemian 'Patella beach' at Minchin Hole extends from +6 to +12 m OD, passing from boulders at the bottom, to gravel in the middle, to sand at the top. The deer-containing Belle Hougue deposit comprises coarse sand and fine gravel with many broken shells, resembling parts of the Minchin Hole beach. It is not necessarily the case that Belle Hougue is a storm beach well above contemporary mean sea-level, but the possibility cannot be ruled out. The inclusion of well preserved deer bones, which do not appear to have been tossed about by the sea, does not necessarily preclude this possibility, as they may have become incorporated by falling from above into the already-existing beach sediments.

Another consideration is tidal range. This is currently high in the south Channel area, with a total range of 11 m at Jersey. This again makes it quite likely that beaches would have been deposited a few metres above mean sea-level.

In general, therefore, storm waves, tidal ranges and other local conditions can combine to produce beaches at different heights relative to 'mean sea-level' (Sutcliffe 1981). However, in specific relation to the question being asked in this study, this is not considered a serious problem. As mentioned above, because of the steepness of the sea-level curves around the times of separation and reconnection (Figs 7 & 8), uncertainties of a few metres make very little difference to the calculated duration of the island's isolation. Insofar as factors such as storms and tides are inclined to place the beach *above* contemporary mean sea-level, the calculations above, based on the assumption that the two were equal, would again bias our estimate of Jersey's isolation to longer intervals.

Further assumptions are implicit in the models of both tectonic stability and instability. First, it is assumed that the height of the isthmus has not been altered by subaerial erosion, particularly during the Weichselian exposure. This effect is likely to have been minimal in view of the granite bedrock (Institute of Geological Sciences 1982). However, if the Eemian sea-floor were higher than today, our estimate of Jersey's isolation would yet again be biased towards a longer time interval, because the sea would have had further to climb before isolating the island and less far to drop before connecting it again.

Second, we have to consider the unfortunate scenario that Jersey has been subject to tectonic displacement even relative to the adjacent sea-floor. Such a situation is possible on the basis of local geology, which indicates Jersey to be a fault-bounded block (Institute of Geological Sciences 1982). This would jeopardize any prospects for detailed reconstruction of Jersey's land–sea history. However, it seems unlikely to be a major effect in view of the approximate coincidence in height of Eemian raised beaches on Jersey and around the Channel coast.

On the basis of all the above assumptions, an 8–10 ka maximum interval for the year-round isolation of Jersey gives an approximate maximum time of isolation of the deer. Note that each of the possible sources of error enumerated above would tend to cause us to overestimate the duration of isolation of the island, so we may say with some confidence that Jersey was isolated during the Eemian for 8–10 ka or less.

In any case, the interval in which the deer became dwarfed was probably shorter than the total time of isolation of the island. Around the time of isolation and reconnection, the sea would have been very shallow, and punctuated by small outcrops of land between Jersey and the mainland. Deer could have moved between island and mainland during this period when the island was formally isolated. However, because of the steepness of the sea-level curves in these periods, the shortening of the time of true isolation was probably small.

A more significant point is the likelihood that the dwarfs had already evolved well before the reconnection of the island. On the assumption that the Belle Hougue beach was deposited some time during the episode of highest sea-level, then the dwarf deer whose bones it contains must have been in existence at the latest by the end of that episode. The maximum duration for the evolution of the deer can then be narrowed to the interval between the isolation of Jersey and the deposition of the beach. According to the oxygen isotope-based curve, a high stand persisted until less than 1 ka before the end of the interglacial (Fig. 8), hardly reducing the 9–10 ka available from the duration of total isolation. However, caution should attach to such detailed deductions because the curve is dated by interpolation and assumes constant sedimentation (N. J. Shackleton, pers. comm. 1989). The pollen-based curve (Fig. 7) indicates a high stand only during sub-stage III, implying that the beach was deposited (and the deer had evolved) by the end of that sub-stage, only 7 ka into the interglacial and *c.* 5.8 ka after isolation. This assumes both that the low-level sub-stage IV sites are correctly pollen dated (Zagwijn 1983), and that Müller's (1974) timescale for the Eemian pollen zones is correct. If so, a figure of around 6 ka emerges as a maximum time available for the dwarfing of the Belle Hougue deer (Lister 1989, 1993*b*).

A further recent development is the suggestion that the Eemian (stage 5e) may itself have included dramatic fluctuations of climate, including several 'stadial' episodes within a generally 'interglacial' period (GRIP 1993; Field *et al.* 1994; McManus *et al.* 1994; Thouveny *et al.* 1994). Evidence on this issue is currently being assessed (Zahn 1994), and it is in any case unclear whether such rapid events would affect global sea-level. According to Cortijo *et al.* (1994), an oceanic cooling event evidenced midway through the Eemian was not accompanied by any significant change in continental ice-volume, implying that sea-levels remained constant. If sea-level did fall during intra-Eemian cool episodes, any resulting episodic reconnection of Jersey with the mainland would serve only to reduce still further the maximum amount of time available for dwarfing of the deer.

Finally, in the unlikely event that the Belle Hougue beach were to prove to date from a pre-Eemian interglacial, the duration of Jersey's isolation would probably still have been only of the order of a few thousand years. Based on oxygen isotope curves, the peaks of earlier interglacials do not appear to have lasted longer than that of stage 5e, and stage 7 as a whole is thought to have had generally lower sea-levels than stage 5e (Shackleton 1987).

The problems encountered in the above argument, and the various assumptions required, graphically illustrate how difficult it is to give a chronology for the isolation and connection of offshore islands, even in an apparently promising case such as Jersey. However, it has proved possible to give, with some degree of confidence, a maximum available duration for Jersey's isolation, and the dwarfing of its deer, in the Last Interglacial.

Evolutionary significance

Origin of the Jersey population

The maintenance of a population of dwarfed red deer on Jersey during the Last Interglacial implies that they were largely or entirely genetically isolated from the normal-sized, mainland population. At the present time, the distance from Jersey to mainland France is about 25 km, and the situation during the Eemian high sea-level was presumably similar. Although red deer are good swimmers, this distance was probably sufficient to keep the populations isolated. Whitehead (1993) indicates that modern *C. elaphus* can swim several kilometres, but 7 km is the furthest distance actually recorded.

How did the deer get to the island in the first place? Sondaar (1977, 1986), reviewing island populations of Pleistocene mammals in general, suggests 'sweepstakes' dispersal (Simpson 1940) as the commonest route. This implies that founder populations of the species arrived on the island once it was already isolated from the mainland. Only very rarely would individuals achieve the crossing, by swimming or rafting, so that once founded, there would be little opportunity for significant introgression with further mainland animals. This mechanism seems plausible in many instances, where the islands were never connected to the mainland during the Pleistocene, and their fossil assemblages are restricted to species which are good swimmers (deer, hippopotamus and elephants) (Sondaar 1977, 1986). On the other hand, for a situation such as Jersey, where the island was connected to the mainland until shortly before its endemic deer are recorded, it seems equally plausible that the dwarfs evolved from full-sized animals already present on the island when it became cut off. The Saalian red deer from La Cotte indicate a potential ancestral population, spanning tens of thousands of years preceding the Eemian isolation.

Genetic versus ecophenotypic change

Once isolated, a process of adaptation to the new habitat began. As indicated by Sondaar (1986), the consistency of modifications seen in island populations of different mammalian species makes these changes unlikely to be mere 'chance' or 'degeneration'. The most obvious modification is size reduction. In the Jersey deer, this descended from $c.$ 200 kg to $c.$ 36 kg in adult males – a reduction of $c.$ 80%. Although red deer are ecophenotypically quite labile, such a reduction lies outside the known bounds of environmentally induced size change, and so must have required genetic modification. The present-day ecophenotypic range is indicated by the animals which were moved from the Scottish highlands (among the smallest European red deer) to the pastures of New Zealand, resulting in an approximate doubling of size from about 125 kg to over 200 kg (Huxley 1932). As suggested by Bonner (1968) for the origin of dwarf forms in general, it is quite likely that the initial size reduction on Jersey was ecophenotypic but that later this was translated into, and augmented by, genetic change through natural selection. This could have involved genetic assimilation – selection for individuals genetically predisposed to show the greatest degree of ecophenotypic stunting.

Marshall & Corrucini (1978) suggested that, theoretically, it might be possible to distinguish phenotypic from genotypic dwarfs on the basis of their dental and skeletal proportions. Phenotypic dwarfs might resemble juveniles, with large heads and teeth relative to body size. Genotypic dwarfs are more likely to resemble miniature adults and show modifications outside the ontogenetic trajectory of the ancestor. While these growth considerations are plausible, especially for the early stages of dwarfing, it could be difficult to distinguish between phenotypically large head and tooth size, and a possibly adaptive, genetic reason for this feature in dwarfed forms. We cannot be certain which of these explanations accounts for the relatively large teeth of the Jersey dwarf deer. The short limbs often encountered in dwarf ungulates (counter to juvenile form and hence likely to be of genetic and adaptive origin) are insufficiently evidenced in the Jersey fossils.

Adaptive aspects of morphology

The adaptive advantages of small body size in island populations of ungulates have been widely discussed (Foster 1964; Thaler 1973; Sondaar 1977; Case 1978; Heaney 1978; Lomolino 1985; Roth 1990, 1992). Generally, it is thought that in a situation of limited food supply, survival is favoured in small individuals with a lower absolute food requirement. Roth (1992) adds that large mammals under food-stressed conditions are likely to be at a selective advantage if they cease growth at a smaller body frame and invest remaining resources in fat storage and reproduction, a prediction confirmed on studies of modern island red deer (Clutton-Brock et al. 1982, 1985). These tendencies would be enhanced by several factors. In non-tropical regions, seasonal variation in plant growth would place stress on herbivore populations in the winter months, and with no possibility of migrating to better feeding grounds, a premium would be placed on frugality. In addition, as suggested by Roth (1992), an overcrowded population may reduce its resource base still further by destroying its own habitat. The bark-stripping activities of deer, sometimes killing trees, are an example of this. Illius & Gordon (1987) show, using the example of red deer on Rhum, that individuals of small body size are better able to survive on the short grass swards produced by overgrazing.

Second, the population sizes of herbivores on small islands are generally too small to support predators. In this situation, one of the advantages of large size – predator defence and avoidance – is removed. Further, the absence of predators would allow the herbivore population to expand to the carrying capacity of the island vegetation, so that food requirement would impose a direct selective pressure.

Third, the absence or lower number of competitor species would reduce niche packing and allow the transition to a smaller body size if metabolic selection directed it. Recently, Maiorana (1990) and Damuth (1993) have provided concrete evidence that 'medium-sized' mammals (about 1 kg) are the most efficient in physiology and energy transfer. Brown et al. (1993) suggested that 100 g was the optimum for converting energy into offspring. These studies may explain why, under the simplified constraints of island conditions, smaller mammals tend to increase in size, while larger ones become reduced.

The contemporary mammalian fauna of the Eemian deer on Jersey would be of considerable interest, in view of the implication that an absence of competitors or predators may have contributed to the dwarfing process. Unfortunately, no information is available. Within the collection from Belle Hougue are a few bones other than the deer. These include some unidentified rodent remains, and bones of hare (*Lepus* sp.). Two larger phalanges and a tarsal bone are labelled '*Hyaena spelaea*', i.e. spotted hyaena *Crocuta crocuta* Erxleben, a species known from the Eemian/Ipswichian of mainland Britain and France (Stuart 1982). However, examination of these remains indicates their identity as sheep. Finally, a fragment of tibia is difficult to identify, but probably also corresponds to sheep. The most likely explanation for these remains is that they do not come from the Eemian beach, but are Holocene contaminants.

In addition to small overall size, the Jersey deer exhibit relatively large teeth. As indicated above, this is a common occurrence in small-bodied populations of mammalian species. Lundholm (1947, cited in Fortelius 1985) even noted relatively large teeth among contemporary insular *versus* mainland red deer. Although this difference of proportion may have functional significance (smaller-bodied animals needing to process relatively more food because of higher metabolic rate), Fortelius (1985) suggests that it might alternatively be simply a non-selected consequence of differing growth patterns. Across species of mammals, tooth size is generally isometric to body size (Fortelius 1985), and the non-isometric large tooth size of populations such as the Jersey dwarfs may be a temporary developmental effect reflecting the relative recency of size reduction (Lister in press).

The shortening of distal limb elements, also common among island dwarfs, is uncertain in the Jersey deer because of small sample size. Such differences are linked to 'low-gear' locomotion in rugged terrain, as opposed to fast running (especially in predator avoidance) over larger areas (Sondaar 1977). The shortening in the Jersey deer, if real, was not as profound as in other dwarfed island ungulates of the Pleistocene, which could reflect either the short time of isolation on Jersey or the relatively undemanding topography of the island (Lister in press).

Finally, the Jersey deer had relatively simpler (and probably smaller) antlers than their mainland contemporaries. Developmentally, this can be seen as a growth effect: within red deer today, there is a general allometric effect whereby larger individuals have relatively larger antlers (Huxley 1931). It is also the case that across the Cervidae, larger species of deer tend to have relatively larger and more complex antlers, and this is linked to a more complex social and reproductive behaviour, with males holding harems and defending them by fighting with other males (Barrette 1977; Clutton-Brock et al. 1980). Whether the smaller antlers of the island deer indicate a simpler, less polygonous mating system is unknown. The antlers are too incomplete to show their full morphology, but the preserved parts give no indication that a new species-recognition pattern, implying speciation from the mainland ancestor, had evolved. This contrasts with the unique antler plans seen in some of the dwarf Pleistocene deer species of Crete, which had been isolated for much longer than the Jersey population (Vos, 1984; Lister in press).

Modern analogues

In addition to the many other fossil examples of dwarfed mammals on islands, there are some modern and historical analogues amongst deer which provide an interesting comparison to the Jersey population. The red deer of the Western Isles of Scotland today are generally smaller than those on the mainland (Whitehead 1993). On Rhum, the modern herd is derived from 19th century introductions, but earlier populations were also of small body size. According to Munro (1549; cited in Nature Conservancy Council 1974), the island of Rhum supported an 'abundance of little deire'. Small red deer also occurred in the Western Isles during Mesolithic times, evidenced by the remains from Oronsay (Grigson & Mellars 1987). Finds from the Scilly Islands, SW of the British mainland, similarly indicate relatively small red deer between about 4000 years BP and Roman times; it is not clear whether these were naturally present or were introduced by human settlers (Turk 1968, 1971; Thomas 1985). None of these animals, however, approaches the degree of dwarfing in the Jersey population.

On the Florida Keys, a chain of islands off the SE coast of the USA, are populations of whitetail deer *Odocoileus virginianus* Zimmerman distinctly smaller than those of mainland North America. The 'Key Deer' *O.v. clavium* occur on 22 different islands, and populations vary considerably in mean size (Klimstra et al. 1978). An average for adult males is 36 kg and 62 cm shoulder height – an intriguing similarity to the Jersey deer although mainland *O. virginianus* are smaller than typical *C. elaphus*. Mainland animals are themselves quite variable in size and generally small in the southern states, but 60–80 kg would be typical for an adult male, indicating a reduction of around 50% on the Florida Keys. Tooth row length of Key Deer is relatively larger, at c. 80% of the mainland value (Maffei et al. 1988), again similar to the situation in the Jersey deer. There are no native predators on the Keys, but nor does food appear to be a limiting factor on the deer, at least at the present time, when human disturbance and accidental kills are an important influence (Klimstra et al. 1978). Hardin et al. (1976) found that behaviourally, Key Deer associate with each other less than in mainland populations.

Another interesting modern analogue is provided by the reindeer of Svalbard (Spitzbergen) *Rangifer tarandus platyrhyncus*, which are of small body size and markedly shorter limb length than their mainland counterparts (Klein et al. 1987). Finally, Case (1978) noted the occurrence of small races of both caribou *Rangifer tarandus* and blacktail deer *Odocoileus hemionus* Rafinesque on several islands off British Columbia, Canada.

Rate of change

As discussed above, a figure of c. 6 ka has been deduced as a maximum duration between the isolation of Jersey in the early Eemian and the deposition of dwarf deer bones at Belle Hougue. This figure is subject to revision because of the various assumptions involved in its calculation, but it would seem very unlikely that the true maximum figure is greater than 10 ka. From a palaeontological perspective, this represents one of the shortest evolutionary transitions to have been resolved in the mammalian fossil record (Lister 1993b). *Cervus elaphus* has existed in Europe since the Cromerian, c. 600 ka BP (Lister 1986), so the c. 6 ka available for the Jersey dwarfing represents only 1% of the species' range. Mainland populations show size fluctuations through this period (Lister 1993a), but never as extreme as in the Jersey dwarfs.

A rapid transition in an isolated population formed the foundation of Mayr's (1963) allopatric speciation model, and of Eldredge & Gould's (1972) punctuated equilibria model which was based upon it. However, these models invoke full speciation by reproductive isolation, and we have no evidence of this in the Jersey deer. After the reconnection of Jersey to the mainland in the Weichselian, the dwarf deer seem to have disappeared. Survival alongside the ancestral large deer would have indicated that speciation had been achieved, although the disappearance of the dwarf does not rule this out as the new species might simply have been outcompeted under mainland conditions. On the other hand, it is at least as likely that the animals were not genetically isolated from the large mainland deer, and succumbed either to individual competition, predation by mainland predators, or conceivably introgression by the larger animals.

Although geologically rapid, the change in the Jersey deer does not require the postulation of any special genetic or ecological processes. Six thousand years probably represented about 2000 generations of the deer, given a probable shortening of generation time as the animals became smaller. Under plausible parameters of selection, this is plenty of time for gradual change on a biological timescale. Gingerich (1983) pointed out that rates of evolution measured over short timescales tend to be higher than the 'averaged' rates perceived over longer spans of time. More recently, Gingerich (1993) has analysed Lister's (1989) data on size change in the Jersey deer, and shown that the observed rate (i.e. the degree of size change in the time available) falls within expected limits based on a survey of mammalian rate distribution, though at the 'fast' end of the range. These high rates tend not to be maintained over longer periods of time, but this does not of course negate the likely importance of such short-term, fast episodes in the history of evolution.

Similar timescales have recently been obtained for other examples of island dwarfing. In the case of the Key Deer, geological evidence indicates that the Florida Keys have been isolated for *c.* 4–5 ka (W. D. Klimstra, pers. comm. 1991). Vartanyan *et al.* (1993) deduced a maximum of 5 ka for the dwarfing of mammoths *Mammuthus primigenius* (Blum.) on Wrangel Island, NE Siberia. Both of these figures are comparable to the 6 ka postulated for the Jersey deer, although in neither of the above cases was size reduction as extreme as on Jersey.

Circumstantial evidence suggests that some introduced populations of island mammals may have suffered size reduction even within historical times. Several islands off the east coast of North America currently support small-bodied populations of feral horse *Equus caballus* L. These include Chincoteague Island (Maryland/Virginia), Okracoke Island (North Carolina) and Sable Island (Nova Scotia). The horses are believed to derive from shipwrecks between the 17th and 19th centuries, but little concrete information is available.

Zeuner (1950) recognized the potential of island populations in general, and the Jersey deer in particular, for estimating evolutionary rates. It is historically interesting to compare the current estimate of 6 ka as a maximum time available for the dwarfing, with the 70 ka quoted by Zeuner, based on his concept of the duration of the Last Interglacial.

Whether or not the Jersey deer had formed a truly separate species, their evolution illustrates the capacity of island populations to produce significant morphological divergence in a short period of time, an important component of the allopatric speciation model. However, we do not know whether the pattern observed on an offshore island can be generalized to the continental masses or the oceans, where most of life's diversity has arisen. 'Islands' of habitat, analogous in some ways, do occur in mainland ecosystems, and in many instances harbour endemic species. However, it would be much more difficult to demonstrate palaeontologically allopatric speciation there, than in the ideally controlled situation of an offshore island, where the timing of evolution in fossil populations can, in principle, be related to periods of known isolation due to sea-level changes. The realization of this potential will depend crucially on advances in the detailed timing of land–sea relations.

I thank the staff of the Jersey Museum and Hougue Bie Museum for access to fossil specimens. Valuable discussion was provided by Dr D. R. Bridgland, Professor A. E. Mourant, Dr V. L. Roth, Dr J. D. Scourse and Dr J. B. Wood. I thank Dr A. J. Sutcliffe for enlightening me about raised beaches. I am grateful to Sylvia Lachter, Nancy Thomas and Henry Bolker for drawing my attention to the insular horse populations of North America.

References

AZZAROLI, A. 1992. The cervid genus *Pseudodama* n.g. in the Villafranchian of Tuscany. *Palaeontographica Italica*, **79**, 1–41.

BARRETTE, C. 1977. Fighting behaviour of muntjacs and the evolution of antlers. *Evolution*, **31**, 169–176.

BERRY, R. J. 1977. *Inheritance and Natural History*. Bloomsbury Books, London.

BONNER, J. T. B. 1968. Size change in development and evolution. *Journal of Paleontology*, **42**, 1–15.

BOUCHUD, J. 1978. Les grands herbivores rissiens de l'Aven II des Abîmes de la Fage à Noailles (Corrèze). *Nouveaux Archives du Museum d'Histoire Naturelle de Lyon*, **16**, 9–39.

BROECKER, W. S. 1992. Upset for Milankovitch theory. *Nature*, **359**, 779–780.

BROWN, J. H., MARQUET, P. A. & TAPER, M. L. 1993. Evolution of body size: consequences of an energetic definition of fitness. *American Naturalist*, **142**, 573–584.

CALLOW, P. & CORNFORD, J. M. (eds) 1986. *La Cotte de St Brelade 1961–1978: Excavations of C. B. M. McBurney*. GeoBooks, Norwich.

CASE, T. J. 1978. A general explanation for insular body size trends in terrestrial vertebrates. *Ecology*, **59**, 1–18.

CHAPPELL, J. & SHACKLETON, N. J. 1986. Oxygen isotopes and sea level. *Nature*, **324**, 137–140.

CLUTTON-BROCK, T. H., ALBON, S. D. & GUINNESS, F. E. 1985. Parental investment and sex difference in juvenile mortality in birds and mammals. *Nature*, **313**, 131–133.

——, —— & HARVEY, P. H. 1980. Antlers, body size and breeding group size in the Cervidae. *Nature*, **285**, 565–567.

——, GUINNESS, F. E. & ALBON, S. D. 1982. *Red Deer*. University of Chicago Press, Chicago.

CORBET, G. B. & HARRIS, S. 1991. *The Handbook of British Mammals*, 3rd edn., Blackwell, Oxford.

CORTIJO, E., DUPLESSY, J. C., LABEYRIE, L., LECLAIRE, H., DUPRAT, J. & VAN WEERING, T. C. E. 1994. Eemian cooling in the Norwegian Sea and North Atlantic ocean preceding continental ice-sheet growth. *Nature*, **372**, 446–449.

DAMUTH, J. 1993. Cope's rule, the island rule and the scaling of mammalian population density. *Nature*, **365**, 748–750.

—— & MACFADDEN, B. J. (eds) 1990. *Body Size in Mammalian Paleobiology: Estimation and Biological Implications*. Cambridge University Press, Cambridge.

DANSGAARD, W., JOHNSEN, S. J., CLAUSEN, H. B. ET AL. 1993. Evidence for general instability of past climate from a 250-kyr ice-core record. *Nature*, **364**, 218–220.

DARWIN, C. 1859. *On the Origin of Species by Natural Selection*. John Murray, London.

DAVIES, K. H. & KEEN, D. H. 1985. The age of the marine Pleistocene deposits on Portland, Dorset. *Proceedings of the Geologists' Association*, **96**, 217–225.

DELPECH, F. & HEINTZ, E. 1976. Artiodactyles: Cervidés. *In:* DE LUMLEY, H. (ed.) *La Prehistoire Française*, CNRS, Paris, 398–404.

ELDREDGE, N. & GOULD, S. J. 1972. Punctuated equilibria: an alternative to phyletic gradualism. *In:* SCHOPF, T. J. M. (ed.) *Models in Paleobiology*. Freeman Cooper, San Fransisco, 82–115.

FIELD, M. H., HUNTLEY, B. & MÜLLER, H. 1994. Eemian climate fluctuations observed in a European pollen record. *Nature*, **371**, 779–783.

FORTELIUS, M. 1985. Ungulate cheek teeth: developmental, functional and evolutionary implications. *Acta Zoologica Fennica*, **180**, 1–76.

FOSTER, J. B. 1964. The evolution of mammals on islands. *Nature*, **202**, 234–235.

GINGERICH, P. D., 1983. Rates of evolution: effect of time and temporal scaling. *Science*, **222**, 159–161.

—— 1993. *In:* MARTIN, R. A. & BARNOSKY, A. D. (eds) *Morphological Change in Quaternary Mammals of North America*. Cambridge University Press, New York, 84–106.

GOULD, S. J. 1975. On the scaling of tooth size in mammals. *American Zoologist*, **15**, 351–62.

GRANT, P. R. 1981. Speciation and the adaptive radiation of Darwin's Finches. *American Scientist*, **69**, 653–663.

GRIGSON, C. & MELLARS, P. 1987. The mammalian remains from the middens. *In:* MELLARS, P. (ed.) *Excavations on Oronsay: Prehistoric Human Ecology on a Small Island*. Edinburgh University Press, Edinburgh, 243–289.

GRIP. 1993. Climate instability during the last interglacial period recorded in the GRIP ice core. *Nature*, **364**, 203–207.

GROOTES, P. M., STUIVER, M., WHITE, J. W. C., JOHNSEN, S. & JOUZEL, J. 1993. Comparison of oxygen isotope records from the GISP2 and GRIP Greenland ice cores. *Nature*, **366**, 552–554.

HARDIN, J. W., SILVY, N. J. & KLIMSTRA, W. D. 1976. Group size and composition of the Florida Key deer. *Journal of Wildlife Management*, **40**, 454–463.

HEANEY, L. R. 1978. Island area and body size of insular mammals: evidence from the tri-colored squirrel (*Callosciurus prevosti*) of Southeast Asia. *Evolution*, **32**, 29–44.

HEPTNER, V. G., NASIMOVICH, A. A. & BANNIKOV, A. G. 1988. *Mammals of the Soviet Union, Volume 1: Artiodactyla and Perissodactyla*. Smithsonian Institution, Washington, DC.

HUXLEY, J. 1932. *Problems of Relative Growth*, Methuen, London.

HUXLEY, J. S. 1931. The relative size of antlers in deer. *Proceedings of the Zoological Society of London*, 1931, 819–864.

HYDROGRAPHER OF THE NAVY. 1989. *The Channel Islands and Adjacent Coast of France*. Admiralty Charts, Taunton, no. **2669**.

ILLIUS, A. W. & GORDON, I. J. 1987. The allometry of food intake in grazing ruminants. *Journal of Animal Ecology*, **56**, 989–999.

INSTITUTE OF GEOLOGICAL SCIENCES. 1982. *Geological Maps of England and Wales. Sheet 2: Jersey - Solid and Drift*. Ordnance Survey, Southampton.

KEEN, D. H. (ed.) 1993. *The Quaternary of Jersey: Field Guide*. Quaternary Research Association, London.

——, HARMON, R. S. & ANDREWS, J. T. 1981. U series and amino acid dates from Jersey. *Nature*, **289**, 162–164.

KLEIN, D. R., MELDGAARD, M. & FANCY, S. G. 1987. Factors determining leg length in *Rangifer tarandus*. *Journal of Mammalogy*, **68**, 642–655.

KLIMSTRA, W. D., HARDIN, J. W. & SILVY, N. J. 1978. *Population Ecology of Key Deer*. National Geographic Society Research Reports, 1969 projects, 313–321.

LISTER, A. M. 1981. *Evolutionary Studies on Pleistocene Deer*. PhD thesis, University of Cambridge.

—— 1984. Evolutionary and ecological origins of British deer. *Proceedings of the Royal Society of Edinburgh*, **82B**, 205–229.

—— 1986. New results on deer from Swanscombe,

and the stratigraphical significance of deer in the Middle and Upper Pleistocene of Europe. *Journal of Archaeological Science*, **13**, 319–338.

—— 1987a. Giant deer and giant red deer from Kent's Cavern, and the status of *Strongyloceros spelaeus* Owen. *Transactions and Proceedings of the Torquay Natural History Society*, **19**, 189–198.

—— 1987b. Diversity and evolution of antler form in Quaternary deer. *In:* WEMMER, C. M. (ed.) *Biology and Management of the Cervidae* Smithsonian Institution, Washington, 81–98.

—— 1989. Rapid dwarfing of red deer on Jersey in the Last Interglacial. *Nature*, **342**, 539–542.

—— 1990. Critical reappraisal of the Middle Pleistocene deer species '*Cervus*' *elaphoides* Kahlke. *Quaternaire*, **1**, 175–192.

—— 1993a. Cervidae, deer. *In:* SINGER, R., GLADFELTER, B. G. & WYMER, J. J. (eds) *The Lower Paleolithic Site at Hoxne, England*. University of Chicago Press, Chicago, 174–190.

—— 1993b. Patterns of evolution in Quaternary mammals. *In:* EDWARDS, D. & LEES, D. (eds) *Evolutionary Patterns and Processes*. Linnean Society Symposium Series, **14**, Academic Press, London, 71–93.

—— 1993c. The dwarf deer of Belle Hougue Cave. *In:* KEEN, D. H. (ed.) *The Quaternary of Jersey: Field Guide*. Quaternary Research Association, London, 21–23.

—— 1995. Dwarfing in island elephants and deer: processes in relation to time of isolation. *In:* MILLER, P. (ed.) *Miniature Vertebrates: the Implications of Small Size*. Zoological Society of London Symposium Series, in press.

LOMOLINO, M. V. 1985. Body size of mammals on islands: the island rule re-examined. *American Naturalist*, **125**, 310–316.

LUNDHOLM, B. 1947. Abstammung und Domestikation des Hauspferdes. *Zoologiska Bidrag fran Uppsala*, **27**, 1–287.

MACARTHUR, R. H. & WILSON, E. O. 1967. *The Theory of Island Biogeography*, Princeton University Press.

MCMANUS, J. F., BOND, G. C., BROECKER, W. S., JOHNSEN, S., LABEYRIE, L. & HIGGINS, S. 1994. High-resolution climate records from the North Atlantic during the last interglacial. *Nature*, **371**, 326–329.

MAFFEI, M. D., KLIMSTRA, W. D. & WILMERS, T. J. 1988. Cranial and mandibular characteristics of the Key deer (*Odocoileus virginianus clavium*). *Journal of Mammalogy*, **69**, 403–407.

MAIORANA, V. C. 1990. Evolutionary strategies and body size in a guild of mammals. *In:* DAMUTH, J. & MACFADDEN, B. J. (eds) *Body Size in Mammalian Paleobiology: Estimation and Biological Implications*. Cambridge University Press, Cambridge, 69–102.

MARETT, R. R. 1912. Further observations on Prehistoric man in Jersey. *Archaeologia*, **63**, 203–230.

—— 1916. The site, fauna and industry of La Cotte de St. Brelade, Jersey. *Archaeologia*, **67**, 75–118.

MARSHALL, L. G. & CORRUCINI, R. S. 1978. Variability, evolutionary rates, and allometry in dwarfing lineages. *Paleobiology*, **4**, 101–119.

MARTINSON, D. G., PISIAS, N. G., HAYS, J. D., IMBRIE, J., MOORE, T. C. & SHACKLETON, N. J. 1987. Age dating and the orbital theory of the ice ages: development of a high-resolution 0 to 300,000-year chronostratigraphy. *Quaternary Research*, **27**, 1–29.

MATTIOLI, S. 1993. Antler conformation in red deer of the Mesola Wood, northern Italy. *Acta Theriologica*, **38**, 443–450

MAYR, E. 1963. *Animal Species and Evolution*. Belknap Press, Cambridge, MA.

MOTTERSHEAD, D. N., GILBERTSON, D. D. & KEEN, D. H. 1987. The raised beaches and shore platforms of Torbay: a re-evaluation. *Proceedings of the Geologists' Association*, **98**, 241–257.

MOURANT, A. E. 1984. The discovery of the Belle Hougue Cave, Jersey. *Annales et Bulletin de la Société Jersiaise*, **23**, 520–524.

MÜLLER, H. 1974. Pollenanalytische Untersuchungen und Jahresschichtungzählungen an der eem-zeitlichen Kieselgur von Bispingen/Luhe. *Geologisches Jahrbuch*, **A21**, 149–169.

NATURE CONSERVANCY COUNCIL. 1974. *Isle of Rhum National Nature Reserve: Reserve Handbook*.

PHILLIPS, L. 1974. Vegetational history of the Ipswichian/Eemian interglacial in Britain and Continental Europe. *New Phytologist*, **73**, 589–604

PREECE, R. C., SCOURSE, J. D., HOUGHTON, S. D., KNUDSEN, K. L. & PENNEY, D. N. 1990. The Pleistocene sea-level and neotectonic history of the eastern Solent, southern England. *Philosophical Transactions of the Royal Society of London*, **B328**, 425–477.

PROCTOR, C. J. & SMART, P. L. 1991. A dated cave sediment record of Pleistocene transgressions on Berry Head, Southwest England. *Journal of Quaternary Science*, **6**, 233–244.

ROTH, V. L. 1990. Insular dwarf elephants: a case study in body mass estimation and ecological inference. *In:* DAMUTH, J. & MACFADDEN, B. J. (eds) *Body Size in Mammalian Paleobiology: Estimation and Biological Implications*. Cambridge University Press, Cambridge, 151–179.

—— 1992. Inferences from allometry and fossils: dwarfing of elephants on islands. *Oxford Surveys in Evolutionary Biology*, **8**, 259–288.

—— 1995. Dwarfism in elephants of the Californian islands. *In:* SHOSHANI, J. & TASSY, P. (eds) *The Proboscidea: Trends in Palaeoecology and Evolution*. Oxford University Press, Oxford, in press.

SCOTT, K. 1986. The large mammal fauna. *In:* CALLOW, P. & CORNFORD, J. M. (eds), *La Cotte de St Brelade 1961–1978: Excavations of C. B. M. McBurney*. GeoBooks, Norwich, 159–183.

SCOTT, K. M. 1987. Allometry and habitat-related adaptations in the postcranial skeleton of Cervidae. *In:* WEMMER, C. M. (ed.) *Biology and Management of the Cervidae*. Smithsonian Institution, Washington, D.C., 65–80.

SHACKLETON, N. J. 1969. The last interglacial in the marine and terrestrial records. *Proceedings of the Royal Society of London*, **B174**, 135–154.

—— 1987. Oxygen isotopes, ice volume and sea level. *Quaternary Science Reviews*, **6**, 183–190.

—— & PISIAS, N. G. 1985. Atmospheric carbon dioxide, orbital forcing, and climate. *Geophysical Monographs*, **32**, 303–317.

SIMPSON, G. G. 1940. Mammals and land bridges. *Journal of the Washington Academy of Science*, **30**, 137–163.

SONDAAR, P. Y. 1977. Insularity and its effect on mammalian evolution. *In:* HECHT, M. K., GOODY, P. C. & HECHT, B. M. (eds), *Major Patterns in Vertebrate Evolution*. Plenum, New York, 671–707.

—— 1986. The island sweepstakes. *Natural History*, **9**, 50–57.

STOCK, C. & FURLONG, E. L. 1928. The Pleistocene elephants of Santa Rosa Island, California. *Science*, **48**, 140–141.

STUART, A. J. 1982. *Pleistocene Vertebrates in the British Isles*. Longman, London.

SUTCLIFFE, A. J. 1960. Joint Mitnor Cave, Buckfastleigh. *Transactions and Proceedings of the Torquay Natural History Society*, **13**, 1–26.

—— 1981. Progress report on excavations in Minchin Hole, Gower. *Quaternary Newsletter*, **33**, 1–17.

—— 1985. *On the Track of Ice Age Mammals*. British Museum (Natural History), London.

—— 1995. Insularity of the British Isles 250 000–30 000 years ago: the mammalian, including human evidence. *This volume*.

——, CURRANT, A. P. & STRINGER, C. B. 1987. Evidence of sea-level change from coastal caves with raised beach deposits, terrestrial faunas and dated stalagmites. *Progress in Oceanography*, **18**, 243–271.

THALER, L. 1973. Nanisme et gigantisme insulaires. *La Recherche*, **4/37**, 741–750.

THOMAS, C. 1985. *Exploration of a Drowned Landscape. Archaeology and History of the Isles of Scilly*. Batsford, London.

THOUVENY, N., DE BEAULIEU, J.-L., BONIFAY, E. ET AL. 1994. Climate variations in Europe over the past 140 kyr deduced from rock magnetism. *Nature*, **371**, 503–506.

TURK, F. A. 1968. Notes on Cornish mammals in prehistoric and historic times, I (Tean, Agnes). *Cornish Archaeology*, **7**, 73–79.

—— 1971. Notes on Cornish mammals in prehistoric and historic times, IV (Nornour). *Cornish Archaeology*, **10**, 79–91.

VAN DEN BERGH, G. D., SONDAAR, P. Y., DE VOS, J. & AZIZ, F. 1995. The proboscideans of the southeast Asian islands. *In:* SHOSHANI, J. & TASSY, P. (eds) *The Proboscidea: Trends in Evolution and Palaeoecology*. Oxford University Press, Oxford, in press.

VARTANYAN, S. L., GARUTT, V. E. & SHER, A. V. 1993. Holocene dwarf mammoths from Wrangel Island in the Siberian Arctic. *Nature*, **362**, 337–340.

VOS, J. DE 1984. The endemic Pleistocene deer of Crete. *Verhandelingen der Koninklijke Nederlandse Akademie van Wetenschappen, Afd. Natuurkunde, Eerste Reeks*, **31**, 1–100.

WEST, R. G. 1972. Relative land–sea-level changes in southeastern England during the Pleistocene. *Philosophical Transactions of the Royal Society of London*, **A272**, 87–98.

—— & SPARKS, B. W. 1960. Coastal interglacial deposits of the English Channel. *Philosophical Transactions of the Royal Society of London*, **B243**, 96–133.

WHITEHEAD, G. K. 1993. *The Whitehead Encyclopaedia of Deer*. Swan Hill, Shrewsbury.

WINOGRAD, I. J., COPLEN, T. B., LANDWEHR, J. M. ET AL. 1992. Continuous 500,000-year climate record from vein calcite in Devil's Hole, Nevada. *Science*, **258**, 255–260.

YALDEN, D. W. 1982. When did the mammal fauna of the British Isles arrive? *Mammal Review*, **12**, 1–57.

ZAGWIJN, W. H. 1983. Sea-level changes in the Netherlands during the Eemian. *Geologie en Mijnbouw*, **62**, 437–450.

ZAHN, R. 1994. Core correlations. *Nature*, **371**, 289–290.

ZEUNER, F. E. 1946. *Cervus elaphus jerseyensis* and other fauna in the 25 ft beach of Belle Hougue Cave, Jersey, C. I. *Bulletin de la Société Jerseiaise*, **14**, 238–254.

—— 1950. *Dating the Past*. 2nd edn, Methuen, London.

Insularity and the Quaternary tree and shrub flora of the British Isles

K. D. BENNETT

Department of Plant Sciences, University of Cambridge, Downing Street, Cambridge CB2 3EA, UK

Abstract: The distribution of trees and shrubs within the British Isles and elsewhere in northern Europe is considered in the light of palaeoecological evidence for Holocene distributions. A small number of species appear to be capable of spreading across any sea-channel in the region, and the overwhelming majority are capable of spreading across channels of 10–100 km extent. Thus, there is little evidence that insularity would have been a significant factor in controlling the tree and shrub flora in the present or previous interglacials, and hence the presence or absence of any species cannot be used as evidence or otherwise for isolation. However, there have been significant losses of tree and shrub species from islands during the late Holocene as a consequence of anthropogenic activities. These may be among the more dramatic prehistoric extinctions yet described.

The region known as the British Isles is a collection of about 5000 islands, dominated by the two large islands of Great Britain and Ireland, lying off the northwest coast of the European continental mainland. At its nearest point, the southeast extremity of Great Britain is about 34 km from northeast France. The archipelago is generally considered as an example of 'continental' islands (Wallace 1880), once connected to the neighbouring landmass, as distinct from 'oceanic islands' which have never had a connection. Globally, islands are well known as sites of high plant endemism, but the British Isles and other high-latitude islands and archipelagoes do not have an endemic flora. In the case of the British Isles, the flora of about 1200 presumed native species (i.e. not introduced as a consequence of human activities) is a sub-set of the European flora. Within the British Isles, individual islands have floras that, with few exceptions, consist of a sub-set of the flora of Great Britain. The tree and shrub component of the flora, however, becomes generally less diverse moving from southeast to northwest, and, with one exception, all occur on Great Britain. The exception, *Arbutus unedo*, occurs in southwest Ireland, probably as a result of long-distance dispersal from western France (Webb 1983), and I shall not discuss it further. The aim of this paper is to address three questions, all potentially answerable from the Quaternary fossil record:

(1) Has separation of Great Britain from the European continent, and separation of islands within the British Isles, affected the spread of trees and shrubs to and within the archipelago?
(2) Are differences between interglacial floras in the British Isles attributable to isolation (or lack of it)?
(3) Have there been either distribution or abundance changes (such as extinctions) that are attributable to the separation of the land area of the British Isles into islands?

Vascular plant nomenclature follows Tutin *et al.* (1964–1980). All ages and durations of time are given in radiocarbon years relative to AD 1950 (BP).

Holocene palaeogeography

Past distribution and dispersal of trees and shrubs to islands cannot be witnessed. We need to infer from palaeogeographic evidence which landmasses were islands, and when, and we then need palaeoecological evidence from those landmasses concerning which species were present, and when. Much of the palaeogeographic information needed is available within the British Isles, but information relevant to the dispersal potential of some species of tree and shrub can be obtained from Scandinavian palaeogeographic history.

During the last full-glacial maximum, sea-levels are normally assumed to have reached about 100 m below present levels (e.g. Kutzbach & Ruddiman 1993). The present 100 m contour of sea-depth suggests that, at the time of the full-glacial maximum, eastern Great Britain would have been connected to the European continent along a broad area from eastern Scotland round to southwestern England. The present islands of the archipelago would have been connected to

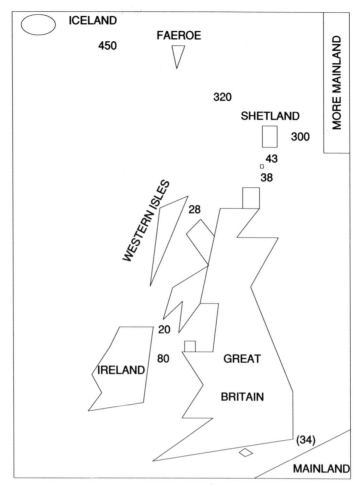

Fig. 1. Quaternary archipelagic features of the British Isles showing main island groups and lengths of significant sea crossings (km). Not to scale.

each other, with the exception of the Western Isles and the St Kilda group (both persisting as separate islands groups off northwest Scotland). Rising sea-levels after the last glacial maximum would have flooded the channels which isolate Shetland (from Orkney), and Ireland (from Great Britain) (Devoy 1985) before the beginning of the Holocene. The present best estimate is that rising sea-levels in the Strait of Dover led to the separation of Great Britain from the European continent at about 8300 years BP (Jelgersma 1979), and isolation for many of the islands of the present archipelago, especially in northwest Scotland, may also have begun well within the Holocene. Porter (1989) points out that 'average' sea-level conditions for the Quaternary are somewhere between those of the present and those of the full-glacial maximum, with a sea-level of perhaps 55 m below present. On this basis, several of the Hebridean islands would have been separated from Great Britain for much of the Quaternary, but Orkney, the Isle of Man and the Isle of Wight would have been connected to Great Britain.

The present state of the archipelago, with significant inter-island distances, is summarized in Fig. 1.

Early Holocene sea-level and Baltic lake-level history in Scandinavia is complex. It seems that by 7000 years BP, something like the present land–sea configuration would have been established, followed by a fall in relative sea-level because of isostatic rebound, particularly in the north. It can be assumed that all the present southern Baltic islands (including Gotland and Öland) would have been isolated throughout the

Holocene after their emergence into the lateglacial Baltic Ice Lake during deglaciation (Eronen 1983).

Significance of the palaeoecological record

The Quaternary palaeoecological record provides insight into the question of insularity and its consequence for the flora in two important areas. Firstly, successful dispersal events over long (kilometre scale) distances are likely to be rare and cannot be readily observed in practice. We are unlikely to be able to tell by observation whether, for example, *Quercus* acorns are dispersed over sea-crossings of 20–30 km, but we might be able to establish that this must have been the case if a fossil record shows that *Quercus* was present on both sides of such a crossing, and must have spread from one to the other. Secondly, there has been much disturbance of the flora as a consequence of human activity. Some of this has been deliberate, and some has been an accidental side-effect of other activities. Disturbance has included both extension and reduction of plant ranges. The fossil record offers the best chance of seeing what plant distributions within the archipelago of the British Isles were really like before anthropogenic effects became significant, and possibly also of seeing consequences of those effects.

Tree and shrub dispersal characteristics

Examination of the native tree and shrub flora and its distribution across the British Isles and other North Atlantic islands, in conjunction with Holocene pollen records from the islands, helps to establish the potential for dispersal across sea-channels of different widths for different species of trees and shrubs. Failure to cross a channel does not, of course, mean that the channel itself was the limiting factor, because there may be some significant difference in the environments on either side, but success provides a key to dispersal potential.

In the context of the Quaternary of the British Isles, we need, ideally, information on the dispersal characteristics of the following trees and shrubs (West 1980): *Abies, Chamaecyparis, Juniperus, Picea, Pinus,* Taxodiaceae, *Taxus, Tsuga, Acer, Alnus, Betula, Carpinus, Corylus, Eucommia, Fagus, Fraxinus, Hippophae, Ilex, Pterocarya, Quercus, Tilia* and *Ulmus*. Of these, *Chamaecyparis*, Taxodiaceae, *Tsuga, Eucommia* and *Pterocarya* no longer occur in Europe, and there is no information on their likely dispersal abilities between islands. There is information available for the remaining taxa, which include most of the important forest-forming taxa of the middle and late Quaternary, and a few additional species that seem never to have been abundant but are likely to have been components of interglacial forest floras.

Dispersal over distances > 100 km

Four European tree and shrub species (*Juniperus communis, Betula pubescens* ssp. *tortuosa, Populus tremula* and *Sorbus aucuparia*) occur on Iceland (Löve 1983). Pollen or macrofossil analyses indicate the presence of all except *Populus tremula* before the arrival of the first Norse settlers (Hallsdóttir 1987, 1990). They are presumed to have arrived at some time during the Holocene rather than as the result of perglacial survival, as is the case for the Icelandic fauna (Buckland *et al*. 1986). *Juniperus communis* occurs on the Faeroes today and throughout the Holocene, and macrofossil remains of *Betula pubescens* have been found there (Jóhansen 1985, 1989), dating from before the arrival of the Norse. *Betula pubescens* ssp. *tortuosa* occurs in Greenland (Böcher *et al*. 1968). Additionally, species of *Salix* are present on all North Atlantic islands. This genus includes species that are high arctic dwarf shrubs and species that are temperate forest trees, and it is not normally possible to separate them on the basis of fossil pollen.

Species in the five genera *Juniperus, Betula, Populus, Salix* and *Sorbus* can be assumed to be capable of spread across any sea-channel in the North Atlantic region.

Dispersal over distances > 10 km

In this group, I include those species that have spread across the sea-channels between Great Britain and the Western Isles, Great Britain and Shetland, Great Britain and Ireland (during the Holocene), to Great Britain from the European continental mainland after about 7000 years BP, or to Gotland from Sweden. Palaeoecological data from the Western Isles is summarized by Wilkins (1984), Bennett *et al*. (1990) and Fossitt (1990), and from Shetland by Bennett *et al*. (1992). The timing of spread of trees and shrubs into Great Britain from the continental mainland, and thence to Ireland, is discussed by Birks (1989), and timing elsewhere in Europe by Huntley & Birks (1983).

At least 12 species have made sea-crossings in excess of 10 km during the Holocene. The palaeoecological record is not able to discriminate between species in some genera, so the total number is almost certainly greater. Thus, *Pinus sylvestris* and *Ilex aquifolium* reached Ireland and the Western Isles; *Taxus baccata, Euonymus europaeus, Rhamnus catharticus, Frangula alnus* and *Cornus sanguinea* reached Ireland; *Corylus avellana, Alnus glutinosa, Quercus* (two possible species) and *Ulmus* (probably only one species) reached Ireland, the Western Isles, and Shetland; and *Fraxinus excelsior* reached Ireland and the Western Isles. *Pinus sylvestris* reached Shetland in a previous interglacial (Birks & Ransom 1969), but is not known to have made the crossing during the Holocene.

There is an additional group of trees which might be capable of sea-crossings this wide, or which are capable, but did not reach the British Isles during the Holocene, despite presence earlier in the Quaternary. *Picea abies* and *Acer platanoides* occur on the Baltic island of Gotland. *Abies alba* occurs on Corsica, an oceanic island in the Mediterranean. This indicates that all three can cross sea-channels of more than 10 km. *Tilia cordata* occurs in Great Britain, the Danish islands and in southern Sweden. It is absent from Gotland today (Pigott 1991), but occurred there during the mid-Holocene (Påhlsson 1977; Svensson 1989). It almost certainly reached Great Britain at about the time of the flooding of the Strait of Dover (Birks 1989). It is absent from Ireland today and is not recorded in any Irish interglacial deposit (Watts 1985). Since *Tilia cordata* managed to cross from southern Sweden to Gotland, it is perhaps unlikely that the Irish Sea has been a barrier to it.

Acer campestre is present today, with presumed native status, on Great Britain and islands such as the Isle of Wight and Anglesey, but not on Ireland, and it has not reached Gotland. It is widely distributed as an introduced species in Ireland (Perring & Walters 1962). It is therefore possible that its absence as a native species is because the crossing to Ireland from west Wales (the species is absent from southern Scotland) was too great. *Acer campestre* produces little pollen and its Holocene record is too weak to establish whether the species reached Great Britain before or after the flooding of the Strait of Dover. Similarly, *Sorbus torminalis* and *Viburnum lantana* are widespread in England and Wales, but are absent from Ireland. The fossil record for both is inadequate to make judgements about relative timings of their spread into Great Britain. It is possible that these three trees are absent from Ireland because the crossing from Wales was too wide (Matthews 1955).

Hippophae rhamnoides is present along the east coast of Great Britain, but absent from all other islands of the British Isles. It is known to have been present during the late-glacial (Godwin 1975), so is likely to have spread across before the Strait of Dover was flooded, or even survived on the continental shelf through the full-glacial. However, it reached Gotland during the early Holocene (Påhlsson 1977; Svensson 1989), and was present in Ireland during previous interglacials (Watts 1985), so it is unlikely to be limited by poor dispersal characteristics.

An interesting case is provided by *Myrica gale*. It reached many islands in the west of the British Isles, including Ireland and the Western Isles, but is absent from Shetland, the Faeroes and Iceland. I suspect that this plant can cope with some long sea-crossings, but Shetland is just too far. The same may be true for *Pinus sylvestris* with the modern land–sea configuration in the northern British Isles.

Fagus sylvatica and *Carpinus betulus* both spread into northern Europe in the late Holocene, reaching Great Britain and southern Sweden at a time when the sea-crossings were certainly flooded. However, it is unclear whether the spread of either of these trees was facilitated by anthropogenic activities (Behre 1988).

Trees and shrubs absent during the Holocene

There appear to be only three tree and shrub species (*Quercus pubescens, Ulmus laevis, Berberis sempervirens*) with European distributions that extend to, or near, the southern shore of the English Channel, but are absent from Great Britain (Jalas & Suominen 1976–1991). It is just possible that any of these might be absent because of inability to cross the 34 km channel, but they might also be limited by factors other than the physical barrier. It is even possible that they were present during the Holocene, but have gone undetected because of poor pollen productivity and dispersal (*Berberis sempervirens*), or confusion with other members of the genus (*Quercus pubescens, Ulmus laevis*).

Anthropogenic influences

Much of the landscape of the British Isles today is unforested. It is clear that in most areas this is the result of anthropogenic forest clearance for agriculture from the Neolithic period onwards, but it has been suggested that parts of the north and west of Scotland and Ireland, including Shetland and the Western Isles, have always been predominantly treeless, except for limited areas of *Betula* scrub (McVean & Ratcliffe 1962;

Table 1. *Tree taxa that have become extinct on islands of the British Isles during the Holocene.*

Island	Taxon
Ireland	*Pinus sylvestris* (m, p)
Clare Island	*Pinus sylvestris* (m, p), *Alnus glutinosa* (m)
Aranmore	*Pinus sylvestris* (m)
Achill	*Pinus sylvestris* (m)
Inishbofin	*Pinus sylvestris* (m, p)
Valencia	*Pinus sylvestris* (p), *Quercus* (m, p), *Corylus avellana* (p)
Skye	*Pinus sylvestris* (m, p)
Mull	*Pinus sylvestris* (m)
Islay	*Rhamnus catharticus* (p), *Viburnum opulus* (p)
Rhum	*Pinus sylvestris* (m)
Uists/Benbecula	*Pinus sylvestris* (p), *Quercus* (p), *Ulmus* (p), *Alnus glutinosa* (p, m, h)
Lewis/Harris	*Pinus sylvestris*? (m), *Alnus glutinosa* (p, m), *Ulmus* (p), *Quercus* (p)
Shetland mainland	*Alnus glutinosa* (p, m), *Quercus* (p)
Yell	*Betula* (m), *Corylus avellana* (h)
Unst	*Betula* (m)
Foula	*Alnus glutinosa* (m), *Betula* (m), *Juniperus* (m, h)
Fetlar	*Juniperus* (h), *Corylus avellana* (h)
Orkney mainland	*Betula* (p), *Corylus avellana* (p)
Isle of Man	*Pinus sylvestris* (m)
Isle of Wight	*Pinus sylvestris* (p)

Evidence for former presence: p, pollen; m, macrofossil; h, historical; ?, possibly not extinct. Data from various sources, including personal observation by K. D. B. and J. A. Fossitt.

Birks & Madsen 1979; Jóhansen 1985). New information relating to former flora and vegetation of these areas has now been provided by pollen analyses of lake sediments (Bennett et al. 1990, 1992, 1993; Fossitt 1990, 1994), macrofossil remains preserved in peat (Birks 1975; Bennett 1984; Wilkins 1984; Fossitt 1990), and historical records (Scott & Palmer 1987; Pankhurst & Mullin 1991). Pollen analyses of Holocene sediment sequences from small lakes in Shetland and the Western Isles, and in northwest Ireland, demonstrate that woodland formerly occurred there, and that the woodland was diverse, with *Quercus*, *Ulmus*, *Alnus glutinosa* and (locally) *Pinus sylvestris*, as well as the expected *Betula* and *Corylus avellana* (Bennett et al. 1990, 1992, 1993; Fossitt 1990, 1994). Most of these trees no longer grow naturally on any of the islands of the far north and west, but other islands, including some that today do have some remnant woodland, have also lost tree and shrub flora (Table 1). The taxa concerned include populations of some of the most abundant and widespread trees and shrubs in the British Isles today, whose ranges are now reduced compared to the early Holocene. These taxa can only indicate a part of the total extinction, as several of them include more than one species, all of which must have become extinct before the genus had disappeared. Additionally, detailed pollen analyses on Islay (unpublished data) have revealed the former occurrence of *Viburnum opulus* and *Rhamnus catharticus*, which indicates that more extensive losses may be discovered when investigations are carried out on other islands, especially in the southern part of the Inner Hebrides.

It is difficult to establish from pollen evidence alone when particular tree and shrub populations were lost from islands, because of problems with the interpretation of small quantities of pollen as an indication of presence or absence during the declining phase of a population's pollen record. However, records of the total of tree and shrub pollen indicate that woodland decline in the northern and western British Isles took place at variable rates, but generally within the last 5000 years, and mostly within the period 4000 BP to 2000 years BP. It is probable that extinction of most of the island populations listed in Table 1 took place within this period, but pollen records of woodland decline vary considerably from place to place within the islands, in timing and rate of decrease. Charcoal records from the same profiles also show large spatial and temporal variation, but generally increase as the woodland declined and the area of blanket peat expanded. Woodland also disappeared from substantial areas of the mainland, particularly in the Scottish Highlands and western Ireland, but there the tree flora has persisted.

The cause of woodland decline and island extinctions may lie with any of three forcing factors. Firstly, variations in the Earth's orbital parameters suggest that insolation has probably decreased during the Holocene, but the oceanic situation of the British Isles makes it unlikely that there has been substantial change in temperatures or precipitation as a consequence (Kutzbach et al. 1993). However, short-term climatic variations due, for example, to volcanic eruptions, cannot be excluded. Secondly, the northern and western islands have experienced a gradual shift from woodland or scrub to blanket peat as the dominant vegetation type. A slight shift in this direction is also seen on the Faeroes (Jóhansen 1985). These islands have a similar climatic regime to the north and west of Great Britain and Ireland, but human occupation is not recorded until about 1200 years BP, suggesting that such a shift may be a consequence of leaching and gradual soil impoverishment over the course of the Holocene. Thirdly, much of the British Isles was occupied by farming people after about 5500 years BP (Megaw & Simpson 1979). Temporal coincidence between increases in charcoal, probably due to human domestic activity (Bennett et al. 1990), and overall decline in woodland strongly suggests that increased activity was responsible for both the loss of woodland vegetation and tree and shrub flora during the late Holocene.

Radiocarbon age determinations on macrofossils establish times when tree and shrub taxa were present on islands. If taxa subsequently become extinct, such dates may provide a *terminus postquem*. They support pollen data by indicating that most of the extinct trees and shrubs persisted until at least 3000 BP to 4000 years BP. Some survived until the historic period (Table 1).

It appears that the arrival of Neolithic farmers in the British Isles after 5500 years BP had a dramatic impact on the native tree and shrub flora, especially on islands, as well as on vegetation. Extinctions of island populations were a visible part of changes that took place as an alien culture introduced crops and domestic animals to a landscape which had developed in their absence. The result was devastating for the native flora and vegetation.

Discussion

The data on modern and past plant distribution, presented above, provide no convincing evidence that any tree or shrub has been limited in its distribution to the British Isles by the existence of water barriers. It should not be forgotten that all trees and shrubs in northern Europe have successfuly dispersed across other barriers (e.g. the Alps, the Danube) during the course of the Holocene. However, within the British Isles there is a certain amount of suspicion that *Acer* spp., *Sorbus torminalis* and *Viburnum lantana* may have been unable to reach Ireland because of the lack of a land connection or a narrow water crossing. *Myrica gale* may have been unable to reach Shetland. It follows that the presence or absence of any individual species cannot be taken to indicate insularity or non-insularity. Webb (1983), in a detailed study of the Irish flora, suggested that about 186 species of gymnosperms and angiosperms may have been excluded from Ireland because of its isolation from Great Britain, but 797 species did make the crossing. This means that about 19% of the potential flora was limited by isolation. The proportion of the tree and shrub flora missing may be higher than this, but the number of tree and shrub species is itself so small that a statistical comparison would be worthless. Whatever the cause of the differences between floras of different interglacials in Great Britain (West 1980), it seems unlikely that isolation was a significant factor: the Strait of Dover is too narrow and too shallow (meaning that it floods late into an interglacial) to provide a barrier. The situation in Ireland may be different.

The dispersal ability of many trees and shrubs is, to human eyes, remarkable. In order to achieve the observed patterns, heavy-seeded fruits such as *Quercus* acorns and *Corylus* nuts are being dispersed across sea-channels of up to 100 km in width with apparent ease. Similar phenomena have been noted with respect to the spread of trees in the Great Lakes region of North America (Davis et al. 1986; Webb 1987). Dispersal events within the British Isles, involving wind-dispersal or carriage by birds, are too rare to have been observed, but too frequent to leave a noticeable lag in the pollen record. This indicates that they might be happening with frequencies of the order of once every 100 years. Given the longevity of woody plants, that might be long enough for one individual of a dioecious species (such as *Quercus* spp.) to be still present as a low number of following colonists arrive, increasing the chance that populations can be established.

The inferred anthropogenic extinctions of the late Holocene are the earliest known example of a type of impact seen later as Polynesians spread to the Pacific islands, or as Europeans occupied the Americas, Africa, Australasia and oceanic islands. The special circumstances which permit the recovery of fossil pollen from lake sediments and macrofossils from extensive peat deposits, as on the islands of the northern and western British Isles, are not widely available globally. It may be that plant extinctions have taken place unseen, as a consequence of human colonizations, more often than has been appreciated hitherto. Attempts to obtain measures of species richness (Johnson & Simberloff 1974; Currie & Paquin 1987; Adams & Woodward 1989) in relation to modern environmental factors cannot be considered complete unless prehistoric anthropogenic losses and modern introductions have been taken into account.

Because the British Isles are an archipelago, there seem to have been regional extinctions of elements of the flora which would not have happened if the same area of land had remained connected as a single landmass. However, this aspect of insularity seems to be the only significant consequence of being an archipelago.

I thank Julie Fossitt for help with Table 1, and Julie Fossitt, Mike Walker and Kathy Willis for helpful comments on earlier versions of this paper.

References

ADAMS, J. M. & WOODWARD, F. I. 1989. Patterns in tree species richness as a test of the glacial extinction hypothesis, *Nature*, **339**, 699–701.

BEHRE, K.-E. 1988. The role of man in European vegetation history. *In:* HUNTLEY, B. & WEBB, T. III (eds) *Handbook of Vegetation Science 7. Vegetation history* Kluwer Academic Publishers, Dordrecht, 633–672.

BENNETT, K. D. 1984. Post-glacial history of *Pinus sylvestris* in the British Isles. *Quaternary Science Reviews*, **3**, 133–155.

——, BOREHAM, S., HILL, K., PACKMAN, S., SHARP, M. J. & SWITSUR, V. R. 1993. Holocene environmental history at Gunnister, north Mainland, Shetland *In:* BIRNIE, J. F., GORDON, J. E., BENNETT, K. D. & HALL, A. M. (eds) *The Quaternary of Shetland: Field Guide*. Quaternary Research Association, Cambridge, 83–98.

——, ——, SHARP, M. J. & SWITSUR, V. R. 1992. Holocene history of environment, vegetation and human settlement on Catta Ness, Lunnasting, Shetland, *Journal of Ecology*, **80**, 241–273.

——, FOSSITT, J. A., SHARP, M. J., SWITSUR, V. R. 1990. Holocene vegetational and environmental history at Loch Lang, South Uist, Western Isles, Scotland. *New Phytologist*, **114**, 281–298.

BIRKS, H. H. 1975. Studies in the vegetational history of Scotland. IV. Pine stumps in Scottish blanket peats. *Philosophical Transactions of the Royal Society of London*, **B270**, 181–226.

BIRKS, H. J. B. 1989. Holocene isochrone maps and patterns of tree-spreading in the British Isles. *Journal of Biogeography*, **16**, 503–540.

—— & MADSEN, B. J. 1979. Flandrian vegetational history of Little Loch Roag, Isle of Lewis, Scotland. *Journal of Ecology*, **67**, 825–842.

—— & RANSOM, M. E. 1969. An interglacial peat at Fugla Ness, Shetland. *New Phytologist*, **66**, 777–796.

BÖCHER, T. W., HOLMEN, K. & JAKOBSEN, K. 1968. *The Flora of Greenland*, P. Haase & Son, Copenhagen.

BUCKLAND, P. C., PERRY, D. W., GÍSLASON, G. M. & DUGMORE, A. J. 1986. The pre-Landnám fauna of Iceland: a palaeontological contribution. *Boreas*, **15**, 173–184.

CURRIE, D. J. & PAQUIN, V. 1987. Large-scale biogeographic patterns of species richness of trees. *Nature*, **329**, 326–327.

DAVIS, M. B., WOODS, K. D., WEBB, S. L. & FUTYMA, R. P. 1986. Dispersal versus climate: expansion of *Fagus* and *Tsuga* into the Upper Great Lakes. *Vegetatio*, **67**, 93–103.

DEVOY, R. J. 1985. The problem of a late Quaternary landbridge between Britain and Ireland. *Quaternary Science Reviews*, **4**, 43–58.

ERONEN, M. 1983. Shore displacement in Finland. *In:* SMITH, D. E. & DAWSON, A. G. (eds) *Shorelines and Isostasy*. Institute of British Geographers, Special Publication, **16**, Academic Press, London, 183–207.

FOSSITT, J. A. 1990. *Holocene Vegetation History of the Western Isles, Scotland*. PhD thesis, University of Cambridge.

—— 1994. Late-glacial and Holocene vegetation history of western Donegal, Ireland. *Biology and Environment. Proceedings of the Royal Irish Academy*, **94B**, 1–31.

GODWIN, H. 1975. *History of the British Flora*, 2nd edn, Cambridge University Press, Cambridge.

HALLSDÓTTIR, M. 1987. *Pollen Analytical Studies of Human Influence on Vegetation in Relation to the Landnám Tephra Layer in Southwest Iceland*. LUNDQUA thesis, **18**.

—— 1990. Studies in the vegetational history of North Iceland. A radiocarbon-dated pollen diagram from Flateyjardalur. *Jökull*, **40**, 67–81.

HUNTLEY, B. & BIRKS, H. J. B. 1983. *An Atlas of Past and Present Pollen Maps for Europe 0–13,000 Years Ago*. Cambridge University Press, Cambridge.

JALAS, J., SUOMINEN, J. 1976–1991. *Atlas Florae Europaeae*. Nine volumes (incomplete) Committee for Mapping the Flora of Europe, Helsinki.

JELGERSMA, S. 1979. Sea-level changes in the North Sea basin *In:* OELE, E., SCHÜTTENHEIM, R. T. E. & WIGGER, A. J. (eds) *The Quaternary History of the North Sea: Symposia Universitatis Upsaliensis Annum Quingentesimum Celebrantis. Acta Universitatis Upsaliensis, Uppsala*, **2**, 233–248.

JÓHANSEN, J. 1985. Studies in the vegetational history of the Faroe and Shetland Islands. *Annales Societatis Scientiarum Faeroensis Supplementum*, XI.

—— 1989. Survey of geology, climate, and vegetational history. *Annales Societatis Scientiarum Faeroensis Supplementum*, XIV, 11–16.

JOHNSON, M. P. & SIMBERLOFF, D. S. 1974. Environmental determinants of island species number in the British Isles. *Journal of Biogeography*, **1**, 149–154.

KUTZBACH, J. E. & RUDDIMAN, W. F. 1993. Model description, external forcing, and surface boundary conditions *In:* WRIGHT, H. E., Jr, KUTZBACH, J. E., WEBB, T., III, RUDDIMAN, W. F., STREET-PERROTT, F. A. & BARTLEIN, P. J. (eds) *Global Climates since the Last Glacial Maximum.* University of Minnesota Press, Minneapolis, 12–23.

— ——, GUETTER, P. J., BEHLING, P. J. & SELIN, R. 1993. Simulated climatic changes: results of the COHMAP climate–model experiments. *In:* WRIGHT, H. E., Jr, KUTZBACH, J. E., WEBB, T., III, RUDDIMAN, W. F., STREET-PERROTT, F. A. & BARTLEIN, P. J. (eds) *Global Climates since the Last Glacial Maximum.* University of Minnesota Press, Minneapolis, 24–93.

LÖVE, A. 1983. *Flora of Iceland.* Almenna Bókafélagid, Reykjavík.

MCVEAN, D. N. & RATCLIFFE, D. A. 1962. *Plant Communities of the Scottish Highlands*, HMSO, Edinburgh.

MATTHEWS, J. R. 1955. *Origin and Distribution of the British Flora.* Hutchinson, London.

MEGAW, J. V. S. & SIMPSON, D. D. A. 1979. *Introduction to British Prehistory*, Leicester University Press, Leicester.

PÅHLSSON, I. 1977. A standard pollen diagram from Lojsta area of central Gotland. *Striae*, **3**.

PANKHURST, R. J. & MULLIN, J. M. 1991. *Flora of the Outer Hebrides*, Natural History Museum Publications, London.

PERRING, F. H. & WALTERS, S. M. 1962. *Atlas of the British Flora.* Nelson, London.

PIGOTT, C. D. 1991. *Tilia cordata* Miller. *Journal of Ecology*, **79**, 1147–1207

PORTER, S. C. 1989. Some geological implications of average Quaternary glacial conditions. *Quaternary Research*, **32**, 245–261.

SCOTT, W. & PALMER, R. 1987. *The Flowering Plants and Ferns of the Shetland Islands*, Shetland Times Ltd, Lerwick.

SVENSSON, N.-O. 1989. *Late Weichselian and Early Holocene Shore Displacement in the Central Baltic, based on Stratigraphical and Morphological Records from Eastern Småland and Gotland, Sweden.* LUNDQUA thesis, **25**.

TUTIN, T. G., HEYWOOD, V. H., BURGES, N. A., MOORE, D. M., VALENTINE, D. H., WALTERS, S. M. & WEBB, D. A. 1964–1980. *Flora Europaea.* Five volumes. Cambridge University Press, Cambridge.

WALLACE, A. R. 1880. *Island life: or, the phenomena and causes of insular faunas and floras, including a revision and attempted solution of the problem of geological climates.* Macmillan, London.

WATTS, W. A. 1985. Quaternary vegetation cycles. *In:* EDWARDS, K. J. & WARREN, W. P. (eds) *The Quaternary History of Ireland*, Academic Press, London, 155–185.

WEBB, D. A. 1983. The flora of Ireland in its European context. *Journal of Life Sciences Royal Dublin Society*, **4**, 143–160.

WEBB, S. L. 1987. Beech range extension and vegetation history: pollen stratigraphy of two Wisconsin lakes. *Ecology*, **68**, 1993–2005.

WEST, R. G. 1980. Pleistocene forest history in East Anglia. *New Phytologist*, **85**, 571–622.

WILKINS, D. A. 1984. The Flandrian woods of Lewis (Scotland). *Journal of Ecology*, **72**, 251–258.

Deglaciation, Earth crustal behaviour and sea-level changes in the determination of insularity: a perspective from Ireland

ROBERT J. N. DEVOY

The Coastal Resources Centre, Department of Geography, University College Cork, Ireland

Abstract: Insularity is defined here in terms of the presence or absence of land-bridge connections between adjacent landmasses. The development of a land-bridge depends upon the existence of a special combination of environmental conditions. These are created through the interaction of climate, sea-level and other physical environmental controls, together with changes in palaeogeography. Such factors operate together to produce conditions suitable for a land-bridge over only relatively short time-scales (10 to 100 years). New palaeoenvironmental data about such factors have emerged in recent years from the Irish Sea and from neighbouring regions. These data, derived from a variety of disciplines, have contributed to earlier discussions about the existence here of land-bridge links in the Late Quaternary. Some of this information, particularly that concerning patterns of Earth crustal behaviour, former ice distributions, deglacial history and sea-level changes, is reviewed. An assessment is given of their significance for possible land-bridge operation at this time.

The identification of former land-bridges forms a particularly complex problem in Quaternary science. A solution is to be seen, in simplest terms, in the study of the interaction of sea-level changes and of Earth crustal behaviour over time.

Devoy (1985) examined the possible land-bridge connections that may have formed between Britain and Ireland during the Late Quaternary. This followed from an earlier (1983) consideration of the patterns of the Post-glacial colonization of Ireland and, in part, of the possible role of land-bridges in plant and animal dispersal (Sleeman *et al.* 1986). The topic of plant colonization had been a particular driving force for the idea of a land-bridge here (Mitchell 1972). Also, the recognition that Ireland lacked many of the plants and animals found in both Britain and mainland Europe suggested the foundering of any such land connections by the early Post-glacial. The 'native' Irish flora (see Coxon & Waldren 1995) lacks up to 30% of the plant species to be found in Britain (Godwin 1975; Webb 1984), with the fauna similarly impoverished (Stuart & Wijngaarden-Bakker 1985; Stuart, 1986, 1995). Unusual distributions of other elements of the biota (namely the Lusitanian flora, the occurrence of *Rhododendron ponticum* in the Middle Pleistocene and of mammals and amphibia (Mitchell 1986; Mitchell & Watts 1970; Coxon & Waldren 1995; Sleeman *et al.* 1986)), further highlight the apparently disjunct and anomalous nature of Ireland's biogeography. Recognition of these characteristics has led in the past to the construction of hypotheses for plant and animal refugia, and of relict floras, as alternative explanations to that of a land-bridge (Godwin 1975).

Discussions of the plant and animal colonization topic by Mitchell (1960, 1963, 1972, 1976) and Synge (1985), emphasize the desirability of a land-bridge connection across the present Irish Sea in the Late Quaternary (Fig. 1). Consideration also extends to the possibility of connections via the Celtic Sea and the eastern Atlantic shelf areas to France and Iberia. Yet in spite of the biogeographical arguments for the operation of such land-bridge connections, and of improvements in the knowledge of plant and animal dynamics in the British Isles (see Stuart & Wijngaarden-Bakker 1985; Sleeman *et al.* 1986), the land-bridge question remained unresolved by the early 1980s.

Subsequently, new biogeographical evidence has been found (Preece *et al.* 1986; Devoy & Sinnott 1993). This re-emphasizes the possibility of the *connected* rather than the *separate* nature of Ireland's links with Britain in the Post-glacial pattern of colonization. Preece *et al.* (1986), in a study of a tufa deposit at Newlands Cross, County Dublin (Fig. 2), identified the early Post-glacial occurrence of *Apodemus sylvaticus* (wood-mouse). The apparent absence of this animal from Ireland until later in the Post-glacial was thought previously to indicate a human origin for the animal's arrival. The previous lack of any early record supported the case for an absence, or the early severance of any

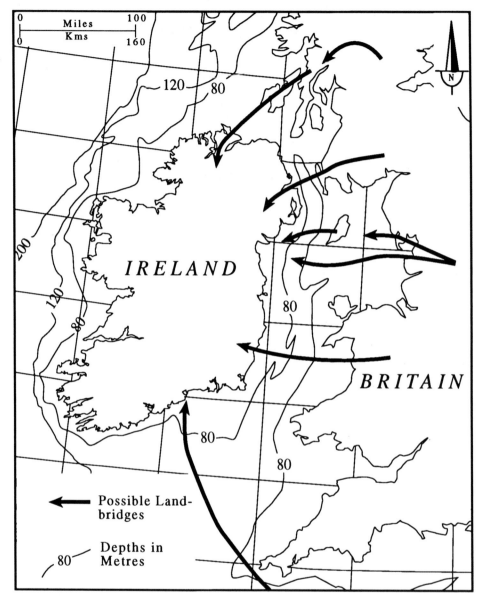

Fig. 1. Approximate locations of postulated Late Quaternary land-bridges between Britain and Ireland.

land connection. It should be noted, however, that the Newlands Cross site also shows the unequivocal presence of Mesolithic people in the area in the early Post-glacial, thereby maintaining a human connection with the wood-mouse record! More significantly, the site has provided good evidence for the timing of the land snail colonization of Ireland. This indicates a broadly synchronous pattern of arrival with that identified in Britain. The sequential order of immigration is very similar on both sides of the Irish Sea (Preece et al. 1986). Given this evidence, then it is difficult to explain the appearance of thermophilous land-snail species in Ireland (e.g. *Discus rotundatus*) without there being a land-bridge connection (Preece et al. 1986).

The value of this faunal and related dating evidence for the land-bridge record is problematic. As with most biogeographical information in this context, the evidence for a possible

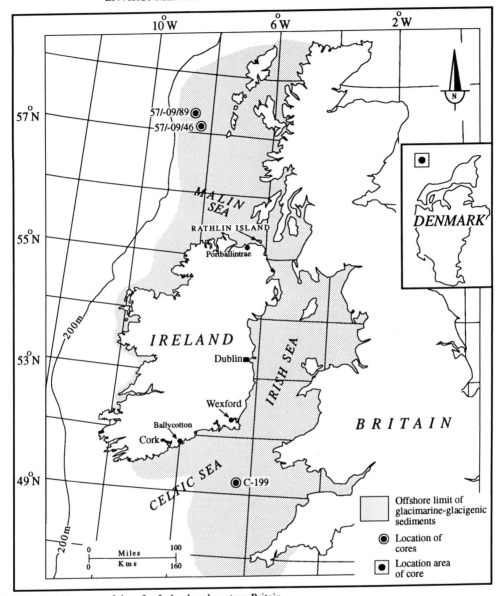

Fig. 2. Location map of data for Ireland and western Britain.

land connection is inferential rather than direct. At Newlands Cross only a few radiocarbon dates have been obtained, several with large standard deviations, which leaves open the possibility of a wide error margin in matching the timing of changes here with those in Britain. This view is particularly relevant given the current knowledge of variations in atmospheric carbon production and other controls governing radiocarbon results (e.g. Zbinden *et al.* 1989). A clearer picture on the comparability of sites and of age determinations needs to be established before firmer conclusions can be drawn from this type of evidence.

The geomorphological and geological record

The proposition of land-bridge operation by Quaternary scientists has stemmed from an appreciation of the interconnections between factors controlling relative sea-level changes and palaeobathymetry (Devoy 1985, 1986, 1989).

A pre-1970s image of the shelf seas separating

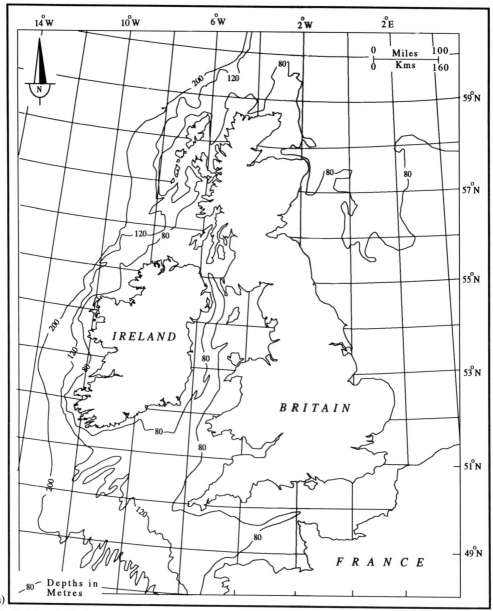

Fig. 3. (a) and (b) Generalized bathymetry of the shelf-seas neighbouring Britain and Ireland (more detailed bathymetric coverages are given by the British Geological Survey Sheets and others).

Britain and Ireland, admittedly simplified, was of a shallow water environment (Fig. 3a) (Mitchell 1960; Shotton 1962). This view was coupled with (1) speculation upon the likely further shallowing effects on palaeobathymetry from former glacial deposits now eroded (Mitchell 1963; Synge 1985) and (2) of ideas of a single, low global 'eustatic' sea-level at the end of the last glacial stage (see Devoy 1987a). Together such ideas fostered an understanding that a land-bridge must have existed across the Irish Sea, or possibly at other locations (Fig. 1) (Flint 1971; Godwin 1975).

This early view has been replaced by a more sophisticated appreciation of the controls operating in land-bridge development (Devoy 1985, 1987a). Survey information of the sea-floor topography around Ireland shows the bathymetry here to be more complex than previously supposed (Devoy 1989; Eyles & McCabe 1989a;

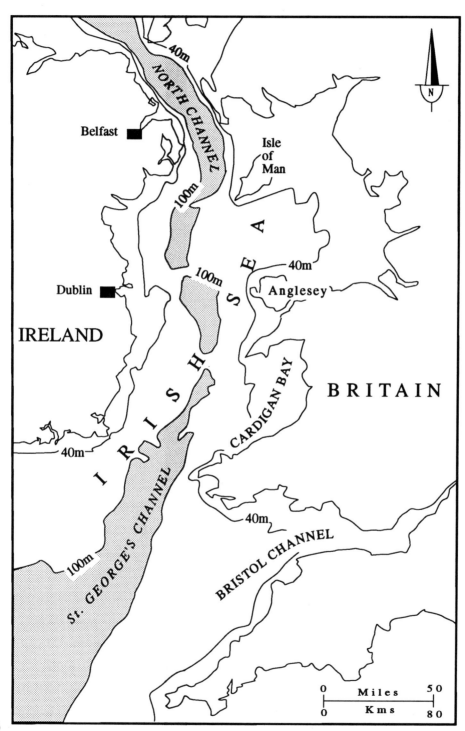

(b)

Wingfield 1989). Both isolated and possibly inter-connected trenches, 'hollows' and 'deeps', with intervening high points at varying elevations, dissect the sea-floor (R. Keary, pers. comm.). Importantly, from a land-bridge viewpoint, a bathymetric 'low' appears to run north to south from the Malin to Celtic Seas, throughout the length of the Irish Sea (Fig. 3b). This bathymetric 'low' is of variable width and is developed to depths of > -80 m to -100 m psl (present sea-level). However, as Fig. 3b illustrates, this is again a generalization and the real sea-floor topography is far from uniform.

This knowledge of the shelf-seas' bathymetry can now be coupled with new detailed sediments and seismic profiling data (British Geological Survey (BGS) bathymetric survey sheets; see e.g. Wingfield (1989) for references). Earlier information (Kidson & Tooley 1977) led to the view that at least parts of this north–south trough, and other neighbouring 'deeps', represent in-filled palaeochannels. The origin of the channels may have been polygenetic. They may have formed from earlier river systems at times of low sea-level and have been modified subsequently by glacial and deglacial processes (Whittington 1977). Wingfield (1989) and others (e.g. Hamblin et al. 1990; Eyles & McCabe 1989a) now hold a different view. Whatever the specific interpretations involved, these data sources together demonstrate that sea-floor areas around Ireland and Britain would have undergone continuous morphological changes throughout the Quaternary. The bathymetry of sea-floor areas is likely to have differed, in some degree, from the present-day appearance. Processes of sediment infilling and redistribution since deglaciation will have reduced the steepness of bottom slope angles and minimized the former depths of negative relief features.

These data, together with other palaeoenvironmental criteria, led Devoy (1985) to conclude that no good physical environmental evidence existed, at that time, to demonstrate the operation of land-bridge connections between Britain and Ireland. If a land-bridge 'had to be' then, on the interpretation of the physical environmental evidence then available, the most plausible position for such a link would have been to the north, from the islands of southwest Scotland to northern Ireland. This would have been at best, 'a disjunct, low and soggy route'. Former islands and possibly even large sandbars may have existed to form 'stepping stones' between the mainlands. This interpretation was purely conjectural and no firm data could be found to verify such a reconstruction.

What more may be added now to the earlier discussions? The Post-glacial Colonization Conference (Devoy et al. 1986) identified the need for the collection and analysis of much more palaeoenvironmental data on the land-bridge issue. These data are now becoming available from varying sources and may be summarized as follows:

1. Earth crustal behaviour and rheology, involving glacio-isostatic modelling of the British Isles and Northwest Europe (e. g. Boulton 1990; Lambeck et al. 1990; Lambeck 1991a, b, 1993a; Boulton et al. 1991; Johnston 1993).
2. Glacier and ice sheet boundaries – their behaviour and reconstructions for Ireland and Britain (e.g. Boulton et al. 1985, 1991; Ehlers et al. 1991; Austin & McCarroll 1992; McCarroll & Harris 1992; Boulton 1993; O'Cofaigh 1993).
3. Offshore mapping and seismic stratigraphy; sediment interpretation, borehole data (namely BGS (Wingfield 1995). (Information in readily accessible and published form is often difficult to obtain, given government and commercial restrictions on such information.)
4. Palaeoenvironmental interpretations of core data from 3 (e.g. Peacock & Harkness 1990; Peacock et al. 1992; Dickson & Whatley 1993; Scourse & Austin 1994; Wingfield 1994).
5. Sea-level data, at regional to global scales (e.g. Devoy 1987b; Tooley & Shennan 1987; Carter et al. 1989, 1993; Fairbanks 1989; Bard et al. 1990a, b; Boulton 1990, 1993; CEC 1990, 1993; Chappell & Polach 1991; Pirazzoli 1991; Tooley 1993; Shennan et al. 1994, 1995; Zong & Tooley 1995).
6. Biota: biogeographical reconstructions (e.g. Preece et al. 1986; Coxon 1993; Devoy & Sinnott 1993).
7. Cave exploration: (a) as repositories of plant and animal information and for sediment dating; (b) for the dating of faunal remains (e.g. A. J. Sutcliffe pers. comm.; P. C. Woodman pers. comm.) in Ireland.
8. Archaeology: early appearances and impacts of people in Ireland (e.g. Woodman 1985).

Despite these new research developments, the studies from both onshore and offshore, highlight how little unequivocal information exists. The gaps in thematic and regional palaeoenvironmental knowledge remain large.

Controls on land-bridge operation

Accurate palaeogeographic reconstructions are dependent upon there being a good understanding of the physical boundary conditions operating in past environments. Essential also is the availability of abundant and good-quality data, together with a clearly defined methodological and conceptual framework for data collection and analysis (Plag et al. 1995). In proving the existence and extent of former land-bridges from now-submerged environments, then these conditions appear almost as an impossibility.

For land-bridge identification two criteria remain important. (1) The use of present-day bathymetry and palaeomorphology, inferred through geomorphological and geological studies, may be quite erroneous in defining the existence of land-bridges. (2) The operation of individual environmental factors and of associated processes of change require treatment not as single parameters but as parts of an integrated complex (see Gregory (1978) and Thorn (1988) for discussion of relevant geomorphological methodologies).

Following from these points a number of factors may be defined for this shelf-seas region as controlling land-bridge existence and the possible time-window for operation:

(a) tectonic changes;
(b) earth rheology (at areal and regional levels);
(c) glaciation – ice presence, limits, shape and deglacial history (regional to far-field scales);
(d) sea-level behaviour (regional to local scales), in both relative and absolute terms;
(e) palaeobathymetric changes – through erosional and depositional histories;
(f) climate change – defining the limiting conditions for the movement of biota, e.g. temperature;
(g) coastal processes (linked with a–e), to define the palaeocoast and the development of coastal structures and landforms;
(h) controls on the biota – taphonomy and ecology.

The role of deglaciation

One of the most controversial of factors affecting the problem of insularity in Ireland and western Britain is that of the glacial and deglacial histories of these regions (e.g. Ehlers et al. 1991).

The application of stratigraphic and sediment studies, amino acid, ^{14}C and other dating techniques (Bowen et al. 1986) has led to the identification of a pattern of rapid ice movements here in the Late Quaternary (Eyles & McCabe 1989a, b; McCabe 1986). Major ice masses over Ireland and the neighbouring shelf seas are seen as undergoing times of wasting and then reforming before c. 30 000 BP (McCabe 1987; Coxon 1993). After an ice maximum at c. 20 000–22 000 BP (Glenavy Stadial) (Fig. 4), ice-wasting began again, accompanied by major ice surges (Eyles & McCabe 1989a, 1991). Ireland may have become ice-free between c. 14 000 and 16 000 BP (see Ehlers et al. (1991) for discussion). Here, the later Nahanagan Stadial was probably of much less significance, in ice-mass terms, than ice associated with the contemporary Loch Lomond Re-advance in Britain (Gray & Coxon 1991).

From sediment facies and stratigraphic work, Eyles, McCabe (Eyles & McCabe 1989a, b, 1991; McCabe & Eyles 1988) and others have suggested the occurrence and importance of glacio-isostatic depression at ice-margins during these times, affecting the present shelf-seas zones (Fig. 4). Sediments formerly interpreted as deriving from processes associated with land-based ice are now identified as glacio-marine. These are seen (e.g. in the Irish Sea region) as having formed in the tidewater margins of glaciers, as marine water again penetrated the isostatically depressed floors of former sea areas. The amount of isostatic depression over Ireland has been placed, arguably, at values as high as >150 m (Eyles & McCabe 1989a, b, 1991) (Fig. 4). These figures have been based upon a number of lines of evidence. These include the recognition of higher level shorelines than formerly identified, arguments for sediments being glacio-marine in origin and the interpretation of shorelines cut into glacial sediments, and also of sedimentary structures (e.g. delta complexes), as graded to former high relative sea-levels (Eyles & McCabe 1989a; Carter 1993; McCabe et al. 1993, in press). More recently, a re-evaluation of these data has occurred and the amount of isostatic uplift is now seen as likely to be less than the published maximal values (A. M. McCabe, pers. comm.).

Following from work in Canada and the Arctic on the deglaciation of isostatically depressed ice-margins (e.g. Andrews, 1982, 1987; Belknap, 1994; England 1994; Stea et al. 1994), Eyles & McCabe (1989a, b) have suggested a relatively rapid end to ice in the Irish Sea region. In their view, early marine flooding under a rising sea-level began a catastrophic feedback in the process of ice-wasting. This

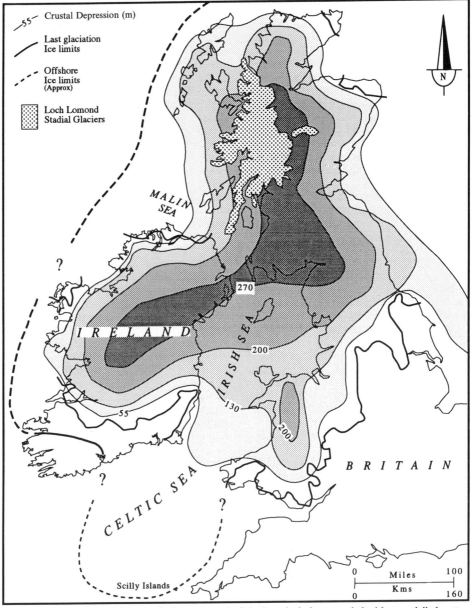

Fig. 4. Regional land and offshore ice limits at the end of the last glaciation, coupled with a modelled pattern of accompanying isostatic crustal depression (sources include Eyles & McCabe (1989a, b, 1991); McCabe (1987), Synge (1981) Warren (1991); Denton & Hughes 1981; Boulton et al. 1977, 1985, 1991.)

resulted in ice surges and finally in deglaciation. Marine incursion may have occurred initially via the topographic 'lows' in the sea-floor. Subsequent to the onset of ice-wasting, the regional sequence of environmental changes would have been of land emergence and relative sea-level fall. Later, with the continued return of water to the ocean basins (Bard et al. 1990a, b), this was followed by a relative sea-level rise from low levels between c. 12 000 and 14 000 years BP (i.e the glacio-eustatic signal now dominant). This interpretation remains consistent with the pattern of environmental changes identified from earlier shoreline studies in the region (Carter 1982; Devoy 1983; Carter et al. 1989, 1993; Synge 1977).

The shelf-seas around Ireland and western Britain are shown to record the occurrence of thick sequences of glacigenic sediments (Pantin & Evans 1984; Ehlers *et al.* 1991; Peacock *et al.* 1992). These are interpreted as being probably glacio-marine in origin, possibly resulting, in part, from conditions of floating ice. This ice may have been in the form of pack, or shelf-ice, developed beyond ice-grounding limits located in the Irish Sea and closer to the land-margins. Along the south coast of Ireland, interpretation of the Irish 'shelly' till as representing glacio-marine facies is also becoming established (Hoare 1991; Warren 1991). This view is strengthened by recent work upon the coastal stratigraphies in this region (O'Cofaigh 1993). Further support is given from the Quaternary sediments found offshore in the Celtic Sea. Here, Pantin & Evans (1984) have identified the existence of glacial sediments, again attributing them to iceberg or other glacio-marine origins. Others (e.g. Scourse *et al.* 1990, 1991), however, discount shelf-ice as a possible explanation for these Celtic Sea glacio-marine sediments. The Celtic Sea sequences may be equated with a Late Quaternary origin (i.e. *c.* 18 000–19 000 years BP), associated possibly with ice expansion southwards from the Irish Sea (Synge 1981, 1985; Scourse 1991*a*,*b*). This ice is seen as affecting the Isles of Scilly and extending further westwards over an isostatically depressed shelf surface (Fig. 4). Possible ice-grounding lines in the Celtic Sea are identified at levels of *c.* −135 m psl (Scourse *et al.* 1991).

Dating of these shelf-sea and neighbouring onshore sedimentary sequences as a whole remains problematic, as does fitting the different areal patterns into the Quaternary event stratigraphy (McCabe 1987; Warren 1991; Coxon 1993; O'Cofaigh 1993). Radiocarbon ages for the Irish 'shelly' till at Ballycotton Bay, County Cork (Vernon 1989; Bowen 1991) at or later than *c.* 17 000 years BP – if correct and at all representative of the wider situation – may support a linked timing of glacio-marine deposition from the south coast of Ireland with that of the Isles of Scilly (Scourse *et al.* 1990, 1991). Accurate radiometric dates from all the sequences, however, are few. Therefore, the errors in defining the timing and the correlation of events are possibly large. More significantly, this widespread recognition of glacio-marine deposition from the Late Quaternary poses fundamental questions about the former palaeogeography and environmental characteristics of the western British Isles at this time – in particular, about the extent and timing of ice-loading throughout the region, glacio-isostatic response patterns, crustal deglacial history and of relative sea-level rise (Boulton *et al.* 1977, 1985, 1991; Boulton 1990).

The recognition of glacio-marine sedimentation, with accompanying arguments for land emergence in the south of the Irish Sea to high levels of +150 m OD (Belfast), is also problematic. The apparent absence south of Dublin of any good evidence for emerged shorelines (Synge 1985) unequivocally dated to the end of the last glacial stage, disputes the existence, or at least the envisaged scale, of a major isostatic depression of the region (Fig. 4). By comparison, glacial reconstructions involving ice in the form of floating ice here would not require the existence of glacio-isostatic loading; however, the presence of land-based ice-masses would. Importantly, in the Isles of Scilly and in neighbouring areas to the south of this region, no evidence has been found of Late Quaternary shorelines above present sea-level. As Scourse *et al.* (1990, 1991) suggest, therefore, any ice loading that took place over these shelf-sea zones at this time must have resulted from relatively thin ice. Shorelines here that might relate to any such former ice loading all occur at elevations below present sea-level.

In southeast Ireland, the implication of a possible marine shoreline at *c.* +70 m OD (Eyles & McCabe 1989*a*, 1991) would surely be that a wider, staircase-type shoreline sequence should exist along this coastline. Other glacio-isostatically loaded regions commonly display such features (Devoy 1987*b*). The absence of Late Quaternary high-level shorelines on the south coast of Ireland, in the light of the identification of glacio-marine sediments here, is thus puzzling. Even if ice palaeogeography prevented, for a time, the formation of shorelines close to the ice front (see Devoy 1983), extensive raised shoreline features should have been formed away from the ice-margins. It is likely that substantial fragments of these, at least, should still be preserved today on present land surfaces. In such situations, the formation of shorelines at former ice-margins will depend on a complex of controls. These will include such factors as the types of coastal processes in operation, times of land exposure to wave activity, and the existence of suitable geological and geomorphological conditions to allow shoreline development to take place. It may be that a suitable combination of these factors, to enable the formation and subsequent preservation of shorelines, did not exist in these south coast and Celtic Sea areas.

Turning to a related question, the issue of the interpretation and origin of glacio-marine sediments in the shelf-seas also needs further study. Do such sediments represent *in situ* glacio-marine deposition, or have they been formed under marine conditions at all? The work of McCarroll & Harris (1992) and Austin & McCarroll (1992) provide other possibilities to the glacio-marine tidewater ice marginal interpretation of Eyles & McCabe (1989a, b, 1991). An interpretation of these glacigenic sediments as equally representative of formation in relatively freshwater environments (e.g. ice-dammed lakes) provides another hypothesis (Warren 1990). However, this 'freshwater formation' scenario of Austin & McCarroll (1992) and others may be of local application only in the Irish Sea region. In a related context, the idea that the deglaciation of the Irish Sea was triggered by a linked climate change and sea-level rise mechanism is also open to question. This mechanism is envisaged as having operated in driving the final deglaciation of the Hudson Bay region (Andrews 1987; Devoy 1987b; England 1994). In the Irish Sea the land surface–bathymetric shapes, the former ice characteristics and other physical boundary controls are likely to have been very different. The two situations may not have been at all analagous (e.g. Wingfield 1992a).

An important lesson from these differing interpretations of the new sediment and stratigraphic data relating to former ice patterns, is that the palaeoenvironmental reconstruction of the region is still far from being precise. Many gaps in the data exist, particularly for the accurate interpretation of offshore sedimentary records. If speculation on regional stratigraphies and palaeoenvironmental interpretations in Ireland and the Irish Sea region is to end, then real progress is still required on the problems of dating events. In this context, the placing of land-ice in the final Late Quaternary glacial stages on the south coast of Ireland would likely change the isostatic picture here and thereby facilitate glacio-marine deposition. But is there good evidence for it? Reconstructed ice limits, ice thicknesses and whether ice-masses formed as grounded ice or were in part floating, would all have a bearing on the operation and timing of regional isostatic recovery and of crustal forebulge operation. If crustal forebulge migration did occur, then the development of this may in turn have had an interactive role with the form of marine water access into the region. Such a role needs precise definition if realistic inferences for former land-bridges are to be made.

Land–sea changes and crustal rebound

Of central importance in any of the controls involving the operation of land-bridges must be factors of land and sea interactions (see fig. 2.5 in Devoy (1990)).

The reconstruction of area to regional scale relative sea-level movements within the British Isles is dependent upon the behaviour of ice masses, glacio-isostasy and of Earth crustal responses to watermass exchanges between the land and oceans (hydro-isostatic changes) (Chappell 1974; Clark et al. 1978; Clark 1980a, b; Devoy 1987a; Mörner 1987). The fundamentals of these controls and of their operation at particular points on the Earth's surface have been defined by Lambeck (1990, 1991a, 1993a), as follows:

The positions of sea-levels at latitude ϕ, longitude λ and time t are observed as:

$$\Delta\zeta(\phi, \lambda: t) = \zeta(\phi, \lambda: t) - \zeta(\phi, \lambda: t_o) \quad (1)$$

where t_0 is present-day sea-level position.

In the absence of vertical tectonic movements then these positions are the sum of four main terms:

$$\Delta\zeta(\phi, \lambda: t) = \Delta\zeta_e(t) + \zeta_r(\phi, \lambda:t) +$$
$$\Delta\zeta_i(\phi, \lambda: t) + \Delta\zeta_w(\phi, \lambda: t) \quad (2)$$

where $\Delta\zeta_e(t)$ = the equivalent sea-level rise or change in ocean volume/ocean surface area; $\Delta\zeta_r$ = sea-level change resulting from changes in gravitational potential, by redistribution of ice and water loads; $\Delta\zeta_i$ = crustal rebound produced by response to glacial loading and unloading, together with additional effects to earth gravitational changes from load redistribution; $\Delta\zeta_w$ = responses to meltwater loading and unloading. Since more than one ice-sheet contributes to sea-level change over time (i.e. ice-sheet $n = 1...N$) at a site, then equation (2) can be rewritten as:

$$\Delta\zeta = \Delta\zeta_e + \sum_{n=1}^{N}\left\{\Delta\zeta_r^{(n)}(\phi, \lambda: t) + \Delta\zeta_i^{(n)}(\phi, \lambda: t) + \Delta\zeta_w^{(n)}(\phi, \lambda: t)\right\} \quad (3)$$

The solution of terms in equations (2) and (3) requires quantitative models of: (a) ice-sheet coverage, growth and decay (this will require accurate quantification of ice thicknesses in space and their changes with time); (b) accurate spatial representation of Earth mantle–crustal structure, viscosity and their subsequent behaviour under changing loads; (c) models of coastal geometry.

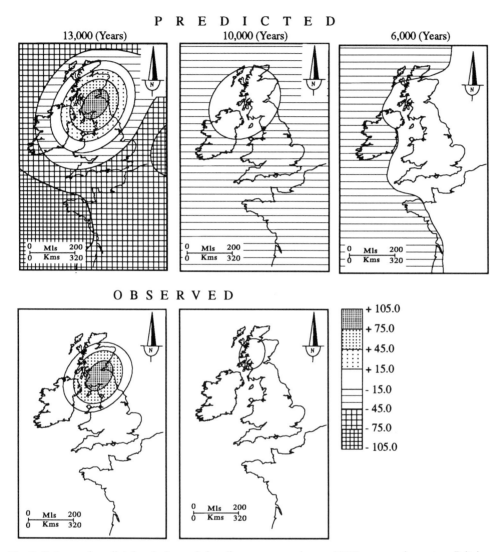

Fig. 5. Patterns of predicted and observed shoreline emergence since c. 13 000 years BP in western Britain and Ireland (sources include Lambeck 1991a, b, 1993a, b).

Regional reconstructions

In northwest Europe, the most recent results of Earth crustal modelling (Lambeck 1991a, b, 1993a; Lambeck et al. 1990) show shoreline patterns for western Britain and Ireland following deglaciation as consistent broadly with earlier regional reconstructions (Flemming 1982; Sissons 1983; Shennan 1983, 1989a, b; CEC 1993). The modelled approaches show rapid land uplift over highland Britain, declining towards the Western Isles and the north of Ireland (Fig. 5).

From the Irish Sea region, a more complex pattern of marginal land uplift is indicated (e.g. from c. 13 000 years BP onwards), though much depends in reconstructions upon the particular Earth model used. Land emergence in the northern Irish Sea is replaced southwards, toward the Celtic Sea and southern Britain, by increasing values of relative land submergence post-16 000 years BP (Fig. 5). Lambeck (1991a, b; 1993a) shows, however, that substantial parts of present-day shallow-water areas in the British Isles (i.e. areas now above c. −45 m psl) such as the Dogger Bank in the southern North Sea, would have been exposed land at c. 13 000 years BP. In the Irish Sea it seems reasonable to assume that the widespread distribution of freshwater peats on the coastal margins (e.g. Tooley 1978, 1982, 1985; Eyles & McCabe 1989a; Kidson & Tooley 1977; Zong 1993; Zong & Tooley 1995) would imply similar land exposures here at this time also. Modelled reconstructions of palaeoshorelines in the St George's Channel area support this view (Lambeck 1993a,b). These may indicate the existence of extensive areas of exposed land in this region at between c. 13 000 and 14 000 years BP: the appearance of the long-hoped-for land-bridge?

In this context, work upon interstadial pollen spectra from long cores (>16 m length) at Clogheenmilcon (Blarney), County Cork, records the occurrence of thermophilous plants here prior to 11 240 ± 90 years BP (uncalibrated ^{14}C age) (Devoy & Sinnott 1993). Above these levels the pollen spectra indicate a return to cold, stadial conditions and later a full Post-glacial pollen sequence. Before this date, results show the development of significant *Pinus*, *Betula*, *Corylus* and *Quercus* pollen values, together with those of other woody plants. The appearance of this pollen assemblage at this time (between c. 11 200 and 12 500 years BP) seems anomalous (e.g. Andrieu et al. 1993). Although there may be no connection with the modelled offshore 'land' areas before c. 13 000 years BP, it is interesting to speculate upon a connection. If the pollen data are interpreted correctly, then are these thermophilous plants here on the south coast of Ireland, at this early time, the result of a southern land connection? If this now-offshore 'land' existed, would such areas have acted as a bridge to the Continent, allowing the early immigration of plants, or could they have acted as a refugia for such plants?

These recently modelled reconstructions (Lambeck 1993b) appear perhaps to be at variance with earlier reconstructions (Lambeck 1991a,b, pers. comm.). Earlier work shows the submergence of the Celtic Sea region below c. −15 to −45 m psl. This would infer the establishment at this time of marine connections northwards, particularly along topographic 'lows', into the Irish Sea. It should be noted, however, that the Irish and Celtic Seas together occupy a clear crustal response transition zone in modelled shoreline behaviour. Further, the modelled reconstructions are not tuned to palaeobathymetric changes resulting from coastal erosion and other geomorphological changes.

As established in the model constraints in these reconstructions (Lambeck 1990, 1991a,b; Lambeck et al. 1990), construction of the driving components to equation (2) form powerful determinants to model accuracy. At the regional scale of northwest Europe some discontinuity exists between predictions for near-ice and far-field sites. Further, the inbuilt error limits on the models' component terms, coupled with their insensitivity to showing other smaller amplitude controls on sea-level behaviour (Devoy 1991), must together increase the error margins in fixing shoreline positions. Important in these models is the truthing of the predictions with empirical data. This is based upon comparisons with relative sea-level index points (Pirazzoli & Grant 1987). Lambeck et al. (1990; also Lambeck 1993a) suggest that, as a whole, the modelled shoreline values for northwest Europe provide a reasonable explanation for the observed data (mainly post-10 000 years BP) for the different regions. Differences with the empirical data do exist though, as instanced for the shoreline reconstructions for western Scotland (e.g. Shennan et al. 1995 in press). In these cases the model accuracy may be improved at particular locations by fitting different rheological parameters and/or ice thickness and behaviour patterns (Peltier 1980, 1987, 1991; Peltier & Andrews 1983). Where discrepancies still remain, then other causes (e.g. tectonic controls) may be involved.

The role of empirical shoreline data

Former shoreline studies, restricted mainly to land-based records, have had a limited power in helping resolve the patterns of land emergence and deglaciation history prior to c. 12 000 years BP (Eyles & McCabe 1989a, 1991; Wingfield 1995). Environments which should contain relevant information for the period following deglacial episodes after c. 20 000 years BP lie offshore. These records have been inaccessible to most shoreline researchers, hitherto restricted by limited physical resources to working within the contemporary shore zone.

Research undertaken after 1985 has, however, extended considerably the available database throughout the region (western Scotland and Ireland) on shoreline heights and ages. Reporting of these data and accompanying reviews are contained in, for example, Carter et al. (1989, 1993), CEC reports (1990, 1993), Shennan et al. (1994, 1995) and Wingfield (1994). In spite of this, the recording of high-level, raised shorelines outside of Scotland is limited. McCabe et al. (1993, 1995) have reported upon marine sequences at +80 − +90 m OD at Portballintrae (Fig. 2), dated to c. 17 000–18 000 years BP. This together with earlier work in the north of Ireland

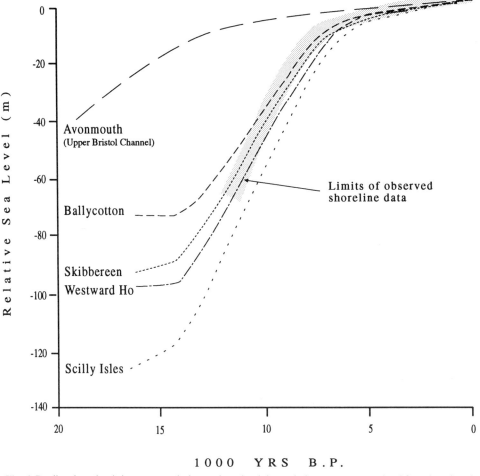

Fig. 6. Predicted sea-level signatures and observed sea-level data relative to present sea-level from locations in the south of Ireland and southwest Britain (sources include Lambeck (1991a, b, 1993a, b; Carter et al. (1989) and CEC (1990, 1993)).

(e.g. McCabe & Eyles 1988), is taken as support for higher values of glacio-isostatic control here than was previously thought to have existed (see also Boulton et al. 1991; Carter 1993).

Collectively, these data still fit with earlier regional interpretations (Carter 1982; Devoy 1983; Sissons 1983; Dawson 1984). In the north of Ireland and Scotland, rapid isostatic emergence took place prior to c. 12 000 years BP. Thereafter, this is overtaken by faster rates of water return to the oceans (glacio-eustacy) until c. 6000 years BP, when glacio-isostatic emergence reasserted itself. Such explanations, as Carter et al. (1989) point out, may be complicated at area levels by other factors, including tectonic and geoid changes. In western Ireland, sea-level index points show no new evidence of isostatic emergence south of Clew Bay (Devoy & Delaney 1995) though studies here have been concentrated primarily on the late Holocene.

In the critical region, incorporating the south of Ireland, southwest England and the Celtic Sea, integrated sea-level and coastal process studies have established more than 35 ^{14}C index points from the mid- to late Holocene and earlier (Carter et al. 1989; CEC 1990, 1993; Healy 1993). These data fit closely the predicted Earth crustal behaviour–shoreline curves of Lambeck (1991a, b, 1993a) (Fig. 6). Whilst this information can help little with early late-glacial shoreline patterns per se, they do suggest that the crustal behavioural assumptions of Lambeck's model for the region are not all wrong. Recent research by Shennan et al. (1994a, b, 1995) in

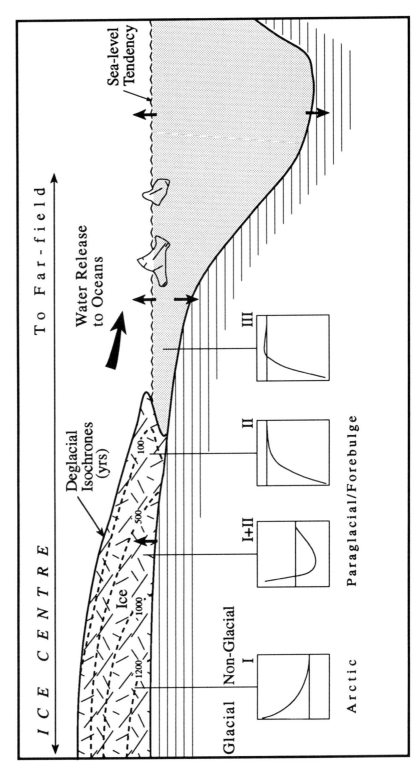

Fig. 7. Generalized trends of expected relative sea-level behaviour, derived from Earth crustal models, from zones of ice loading to far-field locations.

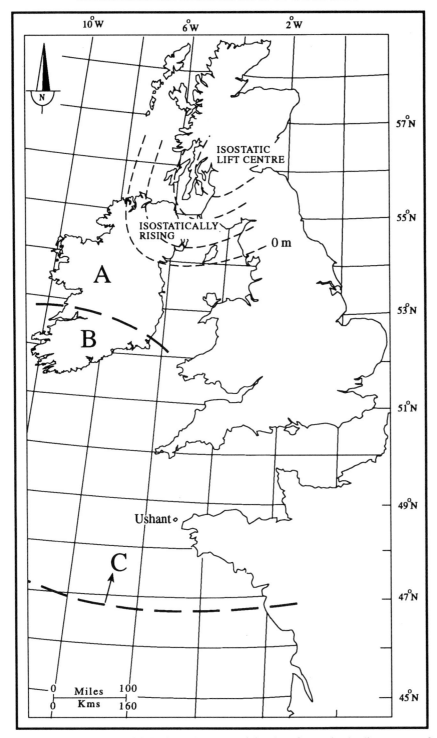

Fig. 8. Patterns of Earth crustal behaviour: **A**, zone peripheral to former ice loading centres of presumed forebulge operation (from Carter 1982; Devoy 1983), **B**, zone of observed Holocene shoreline submergence and late Holocene crustal stability (from Pugh 1981; Carter et al. 1989), **C**, limit of anticipated forebulge peak (from Wingfield 1995).

western Scotland show also that new, precise, sea-level index points and the modelled shoreline patterns are generally conformable here as well. However, differences in the crustal uplift shown by the two approaches do exist and need to be explained.

Evidence of crustal forebulge

The shorelines modelled for the Irish Sea region and southwards (Fig. 6) do not show sea-level signatures characteristic of significant forebulge operation (Devoy 1987a) since the late-glacial period (Fig. 7). Neither do these predictions confirm a definite elevation of land levels sufficient to allow land-bridge existence in this region after c. 10 000 years BP (Fig. 5) (Wingfield 1992b, 1995). The predicted shorelines and empirical data together indicate the submergence of the region in Post-glacial time; as might be expected from a sea-level signature, zone II-type Earth crustal behaviour (see Clark et al. 1978). If this view is correct, then it is useful to define some of the possible implications of these linked shoreline and geophysical reconstructions for the likelihood of forebulge operation in these regions. These imply that: (1) a 'bulge' mechanism as such did not operate in this region (see Boulton (1990) for further discussion); (2) any forebulge collapse had 'peaked' and passed northwards towards ice-loading centres before 10 000–11 000 years BP, or even earlier (see Wingfield 1994, 1995 for further discussion); (3) the zone of any forebulge operation in the late-glacial period (after 12 000 years BP) lay to the north of the south coast of Ireland and the Celtic Sea region.

Different authors (Carter 1982; Devoy 1983; Eyles & McCabe 1989a, 1991; Wingfield 1992b, 1995) have placed the operation of a crustal forebulge zone in different areas (Fig. 8). These range from south of the Celtic Sea to central Ireland and up into the northern Irish Sea. Such zonations have been based only upon interpretations of standard geophysical views (e.g. Daly 1934; Clark et al. 1978) and upon tentative field data (Wingfield 1994, 1995). In comparison, others (e.g. Boulton 1990; Lambeck 1993a) appear not to recognize the operation of a forebulge component in the crustal response to ice loading and shoreline recovery over the British Isles. This reflects the different types of Earth and linked ice reconstruction models envisaged as applicable to northwest Europe by these authors. Over Britain and Ireland the degree of ice loading is seen as relatively small and its effects restricted to the lithosphere. The factors of hydro-isostatic and sediment loading are not always considered fully in these reconstructions (Johnston 1993).

These apparent differences in views upon Earth rheological behaviour are a matter primarily of geophysical investigation. Of central importance in resolving disagreements will be improvements in understanding the appropriate Earth rheological models to apply at local to regional scales of resolution (see Plag et al. 1995, for discussion). Of equal importance, at all spatial scales, is a need for greater accuracy in the reconstruction of ice-mass behaviour, dimensions and chronology.

Offshore shoreline records

The acquisition of additional early (i.e. pre-10 000 years BP) shoreline information to help test the modelled reconstructions of Earth crustal behaviour is essential now. Reliable information of this type is scarce, though some data exist. The sea-bed and related sediment surveys of the northwest European shelf-seas since the 1980s have produced valuable information (e.g. Dickson & Whatley 1993). The analysis of seismic profiles and the production of three-dimensional modelling of sediment structures from the North Sea (e.g. Kay 1993; Boulton 1993) might show the effects of sediment and water loading in inducing a form of 'bulge' operation in this region. Boulton (1993) argues that a complete reconstruction of Late Cenozoic sea-level history is possible from such records.

Less explicable are the results of biostratigraphic and palaeoenvironmental reconstructions from sediment cores in the shelf areas (e.g. Peacock et al. 1992; Rose et al. 1993; Scourse & Austin 1994). These studies indicate early and high elevations for shallow-water environments at different points around the British Isles (Fig. 2). The data, together with the reinterpretation of known sea-bed structures (Wingfield 1994) have led to explanations of their origin through the effects of ice unloading and peripheral forebulge migration and collapse (Wingfield 1992b, 1995; Rose et al. 1993).

A core from the Celtic Sea (BGS vibrocore 51/-07/199; 51° 20′ N 06° 15′ W) shows the development of shallow water, intertidal sediments at a height of c. −123 m psl. A radiocarbon AMS date on a shell (*Spisula elliptica*) at this level gave a calibrated age of 11 165 ± 110 years BP (Scourse & Austin 1994). Comparison with the estimated elevation of the sea surface from the Barbados sea-level curve, showing a glacio- and hydro-eustatic signal (Fairbanks 1989; Bard et al. 1990a,b; see also Shackleton

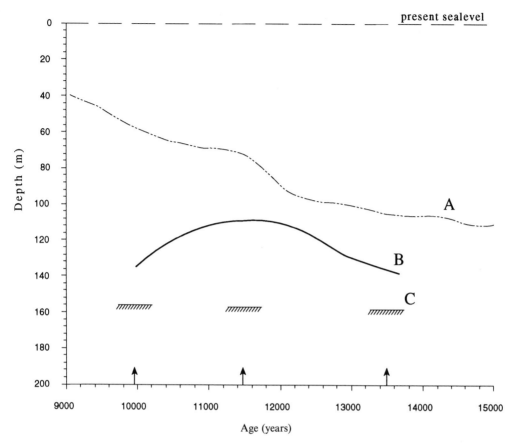

Fig. 9. Relative trend of sea-floor movement from the area of cores 57/-09/89 and 57/-09/46, the Outer Hebrides: A, pattern of ocean water-level movements taken from Barbados, ±2.5 m psl (Fairbanks 1989; Bard et al. 1990a,b); B, sea-floor behaviour for core areas relative to curve A; C, present depth of recorded sediments from core areas from c. 10 000 to 13 500 years BP).

1987; Chappell & Polach 1991), places ocean-surface levels at c. −60 to −70 m psl at this time. Such comparison, although inaccurate, indicates that the sea-floor at the Celtic Sea site has undergone downward movement (subsidence) since that date. Another AMS date a few metres up the same core gives a date of 8425 ± 100 years BP, the palaeoenvironment showing deepening water conditions.

Another location west of the Outer Hebrides (Peacock et al. 1992) (Fig. 2) provides palaeoenvironmental evidence from two vibrocores (BGS cores 57/-09/46 and 57/-09/89) taken from c. −155 m psl. Biostratigraphic work on the contained foraminifera and other faunas in the cores, coupled with sequential dating of the sediments, shows a detailed picture of inferred water-level changes here since c. 15 500 years BP.

Sedimentation is interpreted as beginning in shallow marine environments (c. 30 m water depth) at c. 13 500 years BP. Conditions are taken as representing a proximal glacio-marine and high Arctic environment. Subsequent sedimentation from the late-glacial to the early Postglacial period, records water depth changes under continued glacio-marine conditions. These range from water depths of <40 m at c. 11 500 years BP, deepening to c. 50–60 m by c. 10 000 years BP. Again, comparison with sea-level data from Barbados (Fairbanks 1989; Bard et al. 1990a,b) indicates ocean basin water-levels as much higher at this time. By inference, this sea-floor location has also undergone net subsidence (Fig. 9). As Peacock et al. (1992) point out, the rates of water-depth increase shown in these cores from 13 000 to 10 000 years BP match those from the Barbados sea-level change signal. This is taken to indicate that the core positions

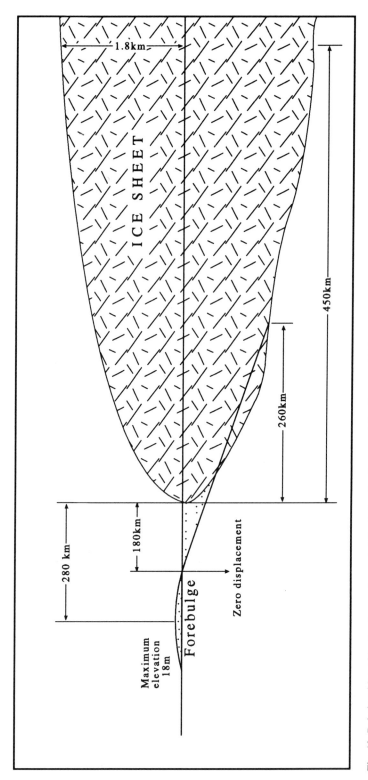

Fig. 10. Relationships of ice loading and expected forebulge development (after Dawson 1992).

show a real post-depositional subsidence trend. The palaeoenvironmental picture drawn by Peacock et al. (1992), for deposition under glacio-marine environments, compares closely with the conditions evidenced in the Celtic Sea (Scourse et al. 1991; Scourse & Austin 1994).

In the same broad region as the cores from the Hebrides, but southeastwards to Rathlin Island (northern Ireland) (Fig. 2) new data from a raised shoreline give a ^{14}C age of c. 12 300 years BP on *Littorina littorea* shells (Carter 1993). The raised beach sequence here is developed at altitudes of between +5 m to +10 m OD (Belfast). This sea-level index point occurs at much higher levels than shown by earlier data in the north of Ireland for this time (Carter 1982). Interpretation of this information serves to underline the positive crustal emergence trend for the region. However, the recovery of what must have been an isostatically depressed crust here at c. 12 000–12 500 years BP (Boulton et al. 1991) has been distinctly differential in form. The Rathlin Island data support possibly a greater degree of land emergence for the region than was formerly thought to have occurred. This contrasts with an apparently real subsidence trend for the contemporaneous shallow water environments c. 150 km westwards, away from ice load centres in Scotland (i.e. cores 57/-09/46 and 57/-09/89, Fig. 2). Explanation for this may fit the idea of peripheral forebulge collapse (see Figs 7 and 10). Equally, it may evidence other causes of subsidence, such as those originating from geoid changes and tectonism.

A third site recently discussed (Rose et al. 1993) is situated in the Danish Sector of the southern North Sea (Fig. 2). Biostratigraphic studies of a vibrocore here show a complex of freshwater and marine sediments at present depths of −62 to −65.5 m psl, developed between 10 500 and 12 200 years BP. Studies indicate marine sediments forming here at relatively high elevations in contrast to the contemporary ocean-basin signal from Barbados. Explanation for the continuity of heights and shallow marine signals at between c. −62 and −65 m psl into the Post-glacial period, may involve the migration and peaking of a crustal forebulge, moving toward the Scandinavian ice centre.

Forebulge operation

These new data and their linking by Wingfield (1992b, 1995) to the existence of a land-bridge between Britain and Ireland brings to the fore the different geophysical arguments for earth crustal behaviour under ice loading. Recognition of the main responses possible to such loading was made by Daly (1934). The concept of a 'peripheral bulge' and of 'forebulge migration' following deglaciation, as popularly adopted subsequently in Quaternary literature (e.g. Lowe & Walker 1984; Dawson 1992), was contrasted with an alternative 'punching' hypothesis. Here, the rheological properties of the Earth's mantle (a uniform mantle viscocity is inferred) were seen as necessitating a much lower degree of peripheral bulge and the subsequent *in situ* dissipation of this with deglaciation. Daly suggested that, at that time, no good evidence existed to support the former operation of the 'bulge' hypothesis. Work from Antarctica (Drewry 1983) indicates, however, the existence here of a possible forebulge zone, but little data exist on the dimensions or behaviour of this feature. In North America, shoreline data are also taken as evidencing regional forebulge operation since deglaciation (see e.g. Devoy 1987a; Gehrels et al. 1993; Josenhans 1994).

Since Daly's work, discussion of Earth rheology and glacio-isostatic crustal rebound has returned repeatedly to these same problems (Mörner 1980, 1987; Sabadini et al. 1991). Cathles (1975, 1980) reviewed the necessary properties of crustal thickness, strength and mantle composition for forebulge operation. Different mantle rheologies and viscosities are shown to explain equally the existence of crustal forebulge. Forebulge operation should result theoretically from mantle rheology states ranging from (1), a relatively fluid upper mantle, or where linear-fluid flow is restricted to an upper mantle channel, to (2), where mantle rheology is non-linear and viscosities vary with depth. The absence of forebulge type crustal behaviour could be expected for an Earth mantle with Newtonian flow characteristics and where fluid viscosity is held constant.

The Earth models of Walcott (1970, 1972, 1973, 1980) and others (e.g. Clark et al. 1978; Clark 1980a), which together have been seminal in linking Earth crustal behaviour to expected shoreline recovery patterns, have all adopted forebulge operation. Here, the Earth's behaviour is modelled as that of a viscoelastic Maxwell-type material. Mantle rheological properties are seen as variable, but with mantle viscosity held uniform with depth. The probable shortfalls in this Earth behaviour are understood (Plag et al. 1995). As illustrated by subsequent research (e.g. Peltier & Andrews 1983; Peltier 1987; Lambeck 1990; Nakada & Lambeck 1991; Sabadini et al. 1991), this view of Earth rheological characteristics is seen as oversimplified.

Model reconstructions may be further complicated by the multiple-loading histories of the Earth's crust by ice and water, the exact nature of which are uncertain (Peltier & Andrews 1983; Peltier 1991). In the Earth models in current use, problems arise from the different Earth environmental assumptions made, i.e. about ice limits, thicknesses, melting histories, different contributions of ice sources, and the amount of water return to the oceans (see Shackleton 1987; Chappell & Polach 1991). Clark et al. (1978) took a medium- to low-level value for water recovery (c. 75.5 m). Subsequent work by Peltier (1987, 1991) and others has allowed for reassessments in these parameters, though discussions on the appropriateness of model fit and accuracy continue.

From the Walcott (1970, 1972) and Clark et al. (1978) modelled predictions of sea-level recovery patterns, the spatial limits of forebulge migration are shown to lie theoretically in a 'transitional behaviour' zone (Fig. 7). This exists between regions of persistent land emergence and falling shoreline elevation in time (zone I) and 'far-field' areas of Earth surface submergence (zone II), which would result from forebulge collapse. In zone II, early shoreline elevations increase in depth (below psl) the further the recording point is located from the former centre of ice loading. In the I/II transition zone shoreline behaviour is represented by initially emergent shorelines and relative sea-level fall. This is followed, at its simplest, by land submergence and relative sea-level rise. This shoreline signature is thought to mark the initial forebulge elevation, followed by the progressive migration and decline in height of the forebulge with time.

Testing of these predicted patterns with empirical shoreline data from the east coast of the USA and Canada (Clark et al. 1978; Clark 1980b; Newman et al. 1980; Bloom, 1983), i.e. data derived before c. 1985 from expected (I/II) transitional areas, does not show a good fit of information (Plag et al. 1995). Therefore, this begs the question as to how realistic is the expected 'forebulge behaviour' of the Earth's crust? Are the different model boundary conditions used wrong, or are the field data wrong, or a combination of these factors? It should be stressed that reliable sea-level index points from times of expected forebulge migration in reconstructions are often lacking. More recently, detailed and integrated research studies from Maine, and northeast and arctic Canada (Gehrels 1993; Gehrels & Belknap 1993; Gehrels et al. 1993; Stea et al. 1994; Belknap 1994; England 1994) have provided much more good-quality shoreline data. These results are again interpreted as evidencing forebulge operation here following ice unloading. However, whether these new data really represent forebulge migration, or possibly a complex and local-scale element of Earth crustal behaviour to which present Earth models are not tuned (Plag et al. 1995), is a question of geophysical debate.

The current differences in Earth modelling approaches are important. The varying environmental boundary conditions used in models, indicating differences between real and predicted data fits, serve together to stress the possible wide error margins that exist in reconstructing former Earth surface elevation changes, at timescales of 1000 years or less. For Earth surface elevation read, 'fixing possible land-bridge existence and limits'.

Discussion

The reintroduction of the forebulge mechanism in the explanation of shoreline and shelf seas palaeoenvironmental data (Wingfield 1992b, 1995) presents a valuable stimulus to the study of these data and to problems of palaeogeographic reconstruction. The idea of a migrating forebulge, resulting in the temporary elevation of Earth surface levels and thereby land-bridge creation, is intuitively appealing. It solves the problem of a now untraceable (in terms of other physical data) land connection. The basis of the interpretation is perhaps more problematic. The data used in construction derive from different environmental regimes, are of varying reliability and accuracy of indicative meaning (see van de Plassche 1986), and come from widely separated shelf-sea areas. Their provenance from shelf-sea environments makes the data difficult to verify. Importantly, accurate height and dating control on the different evidence types is absent. In short, good data are lacking.

Other problems exist. The operation of a migrating forebulge mechanism assumes the existence of a specific Earth rheology. The nature of the Earth's geophysical structure and properties, above the level of generalized parameters, remains in dispute (Sabadini et al. 1991; Plag et al. 1995). Consequently, the construction of possible forebulge dimensions and behaviour over time, based upon Earth models (e.g. Walcott 1970, 1972) which probably have significant inaccuracies in boundary controls and do not compare well with empirical data, is insecure (Fig. 10). Such constructions are likely to have significant margins of error in the elevations and timing of any predicted Earth surface-level changes. These probable errors

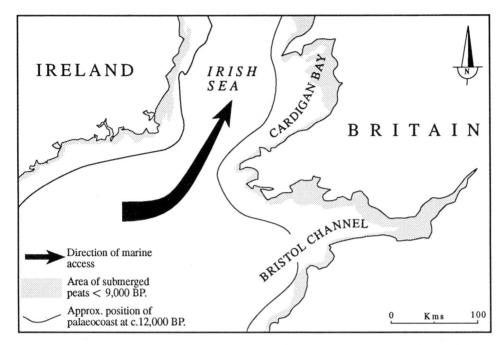

Fig. 11. Approximate shoreline position and inferred direction for marine access into the Celtic and Irish Seas at c. 12 000 years BP (shoreline limits for 12 000 years BP are based upon observed data and the Earth modelled predictions of relative sea-level behaviour).

may be compounded by the method of forebulge construction used. In the case of the method used in connection with the Irish Sea landbridge (Wingfield 1992b, 1995), then this relies upon essentially geometrical relationships. These appear to be derived from generalized Earth models, used to estimate former land-surface elevations. No account is taken in this case of the influences of multiple ice centres on Earth loading and recovery. This ice factor alone may be an important element in fixing the shape and behaviour of any forebulge migration, particularly in a region characterized by a complex deglacial history.

There is good conceptual reason to expect any such migration to be coincident with, and to be completed 'soon' after deglaciation and the major phase of crustal recovery (Andrews 1970; Eyles & McCabe, 1989a; Devoy 1987b). For the Irish Sea and neighbouring regions the timing of deglaciation and of ice types, thickness and limits remain uncertain. It would appear (Ehlers et al. 1991) that the major phases of ice disappearance from these regions is being pushed back beyond 15 000–16 000 years BP. If such interpretations are correct, then the timing of any forebulge movement should consequently occur earlier than that currently envisaged for land-bridge operation (Wingfield 1995).

Lambeck (1990, 1991a, b, 1993a) and others draw specific attention to the complex crustal behaviour that can exist and complicate Earth model reconstructions within, or close to, former ice-margins, such as that of the Irish Sea. Observed shorelines from the northern Irish Sea (Sissons 1983; Jardine 1982; Tooley 1982, 1987; Synge 1977, 1985) show a zone I type crustal emergence behaviour for this region. Both these data and the modelled reconstructions also show, however, that the shoreline and Earth crustal behaviours differ between even small contiguous areas (Figs 5 and 6). This record may reflect a complex regional deglacial history and/or interplay with other Earth structural controls e.g. 'piano key tectonics' (Eyles & McCabe 1989a; Tooley 1978). The recognition of any forebulge operation, given these controls and the coarse scale of resolution of some of the palaeoenvironmental indicators used, may therefore be misleading or wrong, and difficult to test with the simplified land-bridge methodology applied. Equally, such explanation may fortuitously be correct! At this stage, in the context of sparse and poorly constrained data, the forebulge explanation should be viewed cautiously, as Wingfield (1995) suggests.

The application of established geophysical modelling to the regional data (Lambeck 1991a, b, 1993a) does not appear to show the operation of forebulge migration. It is argued that the predictions fit the existing regional data well. Palaeoshoreline reconstructions based on these predictions for c. 12 000 years BP (Lambeck 1991a, b; pers. comm.) show marine access through St George's Channel into the Irish Sea at this time (Fig. 11). This reconstruction would fit other palaeoenvironmental evidence for marine conditions established close to the present shoreline within Dublin Bay by late-glacial times (i.e. 11 000–12 000 years BP) (Penney 1983). However, the most recent model results may also indicate extensive land exposure in this region earlier at c. 13 500 years BP, causing a possible isolation of the Irish Sea from the south. Proponents for such a land-bridge, temporarily closing the southern end to the Irish Sea at this time and later, suggest a marine access to the region via a North Channel route. Presumably such a view needs to establish accurately the possible forebulge and crustal recovery patterns in this northern sector too.

The field evidence reported from Peacock et al. (1992), Scourse & Austin (1994) and others does seem to suggest at first analysis, the existence of real sea-floor subsidence in these locations. Such data may fit with the operation of a forebulge mechanism. However, reliable data points are few. The data may equally simply represent peripheral subsidence following ice unloading. Other sea-level data to support the existence and limits of an expected forebulge zone from regions beyond the former ice-margins e.g. the Bay of Biscay (Wingfield 1994, 1995) appear, however, to be either non-existent (Pirazzoli 1991), or are inconclusive and await verification. The sources of these newly reported offshore data, from close to established major sedimentary basins (graben) (see Banner et al. 1979; Devoy 1989), are perhaps significant. Consideration of a depositionally and structurally controlled subsidence, linked possibly to glacio-isostatic triggers, should be investigated. Daly's (1934) idea of 'hinge' zones and the existence of structurally related stress-induced subsidence may be more real than previously thought.

Examination of these possibilities through the application of established geophysical models seems a logical way forward. The quantitative base for modelling forms the most rigorous approach yet developed to calculating the regional- to point-scale effects of environmental controls on the Earth's crust. It is important to note that, dependent on location with respect to former ice positions and the timing of deglaciation, all the terms in the Earth models (e.g. equation (3)) may be significant in such calculations. As Fjeldskaar (1991) and others (Clark & Lingle 1979; Clark & Primus 1987) show, the geoid term ($\Delta\zeta_r$), for example, should have measurable repercussions on regional water levels. Temporary disequilibrium of the geoid upon ice-melt would cause relative sea-levels to fall, even at long distances from ice margins. Such geoid falls may have contributed in these shelf seas to the shallow-water conditions recorded by the 'anomalous' offshore data.

In spite of the explanatory power of these Earth models they do need to be better constrained by additional, good-quality data and particularly by sea-level index points from before 10 000 years BP. The inclusion of the 'anomalous' offshore data into these established models would be a valuable next step in analysis.

Conclusion

Recent applications of geophysical modelling and the results of other palaeoenvironmental reconstructions for western Britain and Ireland have renewed debate on the operation of a land-bridge. New data, particularly from offshore records, are providing important contributions in testing hypotheses on the environmental development and regional palaeogeography in the Late Quaternary. The environmental data and explanations, however, rarely appear to fit everywhere exactly, or even well. Many reconstructions remain at the level of speculation. The plea for more good data, collected within integrated research programmes continues and remains relevant (Devoy et al. 1986; Plag et al. 1995).

The redefined patterns of deglaciation and of expected Earth crustal recovery involving a forebulge migration mechanism may, for some, provide an Irish Sea land-bridge in late-glacial and even in Post-glacial times. Examination of both the data and concepts leading to this view reveals problems. The development of any land-bridge surface must result from the interaction of a complex of environmental controls. The operation of these controls remains difficult to date accurately and to scale in height and spatial terms. The height error margins in each of the contributory factors used in reconstructions (see equations (1)–(3)) may be measured in metres and collectively in tens of metres. This is particularly true the further back in time one goes. The physical dimensions and the timing of any land-bridge forming in these shallow seas might be accommodated, therefore, within the collective confidence limits of the environmental terms and models used in reconstructing land-bridge development. Even if the ideas underpinning the land-bridge's operation are correct, the existence of the land-bridge must remain a matter of conjecture, at least until greater accuracy can be given to the environmental terms controlling the dimensions of such a feature.

Whilst it is now possible to attempt to reconstruct some of these controlling parameters in quantitative terms (e.g. modelled palaeotidal levels, amplitude of crustal movements), any meaningful modelled accuracy must depend upon a good knowledge of the other environmental terms involved: a pack of cards situation results! Other contributory factors such as the effects of sea-level rise rate (Δ_{rsl}), coastal erosion or sediment budgets, are less dependent, but equally they are more difficult to reconstruct accurately. In these factors much depends upon analogy with present day process observations, and the concept of analogue may be wrong (Thorn 1988).

From a biogeographical viewpoint no new substantial imperative for a land connection has emerged, although the tentative results for the presence of thermophilous trees before c. 11 500 years BP in County Cork are intriguing. However, the existence of low sea-levels and closer shorelines between Britain and Ireland during c. 9500–13 000 years BP may have been enough to facilitate plant and animal movements without the need for a land-bridge. As Preece et al. (1986) infer, passive vectors in plant and animal dispersals may be more important than biogeographers have been able to demonstrate. Other biogeographical problems also remain, even with land-bridge operation. For example, why does the Irish fauna remain impoverished if a land-bridge really operated (Yalden 1982; Stuart & Wijngaarden-Bakker 1985; Stuart 1995)? The geophysical arguments may have revitalized the land-bridge discussion, but the existence of any connections remains, for present, at the level of *possibility* rather than *probability*.

The author wishes to thank Michael Murphy, Department of Geography, University College Cork, for the preparation of figures and also, particularly, Catherine Delaney, Michael Healy, James Scourse, Peter Coxon and Richard Preece for discussion and critical comment.

References

ANDREWS, J. T. 1970. *A Geomorphological Study of Post-glacial Uplift with Particular Reference to Arctic Canada*. Institute of British Geographers, Special Publication, **2**, 1–156.
—— 1982. On the reconstruction of Pleistocene ice sheets: a review. *Quaternary Science Reviews*, **1**, 1–31.
—— 1987. Glaciation and sea-level. *In:* DEVOY, R. J. N. (ed.) *Sea Surface Studies: A Global View*. Croom Helm–Chapman Hall, London, 95–126.
ANDRIEU, V., HUANG, C. C., O'CONNELL, M. & PAUS, A. 1993. Lateglacial vegetation and environment in Ireland: first results from four western sites. *Quaternary Science Reviews*, **12**, 681–705.
AUSTIN, W. E. N. & MCCARROLL, D. 1992. Foraminifera from the Irish Sea glacigenic deposits at Aberdaron, western Lleyn, North Wales: palaeoenvironmental implications. *Journal of Quaternary Science*, **7**, 311–318.
BANNER, F. T., COLLINS, M. B. & MASSIE, K. S. 1979. *The Northwest European Shelf Seas: The Seabed and the Sea in Motion, 1. Geology and Sedimentology*. Elsevier, Amsterdam.
BARD, E., HAMELIN, B. & FAIRBANKS, R. G. 1990a. U-Th ages obtained by mass spectrometry in corals from Barbados: sea-level during the past 130,000 years. *Nature*, **346**, 456–458.
——, ——, —— & ZINDLER, A. 1990b. Calibration of the ^{14}C timescale over the past 30,000 years using mass spectrometric U-Th ages from Barbados corals. *Nature*, **345**, 405–410.
BELKNAP, D. F. 1994. Sea-level variations south east of the Laurentide ice sheet during deglaciation. *In:* WINGFIELD, R. T. R. (ed.) *Glacial Interglacial Sea-level Changes in Four Dimensions: Abstracts*. European Science Foundation, Strasbourg, **4**.
BLOOM, A. L. 1983. Sea-level and coastal morphology of the United States through the late Wisconsin glacial maximum. *In:* WRIGHT, Jr H. E. (ed.) *Late-Quaternary Environments of the United States, Volume 1, The Late Pleistocene*. Longman, London, 215–229.
BOULTON, G. S. 1990. Sedimentation and sea-level changes during glacial cycles and their control on glacimarine facies architecture. *In:* DOWDESWELL, J. A. & SCOURSE, J. D. (eds) *Glacimarine Environments: Processes and Sediments*. Geological Society, London, Special Publication, **53**, 15–52.
—— 1993. Late Cenozoic relative sea-level change and tectonic warping in the southern North Sea basin. *Abstract Supplement 1, Terra Nova*, **5**, 601.
——, JONES, A. S., CLAYTON, K. M. & KENNING, M. J. 1977. A British ice-sheet model and patterns of glacial erosion and deposition in Britain. *In:* SHOTTON, F. W. (ed.) *British Quaternary Studies: Recent Advances*. Clarendon Press, Oxford, 231–246.
——, PEACOCK, J. D. & SUTHERLAND, D. G. 1991. The Quaternary. *In:* CRAIG, G. Y. (ed.) *The Geology of Scotland*. The Geological Society, London, 503–543.
——, SMITH, G. D., JONES, A. S. & NEWSOME, J. 1985. Glacial geology and glaciology of the last mid-latitude ice sheets. *Journal of the Geological Society, London*, **142**, 447–474.

BOWEN, D. Q. 1991. Time and space in the glacial sediment systems of the British Isles. *In:* EHLERS, J., GIBBARD, P. L. & ROSE, J. (eds) *Glacial Deposits in Great Britain and Ireland.* Balkema, Rotterdam, 3–11.

——, ROSE, J., McCABE, A. M. & SUTHERLAND, D. G. 1986. Correlation of Quaternary glaciations in England, Ireland, Scotland and Wales. *Quaternary Science Reviews*, **5**, 299–340.

CARTER, R. W. G. 1982. Sea-level changes in Northern Ireland. *Proceedings of the Geologists' Association*, **93**, 7–23.

—— 1993. Age, origin and significance of the raised gravel barrier at Church Bay, Co. Antrim. *Irish Geography*, **26**, 141–146.

——, DEVOY, R. J. N. & SHAW, J. 1989. Late Holocene sea-levels in Ireland. *Journal of Quaternary Science*, **4**, 7–24.

——, McKENNA, J., ORFORD, J. D. & DEVOY, R. J. N. 1993. *The Irish Sea Coast of Ireland: a Brief Review of Coastal Processes, Erosion and Management.* Irish Sea Forum Seminar Report Number **4**, 3–23.

CATHLES, L. M. 1975. *The Viscosity of the Earth's Mantle.* Princeton University Press, Princeton.

—— 1980. Interpretation of Post-glacial isostatic adjustment phenomena in terms of mantle rheology. *In:* MORNER, N-A. (ed.) *Earth Rheology, Isostasy and Eustasy.* Wiley, New York, 11–43.

CEC (COMMISSION OF THE EUROPEAN COMMUNITIES). 1990. *Investigations of Past and Future European Sea-level Changes and their Impacts.* Contract EV4C-0077 Final Report, Brussels.

—— 1993. *Climate Change, Sea-level Rise and Associated Impacts in Europe.* Contract EPOC-CT90-0015 Final Report, Volume 1, Brussels.

CHAPPELL, J. 1974. Late Quaternary glacio- and hydro-isostasy, on a layered earth. *Quaternary Research*, **4**, 429–440.

—— & POLACH, H. 1991. Post-glacial sea-level rise from the coral record at Huon Peninsula, Papua New Guinea. *Nature*, **349**, 147–149.

CLARK, J. A. 1980a. A numerical model of worldwide sea-level changes on a viscoelastic earth. *In:* MORNER, N-A. (ed.) *Earth Rheology, Isostasy and Eustasy.* Wiley, Chichester, 525–534.

—— 1980b. The reconstruction of the Laurentide ice sheet of North America from sea-level data: methodology and preliminary results. *Journal of Geophysical Research*, **85**, 4307–4323.

—— & LINGLE, C. S. 1979. Predicted relative sea-level changes (18,000 years BP to present) caused by lateglacial retreat of the Antarctic ice sheet. *Quaternary Research*, **11**, 279–298.

—— & PRIMUS, J. A. 1987. Sea-level changes resulting from future retreat of ice sheets: an effect of CO_2 warming of the climate. *In:* TOOLEY, M. J. & SHENNAN, I. (eds) *Sea-level Changes.* Blackwell, Oxford, 35–370.

——, FARRELL, W. E. & PELTIER, W. R. 1978. Global changes in Post-glacial sea-level: a numerical calculation. *Quaternary Research*, **9**, 265–287.

COXON, P. 1993. Irish Pleistocene stratigraphy. *Irish Journal of Earth Sciences*, **12**, 83–105.

—— & WALDREN, S. 1995. The floristic record of Ireland's Pleistocene temperate stages. *This volume.*

DALY, R. A. 1934. *The Changing World of the Ice Age.* Yale University Press, New Haven.

DAWSON, A. G. 1984. Quaternary sea-level changes in western Scotland. *Quaternary Science Reviews*, **3**, 345–368.

—— 1992. *Ice Age Earth: Late Quaternary Geology and Climate.* Routledge, London.

DENTON, G. H. & HUGHES, T. J. (eds). 1981. *The Last Great Ice Sheets.* Wiley, New York.

DEVOY, R. J. N. 1983. Late Quaternary shorelines in Ireland: an assessment of their implications for isostatic land movement and relative sea-level changes. *In:* SMITH, D. E. & DAWSON, A. (eds) *Shorelines and Isostasy.* Academic Press, London, 487–509.

—— 1985. The problems of a Late Quaternary land-bridge between Britain and Ireland. *Quaternary Science Reviews*, **4**, 43–58.

—— 1986. Possible Land-bridges Between Ireland and Britain: a Geological Appraisal. *In:* SLEEMAN, D. P., DEVOY, R. J. N. & WOODMAN, P. C. (eds) *Proceedings of the Postglacial Colonization Conference.* Irish Biogeographical Society, Occasional Publication, **1**, 15–26.

—— 1987a. *Sea Surface Studies: A Global View.* Croom Helm–Chapman Hall, London.

—— 1987b. Sea-level changes during the Holocene: the North Atlantic and Arctic oceans. *In:* DEVOY, R. J. N. (ed.) *Sea Surface Studies: A Global View.* Croom Helm–Chapman Hall, London, 294–347.

—— 1989. The geological and changing physical environment of the Irish Sea basin. *In:* SWEENEY, J. C. (ed.) *The Irish Sea: A Resource at Risk.* Geographical Society of Ireland, Special Publication, **3**, 5–17.

—— 1990. Controls on coastal and sea-level changes and the application of archaeological-historical records to understanding recent patterns of sea-level movement. *In:* McGRAIL, S. (ed.) *Maritime Celts, Frisians and Saxons.* Council for British Archaeology, Research Report, **71**, 17–26.

—— 1991. The study of inferred patterns of Holocene sea-level change from Atlantic and other European coastal margins as a means of testing models of earth crustal behaviour. *In:* SABADINI, R., LAMBECK, K. & BOSCHI, E. (eds) *Glacial Isostasy, Sea-level and Mantle Rheology.* Kluwer Academic Publishers, Dordrecht, 213–236.

——, DELANEY, C., CARTER, R. W. G. & JENNINGS, S. C. 1995. Coastal stratigraphies as indicators of environmental changes upon European Atlantic coasts in the late Holocene. *Journal of Coastal Research, Special Issue* (in press).

—— & SINNOTT, A. 1993. Lateglacial environments in County Cork: patterns and processes of change. *In:* COXON, P. (ed) *Glacial Events.* Irish Association for Quaternary Studies Annual Symposium, 26 November 1993. IQUA, Dublin, 11–13.

—, SLEEMAN, D. P. & WOODMAN, P. C. 1986. Conclusions. *In:* SLEEMAN, D. P., DEVOY, R. J. N. & WOODMAN, P. C. (eds) *Proceedings of the Postglacial Colonization Conference*. Irish Biogeographical Society, Occasional Publication, 1, 86–88.

DICKSON C. & WHATLEY, R. 1993. The biostratigraphy of a Holocene borehole from the Irish Sea. *In:* KEEN, M. (ed.) *Proceedings of the 2nd European Ostracodologists Meeting*, University of Glasgow, 23–27 July 1993. Geological Society of Glasgow & the British Micropalaeontological Society, 90–94.

DREWRY, D. J. 1983. *Antarctica: Glaciological and Geophysical Folio*. Scott Polar Research Institute, University of Cambridge.

EHLERS, J., GIBBARD, P. L. & ROSE, J. 1991. *Glacial Deposits in Great Britain and Ireland*. Balkema, Rotterdam.

ENGLAND, J. H. 1994. Holocene sea-level change in the Canadian high Arctic: questioning the norm. *In:* WINGFIELD, R. T. R. (ed.) *Glacial Interglacial Sea-level Changes in Four Dimensions: Abstracts*. European Science Foundation, Strasbourg, 7.

EYLES, N. & MCCABE. A. M. 1989a. The Late Devensian (<22,000 BP) Irish Sea Basin: the sedimentary record of a collapsed ice sheet margin. *Quaternary Science Reviews*, 8, 307–351.

—— & —— 1989b. Glaciomarine facies within subglacial tunnel valleys: sedimentary record of glacio-isostatic downwarping in the Irish Sea Basin. *Sedimentology*, 36, 431–448.

—— & —— 1991. Glaciomarine deposits of the Irish Sea Basin: the role of glacio-isostatic disequilibrium. *In:* EHLERS, J., GIBBARD, P. L. & ROSE, J. (eds) *Glacial Deposits in Great Britain and Ireland*. Balkema, Rotterdam, 311–331.

FAIRBANKS, R. G. 1989. A 17,000 year glacio-eustatic sea-level record: influence of glacial melting rates on the Younger Dryas event and deep-ocean circulation. *Nature*, 342, 637–642.

FJELDSKAAR, W. 1991. Geoidal-eustatic changes induced by the deglaciation of Fennoscandia. *Quaternary International*, 9, 1–6.

FLEMMING, N. C. 1982. Multiple regression analysis of earth movements and eustatic sea-level change in the United Kingdom in the past 9000 years. *Proceedings of the Geologists' Association*, 93, 113–125

FLINT, R. F. 1971. *Glacial and Quaternary Geology*. Wiley, New York.

GEHRELS, W. R. 1993. *Late Holocene Relative Sea-Levels in Maine: Eustasy, Isostasy, Tidal Range and Tectonics*. PhD thesis. University of Maine, USA.

—— & BELKNAP, D. F. 1993. Neotectonic history of eastern Maine evaluated from historic sea-level data and ^{14}C dates on saltmarsh peats. *Geology*, 21, 615–618.

——, —— & KELLEY, J. T. 1993. Holocene isostasy along the coast of Maine, USA. *In: International Geological Correlation Programme – Project 274. Final Project Meeting, Quaternary Coastal Evolution: Models, Processes and Local to Global Factors*. Oostduinkerke, Belgium, 15–18 September 1993, 38–41.

GODWIN, H. 1975. *History of the British Flora: A Factual Basis for Phytogeography*. 2nd edition. Cambridge University Press.

GRAY, J. M. & COXON, P. 1991. The Loch Lomond stadial glaciation in Britain and Ireland. *In:* EHLERS, J., GIBBARD, P. L. & ROSE, J. (eds) *Glacial Deposits in Great Britain and Ireland*. Balkema, Rotterdam, 89–105.

GREGORY, K. J. 1978. A physical geography equation. *National Geographer*, 12, 13–141.

HAMBLIN, R. J. O., JEFFERY, D. H. & WINGFIELD, R. T. R. 1990. Glacial incisions indicating Middle and Upper Pleistocene ice limits off Britain – two comments and a reply. *Terra Nova*, 2, 382–389.

HEALY, M. 1993. *Sea-level and Associated Coastal Process Changes in Cornwall, southwest England*. PhD thesis, University College Cork, The National University of Ireland.

HOARE, P. G. 1991. Late Midlandian glacial deposits and glaciation in Ireland and the adjacent offshore regions. *In:* EHLERS, J., GIBBARD, P. L. & ROSE, J. (eds) *Glacial Deposits in Great Britain and Ireland*. Balkema, Rotterdam, 69–78.

JARDINE, W. G. 1982. Sea-level changes in Scotland during the last 18,000 years. *Proceedings of the Geologists' Association*, 93, 25–42.

JOHNSTON, P. 1993. The effect of spatially non-uniform water loads on prediction of sea-level change. *Geophysical Journal International*, 114, 615–634.

JOSENHANS, H. 1994. How much and how fast did Post-glacial sea-levels change along the eastern and western Canadian continental margins. *In:* WINGFIELD, R. T. R. (ed.) *Glacial Interglacial Sea-level Changes in Four Dimensions: Abstracts*. European Science Foundation, Strasbourg, 10.

KAY, C. J. 1993. A regional depositional model for the southern North Sea during the late Tertiary and Quaternary. *Abstract Supplement 1, Terra Nova*, 5, 605.

KIDSON, C. & TOOLEY, M. J. 1977. *The Quaternary History of the Irish Sea*. Seel House Press, Liverpool.

LAMBECK, K. 1990. Glacial rebound, sea-level change and mantle viscosity. *Quarterly Journal Royal Astronomical Society*, 31, 1–30.

—— 1991a. A model for Devensian and Flandrian glacial rebound and sea-level change in Scotland. *In:* SABADINI, R., LAMBECK, K. & BOSCHI, E. (eds) *Glacial Isostasy, Sea-level and Mantle Rheology*. Kluwer Academic Publishers, Dordrecht, 33–61.

—— 1991b. Glacial rebound and sea-level change in the British Isles. *Terra Nova*, 3, 379–389.

—— 1993a. Glacial rebound of the British Isles – I: Preliminary model results and II: A high-resolution, high precision model. *Geophysical Journal International*, 115, 941–990.

—— 1993b. Sea-level change and shoreline evolution around Great Britain and the North Sea for the past 20,000 years. *In:* WOODWORTH, W. (ed.) *Sixtieth Anniversary Meeting of the Permanent Service for Mean Sea Level*. Bidston Observatory, Birkenhead, UK, 15–16.

——, JOHNSTON, P. & NAKADA, M. 1990. Holocene

glacial rebound and sea-level change in Northwest Europe. *Geophysical Journal International*, **103**, 451–468.

LOWE, J. J. & WALKER, M. J. C. 1984. *Reconstructing Quaternary Environments*. Longman, London.

MCCABE, A. M. 1986. Glaciomarine facies deposited by retreating tidewater glaciers: an example from the Late Pleistocene of northern Ireland. *Journal of Sedimentary Petrology*, **56**, 880–894.

—— 1987. Quaternary deposits and glacial stratigraphy in Ireland. *Quaternary Science Reviews*, **6**, 259–299.

—— & EYLES, N. 1988. Sedimentology of ice-contact glaciomarine delta, Carey Valley, northern Ireland. *Sedimentary Geology*, **59**, 1–14.

——, BOWEN, D. Q. & PENNEY, D. N. 1993. Glaciomarine facies from the Western Sector of the last British Ice Sheet, Co. Donegal, Ireland. *Quaternary Science Reviews*, **12**, 35–45.

——, CARTER, R. W. G. & HAYNES, J. R. 1995. A shallow marine emergent sequence from the northwest sector of the last British ice sheet, Portballintrae, Northern Ireland. *Marine Geology*, in press.

MCCARROLL, D. & HARRIS, C. 1992. The glacigenic deposits of western Lleyn, north Wales: terrestrial or marine? *Journal of Quaternary Science*, **7**, 19–29.

MITCHELL, G. F. 1960. The Pleistocene history of the Irish Sea. *Advancement of Science*, **17**, 313–325.

—— 1963. Morainic ridges on the floor of the Irish Sea. *Irish Geography*, **4**, 335–344.

—— 1972. The Pleistocene history of the Irish Sea: a second approximation. *Scientific Proceedings of the Royal Dublin Society*, **A4** (13), 181–199.

—— 1976. *The Irish Landscape*. Collins, London.

—— 1986. *The Shell Guide to Reading the Irish Landscape*. Country House Press, Dublin.

—— & WATTS, W. A. 1970. The history of the Ericaceae in Ireland during the Quaternary Epoch. *In:* WALKER, D. & WEST, R. G. (eds) *Studies in the Vegetational History of the British Isles*. Cambridge University Press, 13–21.

MORNER, N-A. 1980. *Earth Rheology, Isostasy and Eustasy*. Wiley, New York.

—— 1987. Models of global sea-level changes. *In:* TOOLEY, M. J. & SHENNAN, I. (eds) *Sea-level Changes*. Blackwell, Oxford, 332–349.

NAKADA, M. & LAMBECK, K. 1991. Late Pleistocene and Holocene sea-level change; evidence for lateral mantle viscosity structure? *In:* SABADINI, R., LAMBECK, K. & BOSCHI, E. (eds) *Glacial Isostasy, Sea-level and Mantle Rheology*. Kluwer Academic Publishers, Dordrecht, 79–94.

NEWMAN, W. S., CINQUEMANI, L. J., PARDI, R. R. & MARCUS, L. F. 1980. Holocene develelling of the United States' east coast. *In:* MORNER, N-A. (ed.) *Earth Rheology, Isostasy and Eustasy*. Wiley, New York, 449–463.

O'COFAIGH, C. 1993. *Sedimentology of Late-Pleistocene Glacimarine and Shallow Marine Deposits from the South Coast of Ireland*. M.Sc. thesis, University of Dublin, Trinity College.

PANTIN, H. M. & EVANS, C. D. R. 1984. The Quaternary History of the central and southwestrn Celtic Sea. *Marine Geology*, **57**, 259–293.

PEACOCK, J. D. & HARKNESS, D. D. 1990. Radiocarbon ages and the full-glacial to Holocene transition in seas adjacent to Scotland and southern Scandinavia: a review. *Transactions of the Royal Society of Edinburgh, Earth Sciences*, **81**, 385–396.

——, AUSTIN, W. E. N., SELBY, I., GRAHAM, D. K., HARLAND, R. & WILKINSON, I. P. 1992. Late Devensian and Flandrian palaeoenvironmental changes on the Scottish continental shelf west of the Outer Hebrides. *Journal of Quaternary Science*, **7**, 145–161.

PELTIER, W. R. 1980. Ice sheets, oceans and the earth's shape. *In:* MORNER, N.-A. (ed.) *Earth Rheology, Isostasy and Eustasy*. Wiley, New York, 45–63.

—— 1987. Mechanisms of relative sea-level change and the geophysical responses to ice – water loading. *In:* DEVOY, R. J. N. (ed.) *Sea Surface Studies: A Global View*. Croom Helm–Chapman Hall, London, 57–94.

—— 1991. The ICE-3G model of Late Pleistocene deglaciation: construction, verification and applications. *In:* SABADINI, R., LAMBECK, K. & BOSCHI, E. (eds) *Glacial Isostasy, Sea-level and Mantle Rheology*. Kluuwer Academic Publishers, Dordrecht, 95–120.

—— & ANDREWS, J. T. 1983. Glacial geology and glacial isostasy of the Hudson Bay region. *In:* SMITH, D. E. & DAWSON, A. G. (eds) *Shorelines and Isostasy*. Academic Press, London, 285–320.

PENNEY, D. N. 1983. *Post-glacial Sediments and Foraminifera at Dundalk, Ireland*. PhD thesis, Trinity College, University of Dublin.

PIRAZZOLI, P. A. 1991. *World Atlas of Holocene Sea-level Changes*. Elsevier, Amsterdam.

—— & GRANT, D. R. 1987. Lithospheric deformation deduced from ancient shorelines. *In:* KASHARA, K. (ed.) *Recent Plate Movements and Deformation*. American Geophysical Union and the Geological Society of America, Washington, Geodynamics Series, **20**, 67–72.

PLAG, H-P., AUSTIN, W. E. N., DEVOY, R. J. N. ET AL. 1995. Late Quaternary sea-level changes and the repercussions of glaciation upon continental shelf areas. *Terra Nova*, in press.

PREECE, R. C., COXON, P. & ROBINSON, J. E. 1986. New biostratigraphic evidence of the Post-glacial colonization of Ireland and for Mesolithic forest disturbance. *Journal of Biogeography*, **13**, 487–509.

PUGH, D. T. 1981. A comparison of recent and historical tides and mean sea-levels off Ireland. *Geophysical Journal of the Royal Astronomical Society*, **71**, 808–815.

ROSE, J., JACKSON, H., HUNT, C. O. & KONRADI, P. 1993. Lateglacial sea-level change in the central part of the southern North Sea. *Abstract Supplement 1, Terra Nova*, **5**, 607–608.

SABADINI, R., LAMBECK, K. & BOSCHI, E. 1991. *Glacial Isostasy, Sea-level and Mantle Rheology*. Kluwer Academic Publishers, Dordrecht.

SCOURSE, J. D. 1991*a*. Glacial deposits of the Isles of

Scilly. *In:* EHLERS, J., GIBBARD, P. L. & ROSE, J. (eds) *Glacial Deposits in Great Britain and Ireland.* Balkema, Rotterdam, 291–300.

—— 1991*b*. Late Pleistocene stratigraphy and palaeobotany of the Isles of Scilly. *Philosophical Transactions of the Royal Society of London,* **B334**, 405–448.

—— & AUSTIN, W. E. N. 1994. A Devensian Lateglacial and Holocene sea-level and water depth record from the central Celtic Sea. *Quaternary Newsletter,* **74**, 22–29.

——, ——, BATEMAN, R. M., CATT, J. A., EVANS, C. D. R., ROBINSON, J. E. & YOUNG, J. R. 1990. Sedimentology and micropalaeontology of glacimarine sediments from the central and southwestern Celtic Sea. *In:* DOWDESWELL, J. A. & SCOURSE, J. D. (eds) *Glacimarine Environments: Processes and Sediments.* Geological Society, London, Special Publication, **53**, 329–347.

——, ROBINSON, E. & EVANS, C. 1991. Glaciation of the central and southwestern Celtic Sea. *In:* EHLERS, J., GIBBARD, P. L. & ROSE, J. (eds) *Glacial Deposits in Great Britain and Ireland.* Balkema, Rotterdam, 301–310.

SHACKLETON, N. J. 1987. Oxygen isotopes, ice volumes and sea-level. *Quaternary Science Reviews,* **6**, 183–190.

SHENNAN, I. 1983. Flandrian and Late Devensian sea-level changes and crustal movements in England and Wales. *In:* SMITH, D. E. & DAWSON, A. G. (eds) *Shorelines and Isostasy.* Academic Press, London, 255–283.

—— 1989*a*. Holocene sea-level changes and crustal movements in the North Sea region: an experiment with regional eustasy. *In:* SCOTT, D. B., PIRAZZOLI, P. A. & HONIG, C. A. (eds) *Quaternary Sea-level Correlation and Applications.* Kluwer Academic Publishers, Dordrecht, 1–26.

—— 1989*b* Holocene crustal movements and sea-level changes in Great Britain. *Journal of Quaternary Science,* **4**, 77–89.

——, INNES, J. B., LONG, A. J. & ZONG, Y. 1994*a*. Late Devensian and Holocene relative sea-level changes at Loch nan Eala, near Arisaig, northwest Scotland. *Journal of Quaternary Science,* **9**, 261–283.

——, ——, —— & —— 1995. Late Devensian and Holocene relative sea-level changes in northwest Scotland: new data to test old models. *Quaternary International,* in press.

SHOTTON, F. W. 1962. The physical background of Britain in the Pleistocene. *Advancement of Science,* **19**, 193–206.

SISSONS, J. B. 1983. Shorelines and isostasy in Scotland. *In:* SMITH, D. E. & DAWSON, A. G. (eds) *Shorelines and Isostasy.* Academic Press, London, 209–226.

SLEEMAN, D. P., DEVOY, R. J. N. & WOODMAN, P. C. 1986. *Proceedings of the Post-glacial Colonization Conference.* University College Cork, 15–16 October, 1983. Irish Biogeographical Society, Occasional Publication, **1**, 1–88.

STEA, R. R., BOYD, R. & FADER, G. B. J. 1994. Glaciation, shorelines and isostasy in Maritime Canada. *In:* WINGFIELD, R. T. R. (ed.) *Glacial Interglacial Sea-level Changes in Four Dimensions: Abstracts.* European Science Foundation, Strasbourg, **21**.

STUART, A. J. 1986. Pleistocene mammals in Ireland (pre-10,000 BP). *In:* SLEEMAN, D. P., DEVOY, R. J. N. & WOODMAN, P. C. (eds) *Proceedings of the Postglacial Colonization Conference.* Irish Biogeographical Society, Occasional Publication, **1**, 28–33.

—— 1995. Insularity and Quaternary vertebrate faunas in Britain and Ireland. *This volume.*

—— & WIJNGAARDEN-BAKKER, VAN L. H. 1985. Quaternary vertebrates. *In:* EDWARDS, K. J. & WARREN, W. P. (eds) *The Quaternary History of Ireland.* Academic Press, London, 221–249.

SYNGE, F. M. 1977. Records of sea-levels during the Late Devensian. *Philosophical Transactions of the Royal Society of London,* **B280**, 211–228.

—— 1981. Quaternary glaciation and changes of sea-level in the south of Ireland. *Geologie en Mijnbouw,* **60**, 305–315.

—— 1985. Coastal evolution. *In:* EDWARDS, K. J. & WARREN, W. P. (eds) *The Quaternary History of Ireland.* Academic Press, London, 115–131.

THORN, C. E. 1988. *Introduction to Theoretical Geomorphology.* Unwin Hyman, London.

TOOLEY, M. J. 1978. *Sea-level Changes in Northwest England.* Clarendon Press, Oxford.

—— 1982. Sea-level changes in northern England. *Proceedings of the Geologists' Association,* **93**, 43–51.

—— 1985. Sea-level changes and coastal morphology in northwest England. *In:* JOHNSON, R. H. (ed.) *The Geomorphology of Northwest England.* Manchester University Press, Manchester, 94–121.

—— 1993, Long-term changes in eustatic sea-level. *In:* WARRICK, R. A., BARROW, E. M. & WIGLEY, T. M. L. (eds) *Climate and Sea-level Change: Observations, Projections and Implications.* Cambridge University Press, 81–107.

—— & SHENNAN, I. 1987. *Sea-level Changes.* Blackwell, Oxford.

VAN DE PLASSCHE, O. 1986. *Sea-level research: A Manual for the Collection and Evaluation of data.* GeoBooks, Norwich.

VERNON, P. 1989. The Late Quaternary History of Ballycotton Bay, Co. Cork. *Irish Quaternary Association Discussion Meeting – Abstracts,* Dublin.

WALCOTT, R. I. 1970. Flexural rigidity, thickness and viscosity of the lithosphere. *Journal of Geophysical Research,* **75**, 3941–3954.

—— 1972. Past sea levels, eustasy and deformation of the earth. *Quaternary Research,* **2**, 1–14.

—— 1973. Structure of the earth from glacio-isostatic rebound. *Annual Review of Earth and Planetary Sciences,* **1**, 15–37.

—— 1980. Rheological models and observational data of glacio-isostatic rebound. *In:* MORNER, N-A. (ed.) *Earth Rheology, Isostasy and Eustasy.* Wiley, New York, 3–10.

WARREN, W. P. 1990. Deltaic deposits in the Carey

valley, Co. Antrim, Ireland: a discussion. *Sedimentary Geology*, **69**, 157-160.
—— 1991. Fenitian (Midlandian) glacial deposits and glaciation in Ireland and the adjacent offshore regions. *In:* EHLERS, J., GIBBARD, P. L. & ROSE, J. (eds) *Glacial Deposits in Great Britain and Ireland.* Balkema, Rotterdam, 79–88.
WEBB, D. A. 1984. The flora of Ireland in its European context. The Boyle Medal Discourse, 1982. *Journal of Life Science Royal Dublin Society*, **4**, 143–160.
WHITTINGTON, R. J. 1977. A late glacial drainage pattern in the Kish Bank area and post glacial sediments in the central Irish Sea. *In:* KIDSON, C. & TOOLEY, M. J. (eds) *The Quaternary History of the Irish Sea.* Seel House Press, Liverpool, 55–68.
WINGFIELD, R. T. R. 1989. Glacial incisions indicating Middle and Upper Pleistocene ice limits off Britain. *Terra Nova*, **1**, 538–548.
—— 1992a. The Late Devensian (< 22,000 BP) Irish Sea Basin: the sedimentary record of a collapsed ice sheet margin – a discussion. *Quaternary Science Reviews*, **11**, 377–378.
—— 1992b. Modelling Holocene sea-levels in the Irish and Celtic Seas. *In: Proceedings of the International Coastal Congress*, Kiel, Germany, 1992. Christian Albrechts Universitat Press, Kiel.
—— 1994. Submarine evidence of palaeosealevels since 20ka BP from the Irish and Celtic Seas. *In:* WINGFIELD, R. T. R. (ed.) *Glacial Interglacial Sea-level Changes in Four Dimensions: Abstracts.* European Science Foundation, Strasbourg, **22**.
—— 1995. A model of sea-levels in the Irish and Celtic Seas during the end-Pleistocene to Holocene transition. *This volume.*
WOODMAN, P. C. 1985. Prehistoric settlement and environment. *In:* EDWARDS, K. J. & WARREN, W. P. (eds) *The Quaternary History of Ireland.* Academic Press, London, 251–278.
YALDEN, D. W. 1982. When did the mammal fauna of the British Isles arrive? *Mammal Review*, **12**, 1–57.
ZBINDEN, H., ANDREE, M., OESCHGER, H., AMMANN, B., LOTTER, A., BONANI, G. & WOLFLI, W. 1989. Atmospheric radiocarbon at the end of the last glacial: an estimate based on AMS radiocarbon dates on terrestrial macrofossils from lake sediments. *Radiocarbon*, **31**, 795–804.
ZONG, Y. 1993. *Flandrian Sea-level Changes and Impacts of Projected Sea-level Rise on the Coastal Lowlands of Morecambe Bay and the Thames Estuary.* PhD thesis, University of Durham.
—— & TOOLEY, M. J. 1995. Holocene sea-level changes and crustal movements in Morecambe Bay, northwest England. *Journal of Quaternary Science*, in press.

A model of sea-levels in the Irish and Celtic seas during the end-Pleistocene to Holocene transition

ROBIN T. R. WINGFIELD

British Geological Survey, Keyworth, Nottingham, NG12 5GG, UK

Abstract: A macroscale model is attempted to accommodate offshore observations indicative of former sea-levels in the Irish and Celtic seas. The indications considered here are those that appear to post-date 12 ka BP. Raised beaches at about present sea-level, and formed in the Last Interglacial, are found about these seas. This shows that controls to produce the later sea-level changes have acted to return to about their conditions in the Last Interglacial, and that later permanent tectonic displacements have not been significant. A simple, geometrically based model is developed of the interaction of: glacio-eustasy, from graphs based on coral-reef studies; glacio-isostatic depression with an annular forebulge of equal volume, both contracting through the interval considered; and hydro-isostasy, as an enhancement of the other effects by up to 20%. The best fit of the empirical submarine data suggests that from 12 to 9 ka BP relative sea-levels varied from 80 m above, and 160 m below present mean sea-level in the two seas. The resultant sea-level graphs for points from Scotland to Ushant compare with published graphs from eastern North America and northern Europe. Derived maps of palaeocoasts show land-bridges from Britain to Ireland and later to the Isle of Man after 11.35 ka BP.

One result of the systematic mapping of the continental shelf of the British Isles in the last 30 years has been a growing dataset of evidence indicative of relatively lower former sea-levels. Systematic offshore surveys have principally been made by the Hydrographic Office and by the British Geological Survey. Evidence indicative of palaeosea-levels discussed will be restricted to that from the Irish and Celtic seas (Fig. 1) and which appears to relate to Post-glacial events. In this area, towards the end of the Last Glacial (Weichselian–Midlandian–Devensian) stage an ice-sheet with a centre in Scotland established a long-lived margin in the present offshore area at about 51°N (Wingfield & Tappin 1991; Tappin et al. 1994), and occasionally extended over 100 km farther south to the present Isles of Scilly at 50°N (Scourse et al. 1990; Scourse 1991). The presently submarine deposits of the deglaciation and Post-glacial events are widespread, complex and thick to the north of 51°N (Jackson et al. in press; Tappin et al. 1994), but largely absent or only locally developed to the south (Evans 1990). These deposits will only be referred to in this paper where they exhibit evidence regarding former sea-levels. It is shown that most of the latest such indications relate directly to present seabed features.

The area of the two seas considered is some 800 km north–south by up to 400 km east–west (Fig. 1). For this reason, a macroscale model is adopted in an attempt to quantify the scattered data in a coherent framework. This model is simplistic and coarse and neglects details such as sea-level fluctuations of a few metres. Thus, since 5 ka BP it takes sea-level as having stabilized at about present mean sea-level (Tooley 1985).

Literature on former sea-levels about the Irish and Celtic seas has concentrated on the onshore data. These comprise two categories relevant to the present study. In the first are the raised beaches from the Last Interglacial at 132 to 120 ka BP (Chen et al. 1991). In the second are three models of relative sea-level changes, which considered only the interval since 20 ka BP or later, by Eyles & McCabe (1989), by Boulton (1990) and by Lambeck (1991a).

Last Interglacial raised beaches

There are widespread raised beaches of this age in Ireland, Wales and southwest England about the Irish and Celtic seas, in the west of Ireland fronting the Atlantic Ocean, and on the English, French and Channel Island coasts of the English Channel (Wright & Muff 1904; Mitchell 1972; Mitchell et al. 1973; Bowen 1977; Synge 1977; Keen 1978; Devoy 1983; Synge 1985; Warren 1985; Preece et al. 1990; Mitchell 1992; Keen 1995). The beaches lie sub-parallel to present mean sea-level (pmsl) at 2 to 8 m higher. Later tectonic disturbance of the original level is cited once (Preece et al. 1990). Over hundreds of kilometres of shoreline the scattered preservation of the Last Interglacial raised beaches at about pmsl implies that, whatever variations of

From Preece, R. C. (ed.), 1995, *Island Britain: a Quaternary perspective*
Geological Society Special Publication No. 96, pp. 209–242

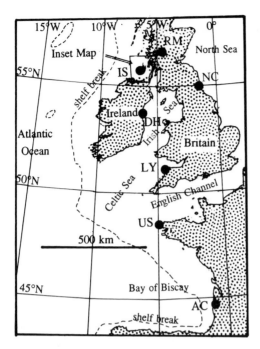

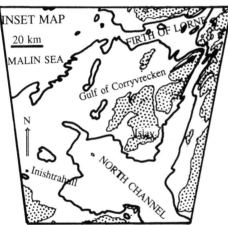

Fig. 1. General map, showing the shelf-seas between Scotland and Spain and fronting the Atlantic Ocean. RM, Rannoch Moor; IS, Islay; NC, Newcastle; DH, Drogheda: LY, Lundy; US, Ushant; AC, Arcachon. The inset map, outlined on the general map, is shown in greater detail to illustrate the deep water route from the North Channel to the Malin Sea through the Gulf of Corryvrecken. The heavy line represents the 60 m isobath. Land areas on both maps are stippled.

sea-level have occurred in the interim, the controls have returned to their settings during the Last Interglacial (K. Lambeck pers. comm. 1990). Evidently permanent tectonic movements, such as those brought about by folding, faulting, subsidence, compaction and salt withdrawal, have not produced significant effects over this period on the scale of modelling here attempted.

Eyles & McCabe (1989)

These authors described three phases of sea-level changes starting from deglaciation of the Late Devensian/Late Midlandian ice-sheet from c. 17 ka BP. Phase 1 was of relative falls of sea-level from pmsl +90 m on Cardigan Bay coasts and +150 m and higher on north Irish Sea coasts. Phase 2 was of rising sea-levels starting in the south with the Phase 1 falls reaching relative levels of below pmsl −55 m on north Celtic Sea coasts during the Pre-Boreal (Stillman 1968), and later northwards until the early Holocene lowstand northeast of the Isle of Man was relatively at least as low as pmsl −40 m (Pantin 1977, 1978). The sea-level rise of Phase 2 continued until it levelled off with the attainment of the range of the present tides at about 5 ka BP. Since that time, Phase 3 was one of quasi-stable sea-level continuing to the present day. The latter part of Phase 2 and Phase 3 follow the sea-level curves established by borehole studies in the coastal lowlands of Lancashire on the eastern margin of the Irish Sea (Tooley 1978). The authors suggested that the phases resulted from 'piano-key tectonics', such that glacio-isostatic depression with ice loading, followed by isostatic recovery with ice unloading, were both effected by differential block movements on pre-existing faults, so as to return to the initial conditions.

Boulton (1990)

This author modelled the interaction of eustatic, isostatic and gravitational effects on relative sea-level near to an ice-sheet during a glacial cycle. Models in the four dimensions of space and time (4-D) were figured for several of the former north European ice-margins, including the southern margin of the British ice-sheet of Late Weichselian age (but not the ice-margin southwest of Scotland in the Irish Sea). The rigid, elastic lithosphere was modelled to respond to glacier growth by increasing depression of the lithosphere beneath the growing ice-sheet with an external 'wave' of uplifted lithosphere, the proglacial bulge, around the periphery of the ice-sheet to a maximum width of 400 km. Deglaciation led to the decay of these effects *in situ* and not to proglacial bulge retreat. In this model worldwide variations in the amount of ice stored on land affected only the eustatic components,

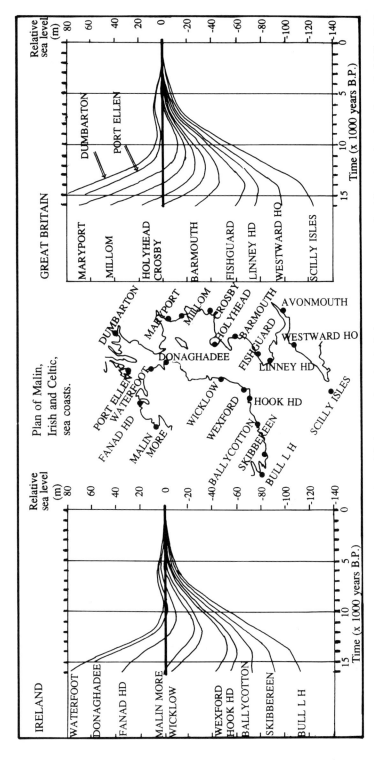

Fig. 2. Sea-level curves from 16 ka BP (after Lambeck 1991a): modelled curves of sea-level relative to present mean sea-level for selected points about the Malin, Irish and Celtic seas. Reprinted with permission from *Terra Nova*, copyright 1991 Blackwell Scientific Publications, Oxford.

which might have been out of phase with the local ice conditions, while the other components might have lagged, or led, ice-sheet growth and decay. Each of the 4-D models of former ice-margins figured had an inner 'isostatic' zone separated by a narrow (less than 100 km wide) hinge region, fixed through time, from an external 'eustatic' zone. Relatively higher sea-levels than psml were modelled solely within the isostatic zones. Relatively lower sea levels than pmsl were modelled to no lower than the eustatic levels in the isostatic zones and eustatic zones, except for the parts of the latter within 500 km of the hinge lines, where lower sea-levels induced by proglacial bulges were modelled, peaking at the glacial maximum (18 ka BP) to no lower than pmsl -160 m. It was inferred that in the Irish Sea area, after Eyles & McCabe (1989), local deglaciation led the eustatic cycle and consequently the hinge line was within the area of the ice-sheet and lay between the Isle of Man and Wales.

Lambeck (1991a)

This author modelled the entire British Isles area from 16 ka BP. The major inputs were all contributed from rheological theory and comprised a near-field produced by the unloading of the British and Fennoscandian ice-sheets and a far-field of eustatic changes and the isostatic effects of distant ice masses. Both the near- and the far-fields had water loading and unloading components. The model was constrained by empirical sea-level studies and was tested to find best-fit values for the geophysical parameters for the lithosphere and the mantle, and also to select the best-fit values for the disposition and volume of the near-field ice-sheets to be modelled from the various published examples. Figure 2 (after Lambeck 1991a) shows graphs of predicted relative sea-level variations about the Malin, Irish and Celtic seas. The three-fold phases of the Eyles & McCabe (1989) model were apparent, but with differing amplitudes in time and space: the changeover from the falling sea-levels of Phase 1 to the rising sea-levels of Phase 2 was lower and earlier in the south, and progressively later and to lesser low levels northwards into the Irish Sea. However, it was notable that the model of Lambeck (1991a) did not accommodate the pmsl -40m relative falls after 10 ka BP in the northeast Irish Sea off Millom reported by Pantin (1977, 1978). In contrast also to Eyles & McCabe (1989) in the Lambeck (1991a) model, Phase 3 was from 7 ka BP, and although it was also a phase of quasi-stability within the range of the present tides generally, relatively higher sea-levels than pmsl were figured to reach a maximum extent and amplitude at 6 ka BP: these higher levels were shown most prominently at Dumbarton and extending into the Irish Sea to Maryport (Fig. 2).

The evidence for these models was from onshore sections. Sub-sea-level data in boreholes were limited. The three models showed differing values for the relatively higher sea-levels. These differences reflected the continuing controversy concerning the levels attained by Lateglacial sea-levels about the Irish Sea. The 'marine minimalists' have suggested no higher than pmsl $+15$ m and the 'marine-maximalists' to higher than $+170$ m (Eyles & Eyles 1984; Thomas & Dackombe 1985; Eyles et al. 1985; Thomas 1985; Eyles & McCabe 1989; McCabe et al. 1990; Warren 1990; Eyles & McCabe 1991; Dackombe & Thomas 1991). Later research has studied several Late Devensian sequences onshore of supposed marine origin. These lie up to pmsl $+50$ m on the shores of NW Wales. It was assessed that the deposits were probably non-marine, and that the included marine faunas were transported from lower altitudes by ice (Austin & McCarroll 1992; McCarroll & Harris 1992; Walden et al. 1992).

The offshore evidence of former sea-levels has only started becoming widely available with the publication from 1983 of the British Geological Survey (BGS) 1 : 250,000 series maps for the UK shelf. Evidence of early Holocene back-beach and beach sediments to lower than pmsl -40 m in the northeast Irish Sea (Pantin 1977, 1978) was incorporated in the Eyles & McCabe (1989) model. Other relatively low Post-glacial sea-levels had been reported in the northern North Sea; with a relative sea-level fall of 180 to 190 m between 16 and 13 ka BP; beach formation at pmsl -160 m, dated at 12.2 ka BP; and beach formation at psml -150 m, dated at 11 ka BP (Rokoengen et al. 1982; Carlsen et al. 1986). Off Scotland, Peacock (1992) reported indications of shallow marine conditions showing relative sea-levels 140 m lower than pmsl at c. 13 ka BP on both the Hebridean shelf and on the North Sea shelf. West of Scotland sea-level had remained some 100 m below pmsl until 10 ka BP. In the north Celtic Sea. J. D. Scourse (pers. comm. 1992) reported shallow sub-tidal, possibly including intertidal, deposits cored at pmsl -123 m and determined by AMS radiocarbon dating at about 11 ka BP. Off Newcastle in the North Sea (Fig. 1), D. Long (pers. comm. 1992) found a deposit including rolled fragments of peat, which is overlain above an erosion surface by shelly, sub-tidal deposits at pmsl -60 m, and gave an AMS radiocarbon date of 9.3 ka BP. The three models described accommodate similar lowstands, but at considerably earlier dates. As Peacock (1992) noted, such lowstand indications from offshore suggest that a radical rethink of Late Quaternary isostatic responses in areas peripheral to the British Isles was needed. This paper presents a simple framework model. This model is geometric. It incorporates features from all the three cited models. Specific former sea-level indications found offshore will be given where appropriate. These include, in the Irish Sea, indications of substantially higher relative sea-levels in Post-glacial, distal glacimarine and boreal marine sediments (Wingfield et al. 1990). The latter findings lend some support to the 'marine maximalist' interpretation of the coastal sequences there. The time-span considered is restricted to the end-Pleistocene to start-Holocene interval.

The model's components and their integration

Changing sea-levels are modelled here relative to the reference surface of pmsl, whether the changes were produced by eustatic changes of water volume or by isostatic effects flexing the lithosphere. Only three components are modelled: glacio-eustasy, glacio-isostasy and hydro-isostasy. These are considered to have operated on medium (greater than 100 km) to worldwide distances. Factors which operated with variations over local (less than 100 km) distances are not considered to be susceptible to the macroscale modelling attempted. Such local factors would include the following:

(i) The gravitational distortion of the potentiometric surface, as the sea-surface, produced at ice-margins (Williams et al. 1993).
(ii) Short-term variations from mean sea-level produced by tides, by waves and by meteorological surges.
(iii) Erosion and deposition, which will have altered the former surface of the land or sea-bed over any interval of time to produce the present morphology. This is assumed to have had zero net effect on a macroscale. It is, however, realized that this assumption is invalid, particularly in the near-shore and coastal zone where both erosion and deposition were concentrated. For example, present extreme cases of inshore processes in the seas considered include the on-going lateral erosion of cliffs up to 30 m high (Thomas & Summers 1983), the build-up and migration of tidal sand ridges to 40 m in height (Kenyon et al. 1981) and the infilling of giant kettle holes as the present enclosed bathymetric deeps with up to 100 m of Holocene deposits (Wingfield 1989, 1990). Where significant changes of morphology are indicated due to net-erosion or net-deposition, allowances are applied.
(iv) Localized subsidence in time and space similar to that of the Lower Thames area in the Holocene (Devoy 1977).

It is intended that each modelled component should be kept as simple as possible. This involves incorporating simplistic assumptions, but allows the formulation of a geometrical model with very few variables. This combined model can then be empirically tested to find best fit values.

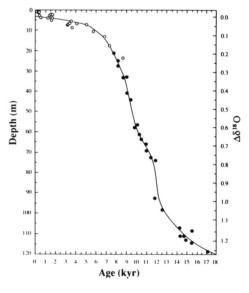

Fig. 3. Worldwide sea-level curve since 17 ka BP based on coral-reef studies (after Fairbanks 1989). Barbados sea-level curve based on radiocarbon-dated A. palmata (solid circles) compared with A. palmata age–depth data (open circles) for four other Caribbean island locations. All radiocarbon ages are corrected for local seawater $\Delta^{14}C$ by subtracting 400 years from the measured radiocarbon ages, but they are not corrected for secular changes in atmospheric ^{14}C levels. The Barbados data are corrected for the estimated mean uplift of 34 cm/ka^{-1}. The right-hand axis of the Barbados sea-level curve (solid line) is scaled to the estimated $\delta^{18}O$ change of mean ocean water. Reprinted with permission from Nature, copyright 1989 Macmillan Magazines Limited.

Modelling glacio-eustasy

Values for glacio-eustasy are taken from Fairbanks (1989, see Fig. 3). This graph was based on the study of palaeocoral-reef surfaces which lay at, or less than 5 m below, contemporary tropical sea-level. Thus for each case the sea-level lowstand included the full hydro-isostatic component of water unloading relative to pmsl. So Fig. 3 is not a 'pure' eustatic graph, but shows a combination of eustatic and hydro-isostatic components that can be separated. In the Irish and Celtic seas a glacio-isostatic component would additionally have been present for much of the time. If the latter component can be isolated, an assessment of the eustatic component can be made and, by reference to Fig. 3, a date assigned.

Modelling glacio-isostasy

This paper concentrates on the submarine evidence of lower sea-levels. In the Celtic Sea, indications of lowstands substantially lower (Pantin & Evans 1984) than any level of Fig. 3 require explanation. Furthermore, in the Irish Sea in the Early Holocene there were sea-level lowstands considerably lower than the points of the graph of Fig. 3 (Pantin 1977, 1978). Sea-levels lower than the eustatic falls were produced in the Boulton (1990) model solely in the forebulge area during the dilation of glacio-isostatic effects. They were not modelled during rebound. The model proposed, on the contrary, envisages a contraction of forebulge effects to account for a phase of substantially relatively lower sea-levels in Post-glacial times into the Early Holocene. Such a model favours the non-preferred shallow-mantle-flow, peripheral bulge concept of Daly (1934). It is not in accord with Daly's preferred 'punching' concept produced by deep-mantle flow. The latter hypothesis was postulated to result in a broad forebulge of negligible amplitude. The debate between the two concepts has continued to the present-day (Fjeldskaar & Cathles 1991). The model used is intended to framework the sea-level, and consequent shoreline, changes in a sector of the wide peripheral area of the last British and Irish ice-sheet in a simple manner. Lambeck (1991b) rheologically modelled sea-level changes in the former central areas of this ice-sheet. He showed a four-dimensional geometry (Fig. 2) that had similarities to the effects to be expected from a contracting forebulge. Lambeck (1991b, p. 36) noted that his model might be less adequate towards the former ice-sheet margins.

Figure 4 illustrates the model concept employed by two cartoon sections, one of an 'interglacial' geometry and one of a 'glacial' geometry. A flat-earth hypothesis is taken. The cartoons show the differing relationships of two surfaces: pmsl (present mean sea-level) and cmsl (contemporary mean sea-level). In cartoon 1, the interglacial situation without local ice loading effects, pmsl and cmsl are near-coincident, horizontal surfaces, which may be separated only by minor glacio-eustatic (+ water loading) effects. In cartoon 2 there is local ice loading. Beyond the influence of this loading, pmsl and cmsl are still both horizontal surfaces, but are separated by a significant, vertical, glacio-eustatic (+ water loading) difference, as Fig. 3. The ice loading depresses the whole lithosphere, of some 100 km thickness for the British Isles (Lambeck 1991b), to produce a conical downwarp. This glacio-isostatic depression is accommodated by lateral displacement of material in the underlying asthenosphere which produces an equivolume forebulge. The latter is taken to form an annular ring of triangular cross-section, with the uplift of the base of the lithosphere produced by the buoyancy of the displaced material (Fig. 4, after Mörner 1991). The lithosphere is considered to be rigid but elastic (Boulton 1990). This geometry models the pre-existing topography of the top of the lithosphere as merely tilted, so as to retain its relationship with the 'surface' of pmsl. This latter surface, henceforth referred to as 'pmsl', reproduces at the surface the new geometry of the base of the lithosphere. Also, as argued by Mörner (1991), the model can be expanded and contracted to mimic the lateral transfer, radially outwards and inwards, of material in the asthenosphere in response to ice loading and unloading. This simple geometry is taken to be adequate to framework macroscale changes. It incorporates numerous simplifications, as described below.

(i) The isostatic effects of lesser ice centres were either negligible, or subsumed into the main ice loading centred on a single point. For Last Glacial ice this neglects the isostatic effects of the lesser ice centres of Leinster, the Lake District, north Wales and Cork–Kerry, about the Irish and Celtic seas. In the British Isles a single point loading is taken of the main Scotland-centred ice at Rannoch Moor (56°40'N, 4°40'W).
(ii) The ice loading focus was stationary on Rannoch Moor (Fig. 1) for the relatively short period considered.
(iii) Only the isostatic effects of one ice centre need to be modelled in the Irish and Celtic

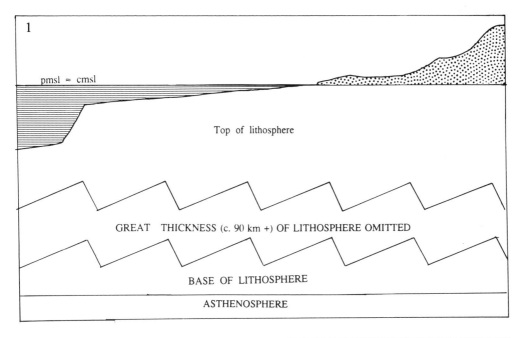

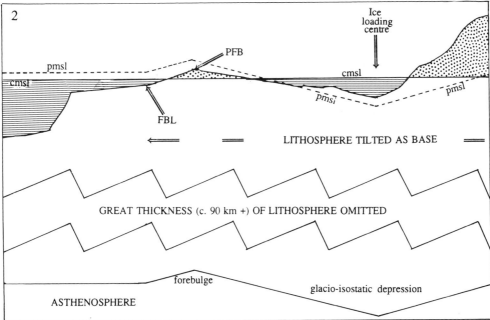

Fig. 4. Interglacial and glacial, sea-level, concepts. Cartoon 1 shows an ocean-to-continent profile during an interglacial stage with a horizontal base to the lithosphere at depth, and the contemporary mean sea-level (cmsl) roughly co-planar with pmsl. Cartoon 2 shows the same section in a glacial stage with local ice loading, and cmsl and pmsl separated by glacio-eustatic sea-level fall. Tilts caused by glacio-isostatic effects at the base of the lithosphere result in differing relations between cmsl, pmsl and the tilted, but otherwise unaltered, topography at the surface of the lithosphere. Points on cartoon 2 of change of tilt on the topography are shown as FBL (forebulge limit) and PFB (peak forebulge). The vertical scale of the cartoons is greatly exaggerated compared to the horizontal scale.

seas. The isostatic effects of other major ice centres, such as Fennoscandia and Iceland, are neglected for the model since 12 ka BP. This assumption would be invalid for similar modelling of the southern North Sea, where the isostatic effects of two ice-sheets (based on Scotland and Fennoscandia) would apply.

(iv) The curvature of the earth may be neglected over the sub-continental sized area modelled.

(v) Following Boulton (1990), the lithosphere is treated as an incompressible whole of constant thickness. This neglects the known variations of chemistry and thickness between continental and oceanic crust, both of which would have been included in the modelled circular area of ice loading effects about Rannoch Moor. For example, when the modelled glacio-isostatic effects had expanded to include the south Celtic Sea, the modelled effects would have included a sector of the northeast Atlantic Ocean (Fig. 1).

To simplify the geometry (Fig. 5) it is further assumed that the slopes of the conic sections, both inwards (from peak forebulge to centre glacio-isostatic depression) and outwards (from peak forebulge to limit forebulge), are the same. These slopes are taken as invariant during expansion and contraction. With this condition, the basic formula which arises from the equality of volumes (Fig. 5) can be solved; first, by eliminating the volume of the cone generated by revolving triangle CDB about CB, which appears on both sides of the equation; then by inspection to establish the following two ratios. (i) The radial dimensions of this model have the relationship of radius glacio-isostatic depression (CD) : radius peak forebulge (FC) : radius limit forebulge (CG) as 1.00 : 1.35 : 1.70. (ii) The amplitudes have the relationship of glacio-isostatic depression (CB) : height peak forebulge (EF) as 1.00 : 0.35.

Modelling hydro-isostasy

Water loading is modelled as an enhancement of the change of water level relative to pmsl produced by the combined modelled glacio-eustatic and glacio-isostatic effects at any point. These changes (dH) may be falls or rises. Since the density of sea-water compared to the average lithospheric density is about 1 : 3, the loading (or unloading) of a thickness of water (x) produces an isostatic enhancement proportional to $x/3$, to which must be added the compensation for that

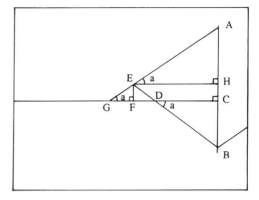

Fig. 5. Model of glacio-isostasy and the derived ratios. Geometry: if AB is the vertical axis of the ice loading and CG is the untilted surface of pmsl, BEG represents the tilting of pmsl by glacio-isostatic effects at depth. Considered as volumes generated by rotation about axis AHCB, the triangles CBD = cone of the glacio-isostatic depression (gid) and CED = annular forebulge (fb). For volumes of CBD and GED rotated about AB to be equal, so volume gid = volume fb:

cone AGC − 2(cone AEH + cone CDB = cone CDB

By inspection:

CD (radius gid) × 1.70 = CG (radius fb)
CB (depth gid) × 0.35 = EF (height fb)

loading (or unloading), as $x/3^2 + x/3^3$ and so on. This gives a formula for the hydro-isostatic enhancement (HIe) of:

$$\text{HIe} = dH(3^{-1} + 3^{-2} + 3^{-3} \text{ to } 3^{-n})K \quad (1)$$

where K is a constant, which was given by Turcotte & Schubert (1982) as 0.5. This gives a relationship of:

$$\text{HIe} = dH \times 0.2 \quad (2)$$

Hydro-isostatic enhancement relates to changes of water-level. Consequently, it must be applied differentially for falls and for rises relative to pmsl: (i) For former sea-levels relatively higher than pmsl, full hydro-isostatic enhancement applied at the position of the present-day coast and seaward, but reduced to zero at the contemporary coast and landward; (ii) for former sea-levels relatively lower than pmsl, full hydro-isostatic enhancement applied at the contemporary coast and seaward, but reduced to zero at the position of the present-day coast and landward.

Figure 6 models, in two dimensions, vertical sections radial to an ice centre to display the

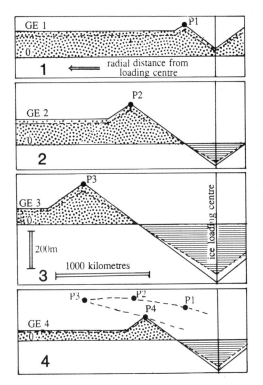

Fig. 6. The modelled locus of the peak forebridge. Sections 1 to 4 show successive time-frames of pmsl surfaces relative to contemporary mean sea-levels (=0). In each section, for pmsl the full lines incorporate full hydro-isostatic enhancement, and dashed lines have no hydro-isostatic enhancement. At the left-hand side of each section, beyond the glacio-isostatic effects pmsl is parallel to cmsl at differing levels (GE1 to GE4) controlled by glacio-eustasy. Progressive positions of the peak forebulge (P) in the sequence of sections define the locus of the forebulge peak through time. The general scales of the vertical and horizontal components are indicated. Dotted areas have pmsl above cmsl and horizontal lined areas have pmsl below cmsl.

general interaction through time of the modelled components for glacio-eustasy, glacio-isostasy and hydro-isostasy. In each section of this figure contemporary sea-level is shown as a fixed baseline, = 0, with the relative positions of pmsl, with and without full hydro-isostatic enhancements, above and below the baseline. (As the differences are relative, it would be equally valid to show pmsl as fixed and the contemporary sea-levels modelled below and above.) Each section has a left-hand side beyond the forebulge limit, where the pmsl lines, with no or full hydro-isostatic enhancement, are parallel to the baseline and relate directly to Fig. 3. Each section has a right-hand side with additional glacio-isostatic distortions, which expand and contract through the time-frames suggested by sections 1 to 4, but remain centred about the centre of ice loading. This simple representation in two dimensions presents a model in four dimensions, since each section represents any radial section at one time-frame, and the sequence of time-frames adds the fourth dimension.

The geometries of Figs 5 and 6 illustrate that, for the macroscale modelling attempted, most contemporary sea-levels in glacial stages were lower than pmsl. This follows since at any time frame the area of the forebulge (of relatively lower sea-levels) in proportion to the area of the glacio-isostatic depression (of relatively higher sea-levels) is as $(1.70^2 - 1.00^2):1.00^2$, a ratio of 1.89:1.00 (Fig. 5). This tendency for more widespread relatively lower levels was further compounded by the effects of glacio-eustasy (Fig. 6), since over the last c. 0.5 Ma, except for minor (5 to 10%) intervals during interglacials, contemporary glacio-eustatic sea-levels (with and without hydro-isostatic enhancement) were always lower than pmsl.

At any time-frame (Fig. 6) the lowest relative sea-level is modelled at the forebulge peak. There all three effects considered reinforce one another. The figure illustrates the predicted locus of the forebulge peak through time.

Indications at the present sea-bed of late, low sea-levels

It is suggested that the locus of the forebulge peak, as identified in Fig. 6, may provide a method of solving this model by the input of offshore indications of extreme lowstands. Lambeck (1991a) had modelled that, over the interval considered here since 12 ka BP, diminishing lowstands should have occurred progressively from south to north through the Celtic and Irish Seas (Fig. 2). Table 1 lists, from south to north, the latest and lowest indications of low sea-levels interpreted from features of the present sea-bed. To describe such features spatially for the modelling, each is referred to by its 'reference latitude' (R.Lat.). Such R.Lat.s lie on the 4°40'W meridian southward of Rannoch Moor, the centre of ice loading modelled. The following figures show arcs of circles centred on Rannoch Moor drawn on equidistant projection maps. In this manner an arc though a feature gives its R.Lat. at its intersection of the 4°40'W meridian.

Table 1. *Indications from south to north through the Celtic and Irish seas of the latest and lowest sea-levels*
All positions are given as distances from Rannoch Moor expressed as equivalent latitudes (R.Lat.s) on the meridian 4°40′W south of Rannoch.

Site	R.Lat. (°′ N)	Present water depth (m)	Indication of relatively lower sea-level	Reference
a	47 30	160	Disposition of tidal sand ridges	Bouysse et al. (1976)
b	48 30	148	Disposition of tidal sand ridges	Pantin & Evans (1984)
c	50 00	130	Disposition of tidal sand ridges	Pantin & Evans (1984)
d	51 10	117	Shallow sub-tidal	J. D. Scourse (pers. comm. 1992)
e	52 43	90	Restricted tidal strait	Mogi (1979)
f	53 20	75	Restricted tidal strait	Mogi (1979)
g	54 00	60	Shallow sub-tidal deposits	D. Long (pers. comm. 1992)
h	55 10	45	Restricted tidal straits	Mogi (1979)

Moribund tidal sand ridges in the Celtic Sea

Figure 7 shows that across the deeper shelf between Brittany and southwest Ireland, numerous sand ridges, up to 200 km long and 12 km wide, have a fan-like disposition. The lineations vary from north–south in the southeast to east–west in the northwest. Table 2 itemizes the statistics of these ridges considered in swathes of 1°R.Lat. width between 47° and 51°N R.Lat.s. The table shows that three factors each diminish northeastwards: the depth of the general sea-floor underlying the ridges, the depths over the ridge crests; and the depths of the terminations of the ridges to the east and north. The sea-floor east and north of the area with sand ridges is a gently sloping rock platform with higher areas formed by the granite outcrops of Haig Fras and the Isles of Scilly and by local shoals and islets of Palaeozoic basement rocks off Ireland and Brittany. In contrast the inter-ridge areas are underlain by sands some 5–10 m thick (Pantin & Evans 1984) over a planed-off surface of pre-Quaternary rocks. The stratigraphy of the sand ridges is three-fold.

(i) The ridges consist of sand with minor gravel. They are up to 50 m thick over planar unconformities on bedrock. Some profiles show cores of pre-existing sand ridges (Pantin & Evans 1984; Belderson et al. 1986).

(ii) Patches of glacigenic deposits are proved on the inter-ridge sands, on ridge flanks and in one case on a ridge crest. Of the 12 patches found, the five northeast of R.Lat. 49°N were attributed to formation by grounded ice (Fig. 7A–D and F), while the seven to the southwest (Fig. 7 E and G–L) were assessed to be of glaci-marine origin (Scourse et al. 1990). Scattered boulders are also reported resting on the sands (Hamilton et al. 1980; Pantin & Evans 1984).

(iii) A gravelly veneer surfaces both the sands and the glacigenic deposits. This was described as a winnowed lag deposit up to 0.2 m thick (Pantin & Evans 1984). This veneer may be lacking across the southernmost tidal sand ridges, where Bouysse et al. (1976) described active sand waves in the areas adjacent to the shelf edge.

Kenyon et al. (1981) considered that the lack of sand waves on the Celtic Sea tidal sand ridges, except in the extreme south, showed that the features were moribund. In contrast they found that active ridges, carrying sand waves, had bases at no more than 40 m water depth, and crestal depths ranging from pmsl −2 to −4 m (in estuaries) to pmsl −15 to −20 m (off Norfolk). All previous authors, when they considered the Celtic Sea tidal sand ridges, inferred that they formed during substantially lower sea-levels with estimates ranging from pmsl −140 m (Pantin & Evans 1984) to pmsl −100 m (Belderson et al. 1986).

The present disposition of the tidal sand ridges of the Celtic Sea shelf is difficult to explain by any of the maximum sea-level falls postulated by previous authors. Pantin & Evans (1984) suggested a maximum fall of some 140 m. Belderson et al. (1986) modelled formation of all the ridges with a 100 m sea-level fall. Scourse et al. (1990) had a maximum fall of 88 m. Yet with the most extreme of these falls the bases of the southernmost ridges, now at 200 m depth would have been at over 60 m contemporary depth and

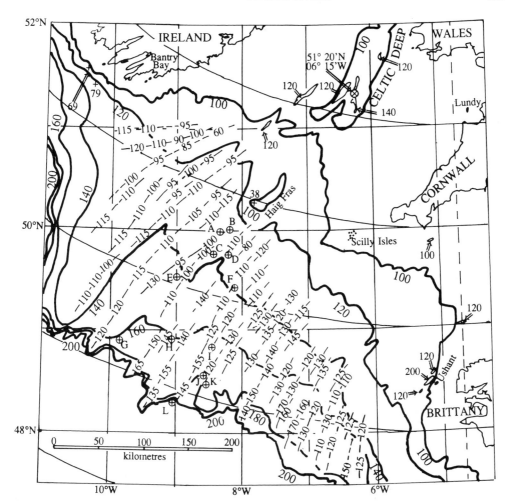

Fig. 7. Tidal sand ridges of the Celtic Sea. All figures are of present-day depths in metres. Bathymetric contours are shown solely between 100 m and 200 m water depths. Figures are shown on the deeper side of contours. Heavy lines with large figures show the general trend of the sea-bed. Dashed lines show the crests of tidal sand ridges, with the least crestal depths in metres shown by small figures along the ridges. Shoreward of the 120 m isobath, fine lines delineate local enclosed bathymetric deeps with depths arrowed in. Plus signs in deep water off Ireland and on Haig Fras, with arrowed least depths in metres, are pinnacle rocks. ⊕ symbols mark the 12 sites (A to L) where cores have been obtained from patches of glacigenic material overlying the sands of the tidal sand ridges of the inter-ridge sands. Depths in metres (uncorrected) are as follows: A, 127; B, 127; C, 151; D, 118; E, 137; F, 144; G, not recorded; H, 147; I, 146; J, 183; K, 173; L, 210. The 4°40′W meridian of Rannoch Moor is shown by a broken line. Arcs are shown about Rannoch for each 1°R.Lat. on this line.

considerably below the level of present-day tidal sand ridge formation. Moreover, the northernmost ridges, if contemporaneous with the others (Belderson *et al.* 1986), with a 100 m fall of sea-level would have largely been above contemporary sea-level. In the framework proposed here it is suggested that the ridges did not form contemporaneously, but in sequence. In this scheme, the southern, now deepest, ridges formed first, and the ones farther north formed progressively later and to lesser present depths. This process is attributed to the northward passage of a diminishing forebulge (Figs 6 and 8).

Four considerations of the dispositions of the tidal sand ridges in the Celtic Sea may allow estimation through the area of the extreme lowstands of sea-level (Table 1; sites a, b and c).

(1) The first consideration is that supply of sediment limited where the tidal sand ridges

Table 2. *Celtic Sea tidal sand ridges.*
Depths in metres below pmsl.

Swathe (R.Lat., °N)	Depth of inter-ridge sea-floor	Depth of ridge terminations to east and north	General crestal depth	Least crestal depth
47–48	165–200	140–170	125–140	120
48–49	135–180	135–150	110–145	100
49–50	125–150	125–140	95–130	80
50–51	110–130	110–125	85–120	60

formed. The sands were deposited previous to tidal sand ridge formation on the Celtic Sea shelf as glacial outwash; either as proximal glacimarine deposits (Scourse et al. 1990), or as subaerial sandur from the ice-margin to the north about the Celtic Deep (Wingfield & Tappin 1991; Tappin et al. 1994) and possibly also from an ice-margin to the east between Cornwall and Ushant (Wingfield 1989). The ice to the north had surged as far south as the northern Scilly Isles about 20 ka BP (Scourse et al. 1990; Scourse 1991). Before 13 ka BP, wastage of the ice-sheets to complete disappearance south of Scotland had occurred (Eyles & McCabe 1989), and the sediment supply had ceased. It is modelled in this paper that later return of the forebulge northwards exposed the shallower parts of the shelf eastwards, and progressively northwards towards Ireland, so that those parts were swept clean of sediment by coastal processes. The sands were concentrated across the deeper parts of the shelf to the west during this interval. In this slightly deeper area, with less than 40 m contemporary water depths, existing moribund remnants of former tidal sand ridges from the previous shallow-water episode acted as cores to a new suite of active tidal sand ridges.

(2) Active tidal sand ridges form sub-parallel to the main tidal streams (Kenyon et al. 1981; Belderson et al. 1986). Consequently, the ridges form sub-parallel one to another and to an adjacent coast, or to coasts. The modelled approach of the peak forebulge from the south first produced north–south aligned coasts west of Brittany and later across the entrance to the present English Channel. Secondly, coastal alignments lay southwest to northeast about the present-day shoal areas with this lineation (then exposed as islands) comprising the southwest salient of the 140 m contour and Haig Fras (Fig. 7). Later still, forebulge passage into the northernmost Celtic Sea produced a west–east aligned coast south of Ireland. At each stage these coastal dispositions are reflected by the disposition of then-active tidal sand ridges, while in deeper waters southwards with declining tidal currents, the ridges with differing alignments were no longer active.

(3) Since tidal sand ridges are formed of loose sediment, they could not have survived through a marine regression as upstanding features, neither could they have formed as subaerial features, such as sand dunes, nor as shoreline features, such as sand spits, and survived drowning (Scourse et al. 1990). Therefore the presence of tidal sand ridges, at any particular present depth, implies that sea-level has not subsequently fallen below that depth. It follows that as relative depths below contemporary sea-level deepened after the passage of the forebulge peak (Fig. 8), tidal sand ridges would have continued to build up to that limiting level while parts of the bedforms still lay in the envelope of 'active tidal sand ridge' depths (i.e. 0 to 40 m). As the waters progressively deepened, activity would have finally ceased, even for the most favourably maintained ridges. Thereafter the only sediment movements would be short-lived agitation in storms (noting that sonar studies suggest that gravel waves are reformed by storms down to depths of at least 60 m in the Irish Sea (Jackson et al. in press)). Such occasional storm action would lead to the degradation of the sand waves associated with the previously active ridges and to winnowing of the immediate sea-bed. The winnowing would have produced the noted top layer of lag sediment. Adjacent to the shelf edge the large oceanic waves appear to be adequate to produce active sand waves at present, and were suggested to have been responsible there for the wholesale destruction of some tidal sand ridges to form sand wave fields and lesser ridges (Bouysse et al. 1976). Alternatively, modern ocean-wave action may produce internal density waves, causing the active sand waves of the shelf-break region (Pantin & Evans 1984).

(4) The occurrence of patches of glacigenic sediments and scattered boulders locally on the sands of the tidal sand ridges imply that these sediments were deposited after the ridges had

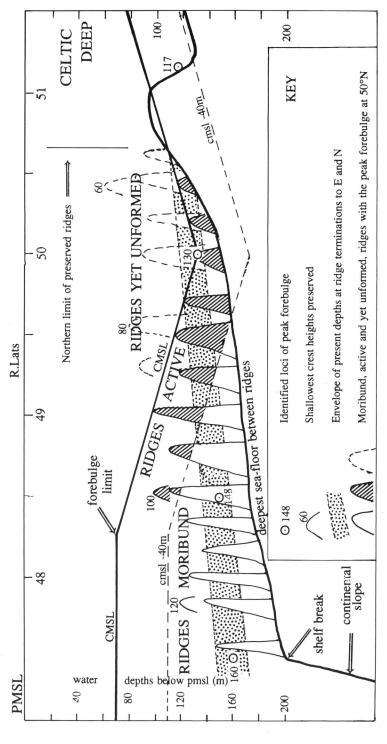

Fig. 8. Cartoon of tidal sand ridge activity in the Celtic Sea with the passage of the forebulge. The section shows a general profile relative to pmsl of the presently moribund tidal sand ridges. The envelope of present depths, in which the ridges terminate to the east and north, is shown dotted, and it is inferred that the locus of the peak forebulge was confined in this envelope. Superimposed on the section is the inferred surface of contemporary mean sea-level (cmsl) to a peak forebulge at pmsl −130 m at R. Lat. 50°00′N controlling the active, moribund and yet-unformed tidal sand ridges.

ceased to be active in that area. Had the tidal sand ridges been active, the glacigenic sediments would have been buried or reworked. Unproven buried patches and boulders may exist. Scourse et al. (1990) suggested that the grounded ice deposits are restricted southwards by the present 135 m depth contour, with glacimarine deposits forming the patches further south in deeper water (Fig. 7). Since grounding of ice was required to produce the northern patches, in the framework model presented, icebergs must have been the glacigenic agent, rather than sea-ice, since keel drafts greater than 40 m would have needed to ground following tidal sand ridge activity. Scourse et al. (1990) identified a northerly provenance, on mineralogical grounds, for these glacigenic deposits, but it is difficult to accept that icebergs from tidewater ice-margins in either the British Isles or northwest Europe could have drifted to the Celtic Sea shelf after the post-12 ka BP date which is put forward here for the formation of the tidal sand ridges. Icebergs from North American sources could have been a possible drift agent after 12 ka BP (Fairbanks 1989; Bond et al. 1992; Alley et al. 1993), but would not fit the mineralogical evidence of provenance.

The following correlations are proposed for the evolution of the Celtic Sea tidal sand ridges and their preservation:

(i) that the maximum sea-level falls with peak forebulge passage corresponded approximately with the termination depths of the tidal sand ridges to the east and north (Figs 7 and 8; Table 1, sites a–c; Table 2);
(ii) that the growth of tidal sand ridges in the still shallow, but progressively deepening, waters after peak forebulge passage northwards (Fig. 8) involved some lateral extension of the ridges onto the bare rock areas that had formerly been exposed to coastal erosion, thus blurring the distinction between the areas that had been above sea-level and those always offshore;
(iii) that after tidal sand ridge formation had ceased in water by then greater than 40 m deep, local iceberg grounding occurred in waters as deep as *c.* 60 m, and iceberg dumping and rainout occurred in deeper waters, followed by winnowing of the sea-bed in these depths during storms.

Applying these correlations to the statistics of each swathe of R.Lat.s (Table 2) leads to two inferences: (a) that present average crestal depths of each swathe are some 40 m shallower than the inferred sea-level fall there with peak forebulge passage; (b) that exceptional least crestal depths in any swathe are some 65 m shallower than the inferred minimum fall (possibly such rare lengths of crest represent parts that suffered minimal subsequent degradation).

Drowned intertidal deposits

Although the rates of relative palaeosea-level change are modelled to have reached extremes of 100 to 160 mm/a (Lambeck 1991*a, b*; Blake 1992; this paper) and may have exceeded the current rate of 1.8 mm/a (Carter 1992; Woodworth 1992) by nearly two orders of magnitude, measured rates of coastal recession in Quaternary deposits are relatively so rapid that the likelihood of the preservation of exposed subaerial and intertidal deposits through a marine transgression is low (Table 3). For example, a point with a spring tide range of 5 m, similar to that at Holyhead on the Irish Sea (Fig. 2), would have remained in the intertidal zone during a 100 mm/a sea-level rise for 50 years (Table 3a). In these 50 years it would have undergone 35 000 semi-diurnal tidal cycles. Furthermore, the point would have remained in the surf zone and exposed to seasonal storm effects for a substantially longer interval.

Only special circumstances would have allowed the preservation of distinctive subaerial and intertidal unlithified deposits through a marine transgression. More coherent deposits, such as peat and tree roots, had only a slightly higher preservation potential. However, recognizable clasts of the latter materials could have survived with minor (*c.* 100 m) lateral displacement (Smith 1972), but with vertical displacements as high as the amplitude of adjacent tidal channels (*c.* 2–5 m). The processes of coastal erosion occur in the intertidal and surf-zone envelope by the lateral retreat of a low-angle ramp (Everts 1987), with the reduction of cliffs by undercutting and talus removal (Williams & Davies 1987). However, distinctive subaerial and intertidal deposits were preserved, in some cases, due to burial by the succeeding shallow marine, and later deeper marine, sediments. Two settings may have enhanced the chances of such preservation in the area of the Irish and Celtic seas during the period considered: (a) where rising sea-level had overtaken the laterally eroding shoreline ramp to bury parts as notches cut into earlier subaerial and backbeach deposits; (b) where a topographically lower area was occupied by either subaerial or intertidal deposits, in which case preservation potential was higher as the sediments were more likely to be buried with

Table 3. *Quantification of rates of rise and of erosion.*
(a) Periods (in years) that a point will remain in the intertidal zone with sea-level rise.

Rise (mm/a)	Tidal range			
	2 m	6 m	10 m	14 m
1	2000	6000	10000	14000
5	400	1200	2000	2800
10	200	600	1000	1400
50	40	120	200	280
100	20	60	100	140
160	13	38	63	88

Note: for semi-diurnal tides of c. 12.5 h cycles between high waters, there are some 700 tidal cycles annually.

(b) Recorded rates of lateral coastal retreat in various sediments and rock types.
Based variously on aerial photographs, ground photographs, maps and surveys dating over several decades, and on engineers' plans (as of lighthouse foundations), mariners' logs and charts, and historical documents (especiallly legal records) dating over centuries.

	Refs	Quoted rate	10^x m/10a $X=$
Loose recent sediments (sands and muds)	Ehlers (1988) Shepard & Wanless (1971)	2 to 3 km in 20 years	3
Permafrost sediments (frozen muds, sands and gravels)	Héquette & Barnes (1990) Ruz et al. (1992)	2 to 18 m/a	2
Semiconsolidated Quaternary sediments and soft Tertiary rocks (friable sands, clays and diamicts)	Steers (1948) Snead (1982)	Spasmodically by up 70 m in extreme storms, average <2 m/a	1
Moderately lithified Mesozoic rocks (chalks, marlstones, mudstones, sandstones and limestones)	Williams & Davies (1987) Davies & Williams (1991)	3 to 47 cm/a, average <10 cm/a localized cliff failures	0
Indurated, strongly lithified, Palaeozoic and older rocks, but including younger igneous rocks	Carter (1991) Shepard & Wanless (1971)	Less than 0.5 m in 100 years. Local collapses of stacks and blocks	−1

the lateral passage of the shoreface ramp.

The preservation potential of deposits in any of these settings is greater where they lie within indented coasts formed by rocks of contrasting lithology which, in consequence, have widely differing retreat rates (Table 3b). Protection in embayments was probably largely responsible for the preservation of the following types of deposits. (a) Freshwater deposits at pmsl −55 m (dated at 11 ka BP (Stillman 1968)) in Bantry Bay, southwest Ireland (Fig. 7); (b) Organic reed-swamp deposits at pmsl −22 m (?Pre-Boreal (Tooley 1985)), proved in a borehole close offshore of Llandudno, north Wales; (c) Interbedded freshwater marsh and shallow marine deposits from pmsl −16m upwards (with dates later than 9.3 ka BP (Huddart et al. 1977; Tooley 1985)) proved in boreholes in the estuaries and coastal lowlands of Lancashire.

Across the broad areas of the present offshore shelf with slight topography, former embayments were generally lacking. The proved cases of intertidal and backbeach deposits recognized in shallow cores (less than 6 m sub-sea-bed) appear to owe their preservation to deposition in places which were, in one way or another, protected by the morphologies associated with former lower sea-levels; as described in the following three paragraphs.

Pantin (1977, 1978) described numerous cores from the Irish Sea east of the Isle of Man (Fig. 9). There, glaciamarine and boreal marine muds, with occasional dropstones and a limited fauna and flora, are overlain with unconformity by temperate intertidal sediments. The latter sediments comprise 0.2 to 3 m thick, variegated intertidal deposits with a distinctive, temperate, low-salinity biota (Pantin 1977, 1978; Table 4). These intertidal deposits form 'Sedimentary Layer 2' (SL2) below the sea-bed, and pass up

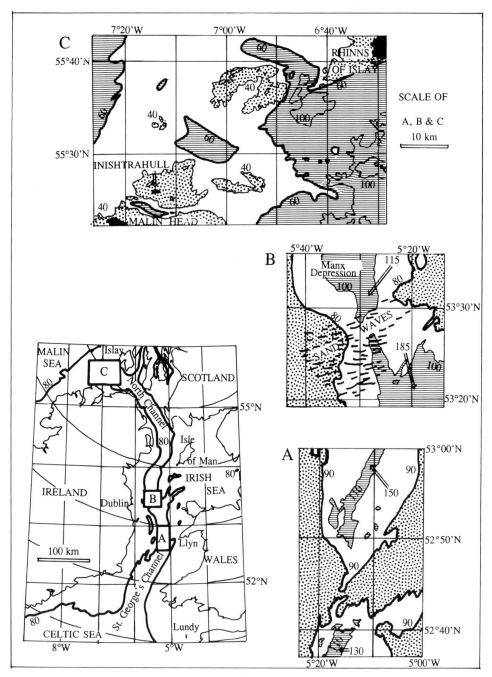

Fig. 9. Former tidal scour cauldrons about restricted straits. All the maps show a limited number of depth contours with the figures in metres below pmsl shown on the deeper sides. On the three detailed maps, with common horizontal scales, the morphologies are emphasized by the use of stipples referring to differing depths on the separate maps. Extreme spot depths are arrowed. The general map from the Celtic Sea to the Malin Sea shows the areas of the detailed maps. Arcs at 1°R.Lat. intervals are centred on Rannoch Moor. Heavy lines show the 80 m isobath. (**A**) St George's Channel: west of the Llyn. (**B**) West Irish Sea: northeast of Dublin. (**C**) North Channel: Malin Head–Inishtrahull–Islay.

Table 4. *Cores of intertidal deposits.*
After Pantin (1977, 1978)

BGS block*	No.	Site†		U/T‡	Depth§	R.Lat. (°'N)		
Isle of Man, east sheet								
55-05/	81	O	1	T	25	54	20	
	109	N	1	T	34	54	18	
	116	M	2	T	32	54	13	
	119	N	3	T	27	54	12	
	122	M	3	U	34	54	10	
	123	N	3	T	34	54	12	
	129	L	3.5	T	37	54	07	
	132	L	4	U	33	54	05	
	134	M	5	U	44	54	03	
	135	J	4	U	41	54	03	
	177			U	52	54	41	Borehole 73/50
	187			U	51	54	39	Borehole 73/69
	193			T	32	54	39	
Lake District, west sheet								
54-04/	29	T	2	T	23	54	22	
	34	T	3	T	25	54	19	
	40	P	2	T	41	54	18	
	46	P	3	T	34	54	14	
	48	O	3	T	25	54	15	
	52	R	5	T	34	54	09	
	59	R	6	T	31	54	05	
	70	N	6	T	46	54	01	
	71	P	7	T	33	53	59	
	72	R	8	T	28	53	58	
	74	O	7	T	40	53	58	
	75	Q	8	T	32	53	57	
	96	S	3	T	28	54	17	
Anglesey, east sheet								
53-05/	68	H	7	U	56	53	51	
	85	L	6	T	48	53	58	
	90	G	5	U	47	53	56	
	92	J	5	U	47	53	59	
	98	G	4	T	48	54	00	

* All blocks are degree rectangles referred to by the southwest corners as degrees N (53) latitude and W -05 and -04) longitude
† Site numbers are those quoted in Pantin (1978)
‡ U, cores showing the base SL2 unconformity on older deposits; T, core terminated in SL2 deposits
§ Depths quoted are in metres below pmsl to the base seen of SL2, the backbeach and intertidal deposits, in each core. As the water depths taken were uncorrected, these levels are accurate to ±3 m (the half-range). Depths from the boreholes are accurate to ±0.5 m

into sub-tidal sediments with evidence of deepening water depths, which comprise 'Sedimentary Layer 1' (SL1) and include the present-day sea-bed sediments. The range of depths of the SL2 deposits (Table 4) is constrained by the present water depths of the east Irish Sea, which range only from 20 to 50 m. This part of the Irish Sea represents a near-land-locked sea which during lower sea-levels would have had the character of a large bay or bight that evidently had preservation potential for intertidal deposits.

In a short core at 51°20'N 06°15'W (Fig. 7), at R.Lat. 51°15'N in the Celtic Deep, a 1 m thickness of shallow sub-tidal, possibly including intertidal, deposits with an abundant diagnostic biota occurs below inner-shelf sediments (Scourse & Austin 1994). These very shallow marine deposits lie at pmsl − 123 m. AMS radiocarbon dates of 11.1 ka BP have been obtained from the shell material. With sea-level reduced by 115–120 m (and less to the north and south as the framework model presented), the Celtic Deep would have formed the extremity of a bay extending 550 km south from the Malin Sea (Figs 7 and 9). This bay would have had the same form, though it was three times longer, as the present-day Bristol Channel between Linney Head and Avonmouth (Fig. 2), and it is probable that a comparable, but an even larger, variation of tidal range would have operated. The present variation of ranges is from 6 m at Linney Head to 14 m at Avonmouth, a half-range variation from pmsl of 4 m. A tidal range allowance is here applied of 6 m to the pmsl −123 m depth of these shallow sub-tidal, and possibly intertidal, deposits to obtain the Table 1 (site d) figure of 117 m for contemporary mean sea-level. The setting of these deposits, both in a bay and in a topographically low area, would have assisted their preservation.

Off Newcastle in the North Sea (Fig. 1), at R.Lat. 54°00′N, D. Long (pers. comm. 1992) found an unfossiliferous sand incorporating rolled peaty fragments, and resting on marine muds. The deposits with peat detritus were overlain, above an erosion surface, by sub-tidal, shelly, muddy sands passing up to the present sea-bed. The erosion surface lies at pmsl −60 m (Table 1, site g) and the shells provided AMS radiocarbon ages of 9.3 ± 0.25 ka BP.

The indication of an Early Holocene, pmsl −60 m, lowstand off Newcastle is consistent with the indications of lowstands then to at least pmsl −53 m (Table 4) in the east Irish Sea, and with other findings at the same R.Lat. 54°N in the west Irish Sea, described below.

Evidence of shallowing has been proved in a number of short cores. A scatter of shallow (less than 6 m long) core samples at sea-bed in the 80 to 110 m present water-depths of the west Irish Sea contain distal, glacimarine and boreal marine muds, with micropalaeontological evidence of deposition in waters notably deeper than the depths of their present setting. These are succeeded upwards by less saline, temperate marine muds, with micropalaeontological evidence of deposition in (sub-tidal) waters notably shallower than the depths of their present-day setting (Wingfield et al. 1990).

In BGS borehole 89/15 at 54°02′N 05°21′W, (R.Lat. 54°00′N), 38 m of temperate marine muds (R. Harland 1990, BGS internal report; Dickson & Whatley in press) were proved below the 92 m deep sea-bed and resting on glacimarine deposits across an erosion surface. There is no significant present-day sedimentation in these deep waters, so it is reasonable to infer that these thick, (?)early Holocene, deposits formed in nearer-shore and substantially shallower conditions, as is indicated by the photophyllic species found among the abundant calcareous nanoplankton from 25 to 37 m sub-sea-bed (117 to 130 m below pmsl).

In the west Irish Sea (Fig. 9) is a broad, flattish plain with 70 to 110 m water depths across thick Post-glacial muds. This mud-belt is very locally interrupted by ten upstanding outcrops, formed by Carboniferous Limestone and Tertiary intrusive rocks patchily mantled by glacial till (Wingfield 1985; James 1990). These outcrops have forms ranging from pyramids, some as little as 100 m across, to ridges, up to 3 km long. They stand from some 5 to 50 m higher than the general sea-bed. Each upstanding feature is surrounded by distinctive scours, which were excavated into the muds to cross-cut the stratification. These scours have the form of moats and hollows of up to 1 km width from a prominence, and descending to between 10 and 55 m deeper than the general sea-bed. The shallowest pyramid, which has the deepest moat, lies at 53°59.2′N 05°34.0′W; it has a minimum depth of pmsl −53 m, lies in an area with an average sea-bed at pmsl −90 m, and the deepest hollow in its moat is at pmsl −145 m. With the postulated extreme sea-level fall of pmsl −60 m at this R.Lat. (Table 1, site g), these upstanding features would have formed shoals and islets in a channel some 25 km wide and up to 50 m deep between Ireland and the Isle of Man. Presumably the scour features formed due to vortices about the shoals and islets, which disturbed the ebb and flow of the tidal currents through this comparatively narrow strait.

Tidal scour cauldrons

Figures 7 and 9 illustrate a number of enclosed bathymetric deeps (Herzer & Bornold 1982) ranging from Ushant to the extreme east of the Irish Sea. Most of these features form the unfilled parts of much larger, boat-shaped, Lateglacial erosion features. Wingfield (1989, 1990) described deeps in the latter setting as giant kettle-holes, and illustrated several examples from the Irish and Celtic Seas (Wingfield 1989, Fig. 8b, c, d, e and h).

An apparently different class of enclosed bathymetric deeps also occurs in this area, concentrated along the waters deeper than 80 m running northwards through St George's Channel, the west Irish Sea and the North Channel, to end southwest of Islay (Fig. 9). These latter deeps are erosional features without significant infills. In plan, they lack the distinctive linear shapes of the deeps described as giant kettle-holes (Wingfield 1989, 1990), forming, in contrast, pairs of irregular, often roughly triangular deeps. The most distinctive feature of these deeps identified is that they are paired: each such pair is separated by a narrow and shallower constriction of the deep waters (Fig. 9A, B and C).

It is suggested that the paired deeps are analogous to the tidal scour cauldrons illustrated by Mogi (1979) in numerous, inter-insular, restricted Japanese straits. The Japanese examples have similar extent, and are also down-cut into both sediments and bedrock, but differ in that they range in depths from comparable depths to more than twice as deep (maximum pmsl −400 m) as the Irish Sea examples (maximum pmsl −185 m). This difference may arise since the Japanese examples of cauldrons lie about present-day restricted straits, and may be inferred to have been excavated over the last 5 ka, since sea-levels stabilized at roughly the present level (Fig. 3); whereas the Irish Sea examples probably formed over a restricted period (? about 200 years) of forebulge-passage-induced sea-level fall to produce restricted straits. Paired deeps on continental shelves ascribed to tidal scour formation have also been described between the Frisian Islands (Ping 1989, pers. comm. 1991) and inside Hainan Island in south China (Chen & Zheng 1987; Shiqing 1989). No analogous features of these dimensions, other than those ascribed to tidal scour action in narrow straits, have been located in the literature.

It is assessed here that three sets of former tidal scour cauldrons identify the late, low sea-levels itemized in Table 1, sites e, f and h. These are as follows.

(1) A strait at R.Lat. 52°43′N (Fig. 9A) which had operated just after an extreme fall of pmsl −90 m, if allowances are made for the very large tidal ranges to have been expected in the deep bays, on either side of the last isthmus here, Subsequent erosion by currents through the strait caused tidal scour cauldrons.

(2) A strait at R.Lat 53°22′N (Fig. 9B) operated over an extreme sea-level fall of some 75 m. Some 15 m general lowering of the surface due to erosion in the strong currents of the former strait is allowed for, and it is noted that a degree of infilling of the northwestern scour cauldron of the strait in the Manx Depression (James 1990) may have occurred. Later sediment movements in this area, continuing to the present day, are shown by the presence of giant sand waves, which are features found throughout the deeper parts of St George's Channel (James & Wingfield 1987; James 1990).

(3) From the North Channel northwestwards to the Malin Sea there would have been three restricted straits if sea-level had fallen at least to pmsl −40 m (Evans 1986). Falls of this size or greater occurred, since each of the restrictions is marked by scour cauldrons (Fig. 9C).

Further north than the area covered by Fig. 9 is a fourth strait, the Gulf of Corryvreckan (Fig. 1), which also has paired tidal scour cauldrons to c. 180 and 135 m below present depths, by which route present-day minimum water depths greater than 60 m extend from the North Channel to the Malin Sea (Evans 1988).

Even an extreme sea-level fall of over 60 m through the area north of Ireland with forebulge passage would not have formed a land-bridge to Scotland across more than one of the four straits at any one time (Devoy 1985). There is a spread of R.Lat.s of the four postulated former straits as follows: 54°45′N (Malin Head–Inishtrahull); 55°00′N (15 km northeast of Inishtrahull); 55°10′N (15 km west-southwest of the Rinns of Islay); and 56°00′N (Gulf of Corryvreckan). While an extreme sea-level fall (of some pmsl −45 m) with postulated peak forebulge passage might have affected one strait, it is modelled that the other straits in this area would have been controlled at that time instant by sea-level falls at least some 10 m less (i.e. to no lower than pmsl −35 m). Table 1, site h takes an average indicated extreme fall of pmsl −45 m for the former strait west-southwest of Islay.

A best-fit solution

If it is accepted that the values listed in Table 1 represent points on the locus of the peak forebulge to be modelled, then the geometry of the framework model (Fig. 7) apply as shown in Table 5. For any point on the locus, its difference in R.Lat. from Rannoch Moor × 1.70/1.35 (= 1.26) gives the difference in R.Lat. to the forebulge limit from Rannoch. By testing trial values of the model's slope (assumed to be invariant), the corresponding levels of the forebulge limit are found for any position of the peak forebulge. These latter values would be for contemporary shorelines below pmsl in each case. They therefore represent eustatic levels with full hydro-isostatic enhancement and relate directly to the graph of Fig. 3, shown in detail in Fig.10A. The best fit is found using three criteria (Table 5): (a) a progressively lessening series of levels without the glacio-isostatic component occurs; thus the values for test slopes of 45 or 50 m/°R.Lat., which vary up and down, are unacceptable, whereas those of lesser slopes than 40 m/1°R.Lat. provide lessening sequences. (b) Ages for peaks of the forebulge at R.Lat. 51.16° and 54.00°N have dates of 11.1 and 9.3 ka BP, respectively, in accordance with the AMS radio-

Table 5. *Finding best-fit dates to points on locus forebulge peak.*

Reference latitudes in decimal degrees					Relative sea-level below pmsl							Corresponding ages (ka BP as Fig. 3)					
PFB	Diff. PFB to R/Moor	×1.26 (diff. FBL to R/Moor)	FBL	Diff PFB and FBL	PFB	Assumed values for model slope (m/°R.Lat.)						Modelled slopes (m/°R.Lat.)					
						25	30	35	40	45	50	33	35	36	37	39
						= test levels of forebulge limits											
47.50	9.17	11.55	45.12	2.38	160	101	89	77	65	53	41	12.35	12.1	12.0	11.7	11.2
48.50	8.17	10.29	46.38	2.12	148	95	84	74	63	53	42	12.2	12.0	11.7	11.5	11.0
50.00	6.67	8.40	48.27	1.73	130	87	78	71	61	52	44	11.9	11.4	11.3	11.1	10.8
51.16	5.50	6.93	49.74	1.42	117	82	74	67	60	53	46	11.4	11.2	11.1*	10.9	10.7
52.72	3.95	4.98	51.69	1.03	90	64	59	54	48	43	38	10.1	10.0	10.0	10.0	9.9
53.33	3.30	4.16	52.51	0.86	75	54	49	45	41	36	32	9.8	9.7	9.7	9.6	9.5
54.00	2.67	3.36	53.31	0.69	60	43	39	36	32	29	26	9.4	9.35	9.3*	9.2	9.1
55.17	1.50	1.89	54.78	0.39	45	35	33	32	29	27	26	9.0	8.9	8.9	8.9	8.85

*Dated levels
PFB, Peak forebulge; FBL, Limit forebulge; Diff, difference; R/Moor, Rannoch Moor at 56°40′N (56.67 decimal degrees)

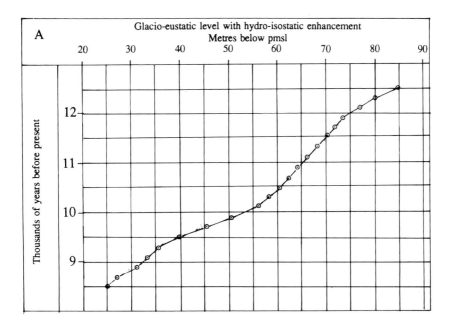

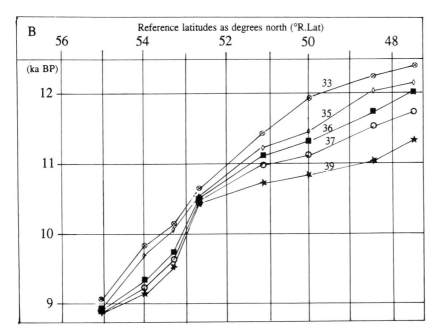

Fig. 10. Graphs of worldwide palaeosea-levels and tested slopes. (**A**) Glacio-eustatic levels with full hydro-isostatic enhancement from Fig. 3 applied in the modelling from 8.5 to 12.5 ka BP. (**B**) Graphic representation of the third block of numbers of Table 5. Values for a test slope of 33 m/1°R.Lat. are shown as ⊗; for 35 m/1°R.Lat. as ◇; for 36 m/1°R.Lat. as ■; for 37 m/1°R.Lat. as ○; and for 39 m/1°R.Lat. as ★. The selected best-fit value for the slope is 36 m/1°R.Lat. This value incorporates full hydro-isostatic enhancement.

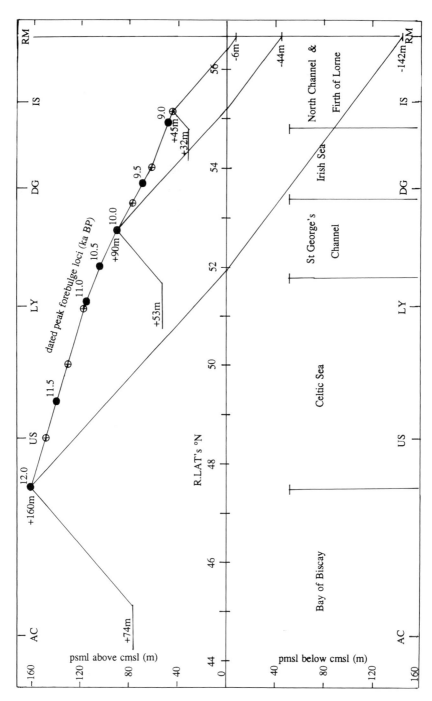

Fig. 11. Sea-level model from 12.0 to 8.9 ka BP from Biscay to Scotland. Conic sections are shown for 12.0, 10.0 and 8.9 ka BP, each of pmsl relatively above and below cmsl (=0). The modelled slopes of pmsl above cmsl are with hydro-isostatic enhancement at 36 m/1°R.Lat., and the slopes of pmsl below cmsl are without hydro-isostatic enhancement at 30 m/1°R.Lat. ⊗ symbols indicate positions of the forebulge peak, other than those indicated by solid circles, identified by the empirical data (Table 1). The coded locations (AC, US, LY, DG, IS and RM) are identified in Fig. 1. The latter are inserted at the position of their R.Lat.s, as are the sea areas.

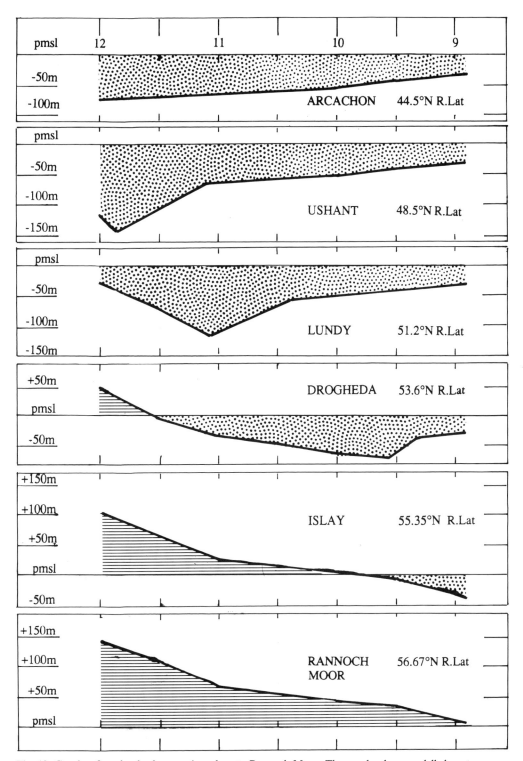

Fig. 12. Graphs of sea-levels changes, Arcachon to Rannoch Moor. The graphs show modelled contemporary sea-level changes relative to pmsl (=0) for the locations plotted in Figs 1 and 11. The solid line on each graph is for values incorporating full hydro-isostatic enhancement for cmsl lower than pmsl, and for values without hydro-isostatic enhancement for cmsl higher than pmsl. The shaded areas show the controls of the positions of palaeoshorelines: dotted for coasts at levels below pmsl and horizontally lined for coasts at levels above pmsl.

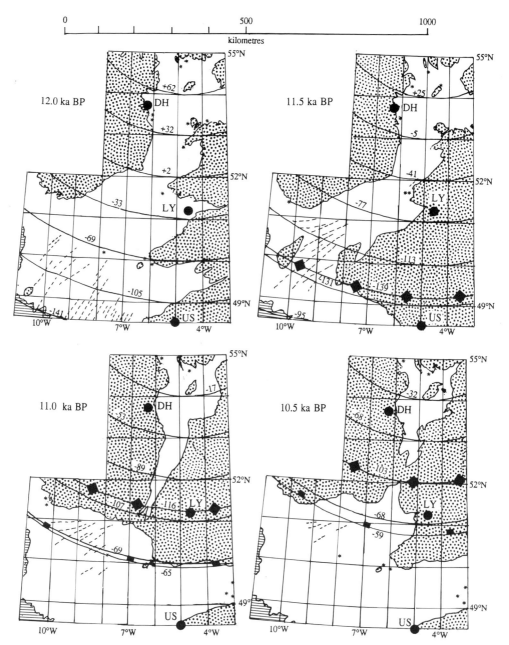

Fig. 13. Maps of palaeocoasts at 500 year intervals. The coded locations (IS, DH, LY and US) are identified in Fig. 1. Arcs at 1°R.Lat. intervals are shown within the forebulge limit on each map with the modelled contemporary sea-level in metres relative to pmsl; values are also shown on the arcs of the peak forebulge and forebulge limit for each map. Areas exterior to the limit forebulge on each map are modelled to have the same contemporary sea-level as that shown for the forebulge limit. Modelled contemporary sea-levels below pmsl incorporate full hydro-isostatic enhancement; levels above pmsl are without the enhancement. Land areas are dotted, shelf-seas are blank, and off-shelf areas of greater than 200 m contemporary water depths are lined. 'Incipient major kettle-holes' show the sites of present-day enclosed bathymetric deeps. These are only shown for the Holocene on maps from 10 ka BP. It has been inferred that ground ice, insulated by overlying sediment, remained at these sites for several thousand years after the onset of temperate conditions before thawing out after the site was submerged by marine transgression with rising sea-level (Wingfield 1989, 1990).

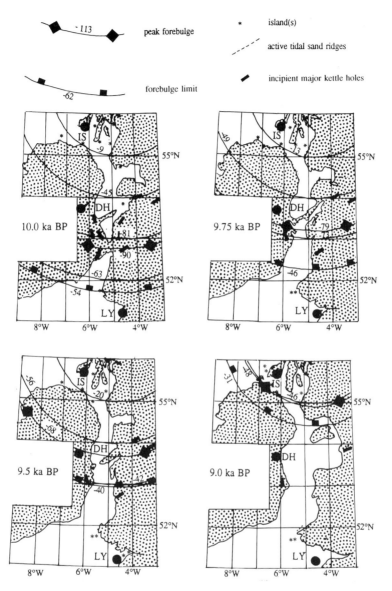

carbon dates cited – this eliminates possible slope values less than 30 m/1°R.Lat.; (c) When tested for small increments of slope change (Table 5, block 3 for 33, 35, 36, 37 and 39 m/ 1°R.Lat. slopes), a smooth curve results when displayed graphically – Fig. 10B shows a graph of these values and emphasizes the chosen best-fit slope of 36 m/1°R.Lat.

As both the peak forebulge and limit forebulge levels analysed would correspond to contemporary shorelines below pmsl, both sets of values incorporate full hydro-isostatic enhancement. The best-fit model slope of 36 m/ 1°R.Lat. would reduce by one sixth to 30 m/ 1°R.Lat. when the hydro-isostatic enhancement is removed. The latter, lesser slope is modelled for contemporary shorelines above pmsl. The two slopes are extremely low: 1 in 3090 with full hydro-isostatic enhancement, and 1 in 3700 without it.

Figure 11 shows the locus of the forebulge with dates derived from Table 5. Conic sections showing the slopes modelled for contemporary relative sea-level, with and without hydro-isostatic enhancement, are shown for 12.0, 10.0 and 8.9 ka BP. This framework model is based on

Table 6. *Modelled land-bridges from Ireland to Britain.*

	R.Lat.(°N), centre of land-bridge	Present position:
	50.8	Cork to west Cornwall, southwest of Celtic Deep
	51.5	Cork to Lundy Island, northeast of Celtic Deep
	52.3	St George's Channel, east of Wexford
	52.7	St. George's Channel, west of Llyn Peninsula
	53.4	Irish Sea, east-northeast of Dublin Bay

FBP below pmsl (m)	Width (km) (x = no connection) 50.8	51.5	52.3	52.7	53.4	Years (BP)	Comments
123 at 50.5°	x	x	x	x	x	11 250	3–5 km wide, 100 km long, strait running west of Celtic Deep from WSW to ENE.
116 at 51.25°	25	x	x	x	x	11 000	Land-bridge shown on Fig. 13 for 11 ka BP
111 at 51.35°	?	10	10	x	x	10 850	The remnant of the land-bridge south of the Celtic Deep at 51.5° survived as sandspits until shortly after the formation of narrow land-bridges southwest and north of western Wales.
103 at 52.0°	x	x	40	x	x	10 500	Wide land-bridge west of Cardigan Bay, shown on Fig. 13 for 10.5 ka BP.
97 at 52.35°	x	x	30—?—30	x	x	10 250	Possibly continuous land-bridge up to 60 km wide, west of whole present Cardigan Bay.
90 at 52.75°	x	x	x	25	x	10 000	Land-bridge west of the Llyn, shown on Fig. 13 for 10 ka BP, included the sites of the later-formed tidal scour cauldrons shown on Fig. 9A.
83 at 53.0°	x	x	x	x	x	9900	No land-bridge between Ireland and Britain. The connection west of the Llyn has submerged to create a narrow strait. The resulting intense tidal scour forming cauldrons (Fig. 9A). Another strait lay to the northwest of Anglesey from St George's Channel to the east Irish Sea south of the Isle of Man (Fig. 13 for 10 ka BP).
79 at 53.2°	x	x	x	x	7	9750	Another land-bridge formed when recession exposed land across the former strait south of the Isle of Man. This 100 km long land-bridge is shown on Fig. 13 for 9.75 ka BP.
72 at 53.5°	x	x	x	x	3	9600	The last land connection to Ireland from Britain has shrunk to a narrow isthmus separating the Manx Depression to the northwest and St George's Channel to the southeast (Fig. 9B and Fig. 13 for 9.5 ka BP).
68 at 53.66°	x	x	x	x	x	9500	No connection. Narrow strait at site of last land-bridge with tidal scour cauldrons shown on Fig. 9B and on Fig. 13 for 9.5 ka BP.

the assumption that the graph of the forebulge retreat was linear between the plotted values, and that intermediate values, as shown, can be interpolated. This is unlikely to have been the case, since rising and falling tendencies (Tooley 1985) caused by lesser cycles were likely. Even neglecting the effects of the possible lesser cycles, the graph of Fig. 11 may indicate that the overall curve reflects a change in the rate of overall forebulge retreat in that the intervals plotted between 12 and 11 ka BP are much larger than from 11 to 9.5 ka BP. It may be speculated that this tendency results from plotting positions along what may merely be part of a large-scale sine curve.

Figure 11 shows that sea-levels relatively above pmsl extended as far south as southern St George's Channel to 52°N at 12.0 ka BP, when they stood greater than 80 m above pmsl in the northernmost Irish Sea. It also shows that by 10.2 ka BP, sea-level above pmsl only occurred north of the Irish Sea; and the limit of the forebulge only as far south as the Irish Sea by 8.9 ka BP. This model allows such conic sections to be drawn for any position of the forebulge peak, so intersects through time of these sections allow graphs of relative sea-level change to be erected for selected points. Figure 12 shows such graphs for the points identified in Figs 1 and 11 for a range of R.Lat.s. between the southern Bay of Biscay and Rannoch Moor.

Figure 13 uses data derived from Figs 11 and 12, combined with the present topography of the area above and below pmsl, to derive the changing palaeocoastlines of the Irish and Celtic seas at 500 year intervals. Allowances are made, as seems appropriate, for the effects of later erosion or sedimentation for each map, but these allowances are relatively minor. The maps are of a small scale and neglect the effects of tides in that they attempt to depict the contemporary mean sea-levels.

Discussion

The proposed model is admitted to be crude and simplistic. Although this allows simple manipulation of the model to incorporate a range of sea-level indications, it incorporates numerous simplistic assumptions. These may be justified for the framework model aimed at. If the model can be shown to give useful insight into the four-dimensional changes that affected former sea-levels, it may be that it merits improvements aimed at eliminating some of these in-built assumptions. Such improvements might take the form of introducing rheological considerations, or the use of a more sophisticated geometry. In addition to the weaknesses of the model, the extreme low sea-level data, used here to constrain the model, are not themselves well constrained and have only two dated levels, with the other dates deduced by reference to Figs 3 and 10A. Considerations touched upon of the probable differing tidal ranges of the former seas tend to suggest that the accuracy of the levels itemized in Table 1 are to ±5 m. Consideration of Fig. 10B may suggest that the framework model presented for dates is constrained to as little as ±250 years.

Taking into account these several sources of error, no attempt is made here to quantify the errors incorporated in the graphs and maps presented. The values shown in both time and space should be regarded as indicative and relative, expressing only a current 'best-fit' solution.

Published sea-level graphs from eastern North America compare with the modelled graphs of Fig. 12. The following North American graphs are for points at increasing distances from the Late Wisconsinan ice-sheet centre at 54°N 75°W east of Hudson Bay (Williams et al. 1993). (a) A graph for the St Lawrence estuary (Praeg et al. 1992) is comparable from 12 to 7 ka BP with that modelled for Islay from 12 to 8.9 ka BP. (b) Graphs for northeast Newfoundland (Shaw & Forbes 1990) and for the Maine coast (Shipp et al. 1991) despite being for points some 1500 km apart, were essentially identical. These graphs from 12.5 to 9 ka BP compare with the graph modelled for Drogheda from 12.0 to 8.9 ka BP. It is noteworthy that these widely separated points in North America lie at about the same distance from the local ice centre. Another point of note is that the post-6 ka BP rise of relative sea-level above pmsl of the St Lawrence estuary graph (Praeg et al. 1992) is not found on either of these graphs. (c) A graph for Sable Island (Amos & Miller 1990) is similar to that modelled for Ushant from 12.0 to 8.9 ka BP, except that the maximum fall of sea-level is shown substantially earlier at 13 ka BP.

Similar comparisons may be made with other sea-level curves: from Canada (e.g. those shown in Fulton 1989, p. 430, fig 5.16); from the Baltic area relative to Fennoscandian ice loading (Raukas 1991, p. 127, fig. 2); and for western Europe (Carter 1992). The later curves were restricted to the Holocene Stage from 10 ka BP, and in the northern areas (north of c. 54°N) had excursions of relative sea-level above pmsl after 7.5 ka BP, as was the case for the St Lawrence estuary graph (Praeg et al. 1992). Although events after 8.9 ka BP are outside the scope of this paper, it may be noted that a re-expansion of the glacio-isostatic model south into the Irish Sea would be required to accommodate the relatively higher sea-levels plotted by Lambeck (1991a) in southwest Scotland and into the northernmost Irish Sea between 6 and 5 ka BP (Fig. 2).

In addition to the evolution of the tidal sand ridges in the Celtic Sea and the later sequential formation and drowning of restricted straits through the Irish Sea, which have been used to constrain the model, the sequence of sea-level changes in the four-dimensional model fits with several other lines of evidence. It agrees with the three-fold phases of major tendencies of relative sea-level change at any point in the models of Eyles & McCabe (1989) and Lambeck (1991a); it also agrees that the 'wave-front' of the change from Phase 1 fall to Phase 2 rise was earlier in the south, and occurred to a lesser extent and later in the north. The dates modelled for the passage of the wave-front are even later than those proposed by Lambeck (1991a) and considerably later than the dates inferred by Eyles & McCabe (1989). There is a problem for both the models of Lambeck (1991a) and the present model in that high sea-levels in the northernmost Irish Sea to post-12 ka BP may conflict with the earliest dates suggested by pollen studies of the freshwater deposits of lowland kettle-holes of this region (Huddart & Tooley 1972).

The model supports the long-running contention that a land-bridge between Britain and Ireland must have existed into the temperate conditions of the Littletonian (Holocene) Stage to explain the Irish biota (Mitchell 1960, 1972, 1990; Preece et al. 1986, and reviews within). Table 6 and Fig. 13 plot that a land-bridge formed from 11.2 ka BP southwest of the Celtic Deep. Over the following 1600 years, until about 9.6 ka BP, a series of land-bridges is indicated to have been progressively exposed and later submerged with forebulge peak passage through the present St George's Channel into the Irish Sea (Figs 7 and 9). The site of each postulated land-bridge is now a wide swathe of flat sea-bed of lessening depths to the north. It is inferred that the flatness of each swathe resulted from phases of lateral coastal erosion, during both the recession to produce the land-bridge and later transgression to submerge it. The postulated northernmost land-bridge, running southwest from the Isle of Man towards Dublin (Figs 9B and Table 6), is inferred to have stood some 15 m higher in its mid-parts that its present psml −75 m maximum depth. A loss of height of this magnitude during submergence by the rapid lateral coastal erosion of Quaternary sediments (Table 3b) and the subsequent formation of tidal scour cauldrons across the initially narrow strait formed by 9.5 ka BP is modelled (Fig. 13). This model does not suggest that land-bridges persisted after 9.6 ka BP. This is not consistent with the detailed biological conclusions (Mitchell 1960, 1972, 1990; Preece et al. 1986), which required a land-bridge persisting for another 2 ka. But the survival of a land-bridge into the early Holocene would remove the major stumbling block of requiring glacial refugia in Ireland through the worst of the Lateglacial climatic conditions (Preece et al. 1986). The model suggests that a land-bridge existed in temperate, Early Holocene conditions for some 600 years – from 10.2 (Mörner 1976) to 9.6 ka BP – which ranged up to 60 km in width (Table 6). Before the appearance of this land-bridge, modelled at c. 11.2 ka BP, the fauna of the Lateglacial Interstadial, including the Irish giant deer Megaloceros giganteus, would have needed to cross from Britain to Ireland where a wide, deep channel is modelled (Fig. 13). Passage of a restricted fauna rafted on sea-ice, as occurs occasionally with Arctic foxes and polar bears to the present day across the 300 km wide Denmark Strait from Greenland to Iceland, is a possibility. Examples of presumed migrations of megafauna across sea-ice or open straits between isolated, North American glacial refugia in the Late Wisconsinan were cited by Pielou (1991). By 11.25 ka BP the Celtic Sea (Table 6) had shrunk to a 100 km long strait, of 2 km minimum width, through 51°10′N 7°30′W, before further recession produced a land-bridge to Ireland. Grounded sea-ice in the strait might have restricted the formation of characteristic tidal scour features. Minor enclosed deeps to the ENE and WSW of 51°10′N 7°30′W (Lawson et al. 1988, Fig. 7) may have been formed during the closure of this strait; alternatively some of the entirely infilled incisions (Delantey 1980), 80 to 170 m deep, of various Quaternary ages along an ENE to WSW aligned tract through 51°40′N 07°30′W may be related to this event. Had tidal scour cauldrons formed, their later infill might have been expected during the transgressive submergence of the land-bridge.

By 10.0 ka BP another land-bridge is modelled (Fig. 13) as forming from Britain to the Isle of Man. This land-bridge is shown as persisting to 9.0 ka BP. Its transgression would have pre-dated by some 500 years the submergence of the final, Post-glacial land-bridge from Britain to the mainland of Europe, which Jelgersma (1979) placed between East Anglia and the northern Netherlands until some 8.5 ka BP.

Whereas the offshore evidence cannot by its nature resolve the controversy regarding the levels obtained by higher sea-levels about the Irish Sea, the model presented shows levels at 12.0 ka BP up to nearly pmsl +85 m and accords with both Lambeck (1991a) and the marine-maximalist school. As noted, numerous short (less than 6 m long) sea-bed cores in the west Irish Sea proved distal, glacimarine and boreal marine muds with micropalaeontological evidence of deposition in waters notably deeper than the depths of their present setting (Harland & Gregory, internal BGS reports 1986, noted in Wingfield et al. 1990). Since the depths of water in which these cores were taken range from 80 to 110 m, 'notably deeper' may imply water depths some 50% greater. These findings support the concept of relatively high, Lateglacial/early Holocene, sea-levels in the Irish Sea (Eyles &

McCabe 1989), but do not quantify the levels attained.

The glacio-isostatic effects modelled as retreating from the Bay of Biscay to the North Channel from 12.0 to 8.9 ka BP (Fig. 11) would have 'lagged' by several thousand years the waning of the ice-sheet over most of the region. Eyles & McCabe (1989) showed that glacial ice was absent from the region of the Irish and Celtic Seas by 13 ka BP, when ice was confined within Scotland. All the glacio-isostatic effects modelled here after 12.0 ka BP took place after glacier disappearance south of Islay. In an analogous manner, graphs for coastal points in eastern North America (Shaw & Forbes 1990; Shipp et al. 1991; Amos & Miller 1990; Praeg et al. 1992) applied to areas from which the Wisconsinan ice-sheet was absent since 11 ka BP (Fulton 1989). The continuing glacio-isostatic rebound at present in both Scotland and Fennoscandia, respectively some 10 and 8 ka after the last presence of ice-sheets, may also exhibit such a lag. The suggested magnitude of the lag to be expected from glacial loading allows the speculation that the resurgence of an ice-sheet in western Scotland in the Younger Dryas from c. 12.8 to 11.7 ka BP (Taylor et al. 1993) is reflected as a re-expansion of the glacio-isostatic effects much later, to produce the relatively high sea-levels in southwest Scotland between 7 and 5 ka BP. Although beyond the scope of the present paper, it may be noted that a re-expansion of the modelled glacio-isostatic effects, but about a different ice loading centre south of Rannoch Moor and with lesser eustatic levels by those dates, could have produced a peak forebulge of no lower than pmsl −25 m at about 54°N. Such a fall of sea-level would not have produced a further land-bridge to the Isle of Man (least depth 35 m), but might account for repeated apparent lowering of sea-level reported by Pantin (1977; pp. 50–51) in the northeast Irish Sea in the uppermost (SL1) strata.

Conclusions

Cyclically acting sea-level controls

There is widespread evidence from the coasts of Ireland and Britain that the sea-levels of the Last Interglacial lay at, or just above, present mean sea-level. This implies that the major sea-level changes since the Last Interglacial were affected by controls which acted cyclically and have restabilized at about their initial conditions. The cyclic controls identified are glacio-eustasy, glacio-isostasy and hydro-isostasy.

The extent of relatively lower, glacial stage sea-levels

A simple geometrically based model of the three controls is described with fixed ratios, which allow its manipulation in four dimensions to accommodate empirical sea-level data. The model shows that at any time instant, other than in relatively short interglacials, areas with relatively lower sea-levels were extensive. Only in comparatively restricted areas, where glacio-isostatic depression exceeded glacio-eustatic sea-level fall, did relatively higher sea-levels occur.

The locus of the modelled peak forebulge

It is modelled that, where the three principal components reinforced each other at the peak forebulge, the lowest sea-level occurred at any one time. Such points are identified from features of the present sea-bed through the Celtic and Irish Seas, and are used to solve the model. This best-fit solution is a slope for the modelled conic sections of 1:3090 with full hydro-isostatic enhancement, giving a slope of 1:3700 without hydro-isostatic enhancement. This solution also gives dates and allows presentation of the model in the dimensions of space and time.

A best-fit solution to the model

The best-fit solution to the model is presented as a series of graphs and maps showing former sea-levels relatively above and below the pmsl reference surface between 12.0 and 8.9 ka BP.

The maps, of palaeocoasts at 500 year intervals for the two seas, show the following main features:

(i) Tidal sand ridges formed in less than 40 m water depths with retreating forebulge passage through the Celtic Sea. The maxima of these sea-level falls decreased from pmsl −160 m in the south to pmsl −130 m in the north. Concurrently, the relatively higher sea-levels in the Irish Sea decreased from pmsl +83 m.

(ii) A land-bridge formed to Ireland across the Celtic Sea southwest of the Celtic Deep after 11.2 ka BP, and transferred to northeast of the Celtic Deep by 10.85 ka BP, by which time very few of the tidal sand ridges in the Celtic Sea were still active, and relatively higher sea-levels in the Irish Sea were restricted to north of the Isle of Man. The last land-bridge is modelled as submerging by 9.6 ka BP, with the last connection southwest of the Isle of Man in the Irish Sea (Table 6).

(iii) Narrow straits with strong tidal streams are modelled as progressively forming with the passage of the peak forebulge northwards between Ireland and Wales, and later between Ireland and Scotland. Although there are abundant features below the present deep waters of the west Irish Sea of substantially shallower conditions, it is in the shallower east Irish Sea that exposure to produce subaerial and intertidal conditions occurred, and a land-bridge extended to the Isle of Man from Britain.

Comparisons of the modelled graphs with published graphs

The graphs produced by the model compare with sea-level graphs published from eastern North America, from Baltica and from western Europe of changes over the Holocene interval. Whereas it is recognized that because a model fits the data (as it should have been constrained to do) it is not necessarily valid, the comparability of the modelled graphs with published graphs may suggest that the model, for all its recognized crudity, merits further study with the aim of improving it.

The apparent lag of the glacio-isostatic effects

The modelled glacio-isostatic effects are later than the removal of the ice load in this area by several thousand years, as also appears to have been the case in North America and Fennoscandia.

Although the model presented is the sole responsibility of the author, he is grateful for discussions over many years with colleagues concerning the former sea-levels of the two seas and their surrounding lands, in particular with G. S. Boulton, R. V. Dackombe, R. J. N. Devoy, C. D. R. Evans, R. Harland, K. Lambeck, D. Long, A. M. McCabe, G. F. Mitchell, H. M. Pantin and J. D. Scourse. Particular gratitude is due to D. Long and J. D. Scourse for permission to quote their unpublished and crucial findings off Newcastle and in the Celtic Deep, respectively. The basis of the work from which the model stems was provided by offshore surveys run by the British Geological Survey and by the Hydrographic Office, and the author acknowledges his thanks to both and his appreciation of the opportunity to participate in, or liaise on the study of the data from, operations by these organizations over three decades. Thanks are further due to BGS reviewers R. S. Arthurton, P. S. Balson, E. R. Shephard-Thorn and R. D. Walshaw. Constructive criticisms from my two external referees are acknowledged to have greatly improved the paper. Figures 2 and 3 are reproduced with copyright permissions, which are acknowledged with gratitude in the respective figure-captions. The paper is published with the permission of the Director of the British Geological Survey (NERC).

References

ALLEY, R. B., MEESE, D. A., SHUMAN, C. A. ET AL. 1993. Abrupt increase in Greenland snow accumulation at the end of the Younger Dryas event. *Nature*, **362**, 527–529.

AMOS, C. L. & MILLER, A. A. L. 1990. The Quaternary stratigraphy of southwest Sable Island Bank. *Geological Society of America Bulletin*, **102**, 915–934.

AUSTIN, W. E. N. & MCCARROLL, D. 1992. Foraminifera from the Irish Sea glacigenic deposits at Aberdaron, western Lleyn, North Wales: palaeoenvironmental implications. *Journal of Quaternary Science*, **7**, 311–317.

BELDERSON, R. H., PINGREE, R. D. & GRIFFITH, D. K. 1986. Low sea-level tidal origin of Celtic Sea sand banks – Evidence from numerical modelling of M2 tidal streams. *Marine Geology*, **73**, 99–108.

BLAKE, W. 1992. Holocene emergence at Cape Herschel, east-central Ellesmere Island, Arctic Canada: implications for ice-sheet configuration. *Canadian Journal of Earth Sciences*, **29**, 1958–1980.

BOND, G., HEINRICH, H., BROECKER, W. ET AL. 1992. Evidence of massive discharges of icebergs into the North Atlantic ocean during the last glacial period. *Nature*, **360**, 245–249.

BOULTON, G. S. 1990. Sedimentation and sea level changes during glacial cycles and their control on glacimarine facies architecture. *In:* DOWDESWELL, J. A. & SCOURSE, J. D. (eds) *Glacimarine Environments: Processes and Sediments*. Geological Society, London, Special Publication, **53**, 15–52.

BOUYSSE, P., HORN, R., LAPIERRE, F. & LE LANN, F. 1976. Étude des grands bancs de sable du Sud-est de la mer Celtique (Great sand banks of the southeastern Celtic Sea). *Marine Geology*, **20**, 251–275 (in French).

BOWEN, D. Q. 1977. The coast of Wales. *In:* KIDSON, C. & TOOLEY, M. J. (eds) The *Quaternary History of the Irish Sea. Geological Journal*, Special Issue, 7, Seel House Press, Liverpool, 223–256.

CARLSEN, R., LØKEN, T. & ROALDSET, E. 1986. Late Weichselian transgression, erosion and sedimentation at Gullfaks, northern North Sea. *In:* SUMMERHAYES, C. P. & SHACKLETON, N. J. (eds) *North Atlantic Palaeoceanography*. Geological Society, London, Special Publication, **21**, 145–152

CARTER, R. W. G. 1991. *Shifting Sands: A Study of the Coast of Northern Ireland from Magilligan to Larne*. Countryside and Wildlife Research Series, **2**, HMSO, Belfast, 1–49.

—— 1992. Sea-level changes: past, present and future. *Quaternary Proceedings*, **2**, 111–132.

CHEN, J. & ZHENG, X. 1987. Sources of inner-shelf sediment in the northern South China Sea and the controlling factors. *Acta Oceanologica Sinica*, **6**, 589–598.

CHEN, J. H., CURREN, H. A., WHITE B. & WASSERBURG, G. J. 1991. Precise chronology of the last interglacial period. ^{234}U-^{230}Th data from fossil coral reefs in the Bahamas. *Geological Society of America Bulletin*, **103**, 82–97.

DACKOMBE, R. V. & THOMAS, G. S. P. 1991. Glacial deposits and Quaternary stratigraphy of the Isle of Man. *In:* EHLERS, J., GIBBARD, P. L. & ROSE, J. (eds) *Glacial Deposits of Great Britain and Ireland*. Balkema, Rotterdam, 333–344.

DALY, R. A. 1934. *The Changing World of the Ice Age*. Yale University Press, Yale.

DAVIES, P. & WILLIAMS, A. T. 1991. Sediment supply from solid geology cliffs into the intertidal zone of the Severn Estuary/Inner Bristol Channel, UK. *In:* ELLIOTT, M. & DUCROTOY, J-P. (eds) *Estuaries and Coasts: Spacial and Temporal Intercomparisons*. ECSA 19 Symposium, Olsen & Olsen, 17–24.

DELANTEY, L. J. 1980. *The Geology of the North Celtic Sea*. PhD thesis, University College of Wales, Aberystwyth.

DEVOY, R. J. N. 1977. Flandrian sea level changes in the Thames Estuary and the implications for land subsidence in England and Wales. *Nature*, **270**, 712–715.

—— 1983. Late Quaternary shorelines in Ireland; an assessment of their implications for isostatic land movement and relative sea-level changes. *In:* SMITH, D. E. & DAWSON, A. G. (eds) *Shorelines and Isostasy*. Institute of British Geographers, Special Publication, **16**, Academic Press, London, 227–254.

—— 1985. The problem of a Late Quaternary landbridge between Britain and Ireland. *Quaternary Science Reviews*, **4**, 43–58.

DICKSON, C. & WHATLEY, R. 1995. The biostratigraphy of a Holocene borehole from the Irish Sea. *In:* KEEN, M. (ed.) *Proceedings of the Second European Ostracodologists Meeting*, University of Glasgow Press, Glasgow, in press.

EHLERS, J. 1988. *The Morphodynamics of the Wadden Sea*. Balkema, Rotterdam.

EVANS, C. D. R. 1990. *United Kingdom Offshore Regional Report: the Geology of the Western English Channel and its Western Approaches*. HMSO, London; for the British Geological Survey.

EVANS, D. 1986. *Malin, Sheet 55°N 08°W, Sea Bed sediments and Quaternary Geology*, British Geological Survey, 1:250,000 series.

—— 1988. *Tiree, Sheet 56°N 08°W, including part of Argyll, Sheet 06N 06W, Sea Bed Sediments*, British Geological Survey, 1:250,000 series.

EVERTS, C. H. 1987. Continental shelf evolution in response to a rise in sea level. *In:* NUMMEDAL, D., PILKEY, O. H. & HOWARD, J. D. (eds) *Sea-Level Fluctuation and Coastal Evolution*. Society of Economic Paleontologists and Mineralogists, Tulsa, Oklahoma, Special Publication, **41**, 49–57.

EYLES, C. H. & EYLES, N. 1984. Glaciomarine sediments of the Isle of Man as a key to late Pleistocene stratigraphic investigations in the Irish Sea Basin. *Geology*, **12**, 359–364.

EYLES, N. & MCCABE, A. M. 1989. The Late Devensian (<22,000 BP) Irish Sea Basin: the stratigraphic record of a collapsed ice-sheet margin. *Quaternary Science Reviews*, **8**, 307–351.

—— & —— 1991. Glaciomarine deposits of the Irish Sea Basin: The role of glacio-isostatic disequilibrium. *In:* EHLERS, J., GIBBARD, P. L. & ROSE, J. (eds) *Glacial Deposits in Great Britain and Ireland*. Balkema, Rotterdam, 311–331.

——, EYLES, C. H. & MCCABE, A. M. 1985. Reply to comment by THOMAS, G. S. P. & DACKOMBE, R. V. on 'Glaciomarine sediments of the Isle of Man as a key to Late Pleistocene stratigraphic investigations in the Irish Sea Basin'. *Geology*, **13**, 446–447.

FAIRBANKS, R. G. 1989. A 17,000-year glacio-eustatic sea-level record: influence of glacial melting rates on the Younger Dryas event and deep ocean circulation. *Nature*, **342**, 637–642.

FJELDSKAAR, W. & CATHLES, L. 1991. Rheology of mantle and lithosphere inferred from post-glacial uplift in Fennoscandia. *In:* SABADINI, R., LAMBECK, K. & BOSCHI, E. (eds) *Glacial Isostasy, Sea-level and Mantle Rheology*. Proceedings of the NATO Advanced Research Workshop on Glacial Isostasy, Sea-level, and Mantle Rheology, Erice, Italy, July 17–August 4, 1990. Kluwer Academic Publishers, Dordrecht, 1–19.

FULTON, R. J. (ed.) 1989. *Quaternary Geology of Canada and Greenland*. Canadian Government Publishing Centre, Ottawa (Geological Society of North America, Geology of North America, K-1)

HAMILTON, D., SOMERVILLE, J. H. & STANFORD, P. H. 1980. Bottom currents and shelf sediments, southwest of Britain. *Sedimentary Geology*, **26**, 115–138.

HÉQUETTE, A. & BARNES, P. W. 1990. Coastal retreat and shoreface profile variations in the Canadian Beaufort Sea. *In:* HILL, P. R. (ed.) The Beaufort Sea Coastal Zone. *Marine Geology*, **91** (Special Section), 113–132.

HERZER, R. H. & BORNHOLD, B. D. 1982. Glaciation and post-glacial history of the continental shelf off southwestern Vancouver Island, British Columbia. *Marine Geology*, **48**, 285–319.

HOLLIN, J. T., SMITH, F. L., RENOUF, J. T. & JENKINS, D. G. 1993. Sea-cave temperature measurements and amino acid geochronology of British Late Pleistocene sea stands. *Journal of Quaternary Science*, **8**, 359–364.

HUDDART, D. & TOOLEY, M. J. (eds). 1972. *The Cumberland Lowland, 26–28 May 1972 Handbook*. Quaternary Research Association, two volumes.

——, —— & CARTER, P. A. 1977. The coasts of

north-west England. *In:* KIDSON, C. & TOOLEY, M. J. (eds) *The Quaternary History of the Irish Sea, Geological Journal*, Special Issue, **7**, Seel House Press, Liverpool, 119–154.

JACKSON, D. I., JACKSON, A. A., WINGFIELD, R. T. R., EVANS, D., BARNES, R. P. & ARTHUR, M. J. 1995. *United Kingdom offshore regional report: the geology of the Irish Sea.* HMSO, London, for British Geological Survey, in press.

JAMES, J. W. C. 1990. *Anglesey, Sheet 53°N-06°W, including part of Dublin, Sheet 53°N-08°W, Sea Bed Sediments*, British Geological Survey and Geological Survey of Ireland, 1:250,000 series.

—— & WINGFIELD, R. T. R. 1987. *Cardigan Bay, Sheet 52°N 06°W, including parts of Waterford, Sheet 52°N 08°W, and Mid-Wales & Marches, Sheet 52°N 04°W, Sea Bed Sediments.* British Geological Survey and Geological Survey of Ireland, 1:250,000 series.

JELGERSMA, S. 1979. Sea level changes in the North Sea basin. *In:* OELE, E., SCHÜTTENHELM, R. T. E. & WIGGERS, A. J. (eds) *The Quaternary History of the North Sea. Acta Universitatis Upsaliensis, Symposia Universitatis Upsaliensis Annum Quingentesimum Celebrantis*, Uppsala, **2**, 233–248.

KEEN, D. H. 1978. *The Pleistocene Deposits of the Channel Islands.* Report of the Institute of Geological Sciences, **78/26**. London, HMSO.

—— 1995. Raised beaches and sea-levels in the English Channel in the Middle and Late Pleistocene: problems of interpretation and implications for the isolation of the British Isles. *This volume.*

KENYON, N. H., BELDERSON, R. H., STRIDE, A. H. & JOHNSON, M. A. 1981. Offshore tidal sand-banks as indicators of net sand transport and as potential deposits. *Special Publication, International Association of Sedimentologists*, **5**, 257–268.

LAMBECK, K. 1991a. Glacial rebound and sea-level change in the British Isles. *Terra Nova*, **3**, 379–389.

—— 1991b. A model of Devensian and Flandrian glacial rebound and sea-level change in Scotland. *In:* SABADINI, R., LAMBECK, K. & BOSCHI, E. (eds) *Glacial Isostasy, Sea-level and Mantle Rheology.* Proceedings of the NATO Advanced Workshop on Glacial Isostasy, Sea-level, and Mantle Rheology, Erice, Italy, July 24–August 4, 1990, Kluwer Academic Publishers, Dordrecht, 33–61.

LAWSON, M. J., JAMES, J. W. C. & WINGFIELD, R. T. R. 1988. *Nymphe Bank, Sheet 51°N-08°W, including part of Waterford, Sheet 52°N-08°W, Sea Bed Sediments.* British Geological Survey and Geological Survey of Ireland, 1:250,000 series.

MCCABE, A. M., EYLES, N., HAYNES, J. R. & BOWEN, D. Q. 1990. Biofacies and sediments in an emergent Late Pleistocene glaciomarine sequence, Skerries, east central Ireland. *Marine Geology*, **94**, 23–36.

MCCARROLL, D. & HARRIS, C. 1992. The glacigenic deposits of western Lleyn, north Wales: terrestrial or marine? *Journal of Quaternary Science*, **7**, 19–29.

MITCHELL, G. F. 1960. The Pleistocene history of the Irish Sea. *Advancement of Science*, **17**, 313–325.

—— 1972. The Pleistocene history of the Irish Sea: second approximation. *The Scientific Proceedings of the Royal Dublin Society*, **4A**, 181–199.

—— 1990. *The Shell Guide to Reading the Irish Landscape* (incorporating *The Irish Landscape*). Criterion Press, Dublin.

—— 1992. Notes on a raised beach between two diamicts, Beginish Island, Valencia Harbour, County Kerry. *Irish Journal of Earth Sciences*, **11**, 151–163.

——, PENNY, L. F., SHOTTON, F. W. & WEST, R. G. 1973. *A Correlation of Quaternary Deposits in the British Isles.* Geological Society of London, Special Publication **4**.

MOGI, A. 1979. *An Atlas of the Sea Floor around Japan, Aspects of the Submarine Geomorphology.* University of Tokyo Press, Tokyo.

MÖRNER, N-A. 1976. The Pleistocene/Holocene boundary: proposed boundary stratotype in Gothenburg, Sweden. *Boreas*, **5**, 193–275.

—— 1991. Course and origin of the Fennoscandian uplift: the case for two separate mechanisms. *Terra Nova*, **3**, 408–413.

PANTIN, H. M. 1977. Quaternary sediments from the northern Irish Sea. *In:* KIDSON, C. & TOOLEY, M. J. (eds) *The Quaternary History of the Irish Sea. Geological Journal*, Special Issue, **7**, Seel House Press, Liverpool, 27–54.

—— 1978. Quaternary sediments from the northeast Irish Sea, Isle of Man to Cumbria. *Bulletin of the Geological Survey of Great Britain*, **64**, 1–43.

—— & EVANS, C. D. R. 1984. The Quaternary history of the central and southwestern Celtic Sea. *Marine Geology*, **57**, 259–293.

PEACOCK, J. D. 1992. The marine palaeoenvironment 15–9 ka BP in shallow UK seas – a progress report. *Quaternary Newsletter*, **67**, 52–61.

PIELOU, E. C. 1991. *After the Ice Age: the Return of Life to Glaciated North America.* University of Chicago Press, Chicago.

PING, S. L. 1989. Cyclic morphologic changes of the ebb-tidal delta, Texel Inlet, The Netherlands. *Geologie en Mijnbouw*, **68**, 35–48.

PRAEG, D., D'ANGLEJAN, B. & SYVITSKI, J. P. M. 1992. Seismostratigraphy of the middle St Lawrence Estuary: a Late Quaternary glacial marine to estuarine depositional/erosional record. *Géographie physique et Quaternaire*, **46**, 133–150.

PREECE, R. C., COXON, P. & ROBINSON, J. E. 1986. New biostratigraphic evidence of Post-glacial colonization of Ireland and for Mesolithic forest disturbance. *Journal of Biogeography*, **13**, 487–509.

——, SCOURSE, J. D., HOUGHTON, S., KNUDSEN, K. L. & PENNEY, D. N. 1990. The Pleistocene sea-level and neotectonic history of the eastern Solent, southern England. *Philosophical Transactions of the Royal Society, London B*, **328**, 425–477.

RAUKAS, A. 1991. Transgressions of the Baltic Sea and the peculiarities of the formation of transgressive coastal deposits. *Quaternaire*, **2**, 126–130.

ROKOENGEN, K., LØFALDI, M., RISE, L., LØKEN, T. &

CARLSEN, R. 1982. Description and dating of a submerged beach in the northern North Sea. *Marine Geology*, **50**, M21–M28.

RUZ, M-H., HÉQUETTE, A. & HILL, P. R. 1992. A model of coastal evolution in a transgressed thermokarst topography, Canadian Beaufort Sea. *Marine Geology*, **106**, 251–278.

SCOURSE, J. D. 1991. Late Pleistocene stratigraphy and palaeobotany of the Isles of Scilly. *Philosophical Transactions of the Royal Society, London, B*, **334**, 405–448.

—— & AUSTIN, W. E. N. 1994. A Devensian Lateglacial and Holocene sea-level record from the Central Celtic Sea. *Quaternary Newsletter*, **74**, 20–29.

——, ——, BATEMAN, R. M., CATT, J. A., EVANS, C. D. R., ROBINSON, J. E. & YOUNG, J. R. 1990. Sedimentology and micropalaeontology of glacimarine sediments from the Central and Southwestern Celtic Sea. *In:* DOWDESWELL, J.A. & SCOURSE, J. D. (eds) *Glacimarine Environments: Processes and Sediments*. Geological Society, London, Special Publication, **53**, 329–347.

SHAW, J. & FORBES, D. L. 1990. Relative sea-level change and coastal response, north-east Newfoundland. *Journal of Coastal Research*, **6**, 641–660.

SHEPARD, F. P. & WANLESS, H. R. 1971. *Our Changing Coastlines*. McGraw-Hill, New York.

SHIPP, R. C., BELKNAP, D. F. & KELLEY, J. T. 1991. Seismic-stratigraphic and geomorphic evidence for a post-glacial sea-level lowstand in the northern Gulf of Maine. *Journal of Coastal Research*, **7**, 341–364.

SHIQING, F. 1989. The marine geological environment of Qiongzhou Strait. *In:* WHYTE, P. *ET AL.* (eds) *The Palaeoenvironment of East Asia from the Mid-Tertiary, Proceedings of the Second Conference, Volume 1: Geology, Sea Level Changes, Palaeoclimatology and Palaeobotany*. Centre of Asian Studies, University of Hong Kong, Hong Kong, 265–266.

SMITH, N. D. 1972. Flume experiments on the durability of mud clasts. *Journal of Sedimentary Petrology*, **6**, 35–47.

SNEAD, R. E. 1982. *Coastal Landforms and Surface Features: a Photographic Atlas and Glossary*. Hutchinson Ross Publishing Company, Stroudsburg, Pennsylvania.

STEERS, J. A. 1948. *The Coastline of England and Wales*. Cambridge University Press, Cambridge.

STILLMAN, C. J. 1968. The post glacial change in sea level in southwestern Ireland: new evidence from fresh-water deposits on the floor of Bantry Bay. *The Scientific Proceedings of the Royal Dublin Society*, **A3**, 125–127.

SYNGE, F. M. 1977. The coasts of Leinster (Ireland). *In:* KIDSON, C. & TOOLEY, M. J. (eds) *The Quaternary History of the Irish Sea, Geological Journal*, Special Issue, **7**, Seel House Press, Liverpool, 199–211.

—— 1985. Coastal evolution. *In:* EDWARDS, K. J. & WARREN, W. P. (eds) *The Quaternary History of Ireland*. Academic Press, London, 115–131.

TAPPIN, D. R., CHADWICK, R. A, JACKSON, A. A., WINGFIELD, R. T. R. & SMITH, N. J. P. 1994. *United Kingdom offshore regional report: the geology of Cardigan Bay and the Bristol Channel*. HMSO, London, for British Geological Survey.

TAYLOR, K. C., LAMOREY, G. W. DOYLE, G. A. *ET AL.* 1993. The 'flickering switch' of late Pleistocene climatic change. *Nature*, **361**, 432–436.

THOMAS, G. S. P. 1985. The Quaternary of the northern Irish Sea Basin. *In:* JOHNSON, R. H. (ed.) *The Geomorphology of North-West England*, Manchester University Press, Manchester, 143–158.

—— & DACKOMBE, R. V. 1985. Comment on 'Glaciomarine sediments of the Isle of Man as a key to late Pleistocene stratigraphic investigations in the Irish Sea Basin'. *Geology*, **12**, 445–446.

—— & SUMMERS, A. J. 1983. The Quaternary stratigraphy between Blackwater Harbour and Tinnaberna, County Wexford. *Journal of Earth Sciences of the Royal Dublin Society*, **5**, 121–134.

TOOLEY, M. J. 1978. *Sea-Level Changes: North-West England during the Flandrian Stage*. Clarendon Press, Oxford.

—— 1985. Sea-level changes and coastal morphology in north-west England. *In:* JOHNSON, R. H. (ed.) *The Geomorphology of North-West England*, Manchester University Press, Manchester, 94–121.

TURCOTTE, D. L. & SCHUBERT, G. 1982. *Geodynamics: Applications of Continuum Physics to Geological Problems*. John Wiley & Sons, New York.

WALDEN, J., SMITH, J. P. & DACKOMBE, R. V. 1992. Mineral magnetic analyses as a means of lithostratigraphic correlation and provenance indication of glacial diamicts: intra- and inter-unit variation. *Journal of Quaternary Science*, **7**(3), 257–270.

WARREN, W. P. 1985. Stratigraphy. *In:* EDWARDS, K. J. & WARREN, W. P. (eds) *The Quaternary History of Ireland*. Academic Press, London, 39–65.

—— 1990. Deltaic deposits in the Carey Valley, Co. Antrim, Ireland. *Sedimentary Geology*, **69**, 157–160.

WILLIAMS, A. T. & DAVIES, P. 1987. Rates and mechanisms of coastal cliff erosion. *In:* KRAUS, N. C. (ed.) *Coastal Sediments '87*., American Society of Civil Engineers, 1855–1870.

WILLIAMS, M. A. J., DUNKERLEY, D. L., DE, DECKKER, P., KERSHAW, A. P. & STOKES, T. J. 1993. *Quaternary Environments*. Edward Arnold, London.

WINGFIELD, R. T. R. 1985. *Isle of Man, Sheet 54°N 06°W, including part of Ulster, Sheet 54°N 08°W, Sea Bed Sediments and Quaternary Geology*. British Geological Survey, 1:250,000 series.

—— 1989. Glacial incisions indicating Middle and Upper Pleistocene ice limits off Britain. *Terra Nova*, **1**, 538–548.

—— 1990. The origin of major incisions within the Pleistocene deposits of the North Sea. *Marine Geology*, **91**, 31–52.

—— & TAPPIN, D. R. 1991. North Celtic Sea

including parts of 1:250,000 sheets: Nymphe Bank, 51°N 08°W; Lundy, 51°N 06°W; Labadie Bank, 50°N 10°W; Haig Fras, 50°N 08°W; and Land's End, 50°N 06°W. Quaternary Geology. British Geological Survey and Geological Survey of Ireland, 1:250,000 series.

——, HESSION, M. A. I. & WHITTINGTON, R. J. 1990. *Anglesey, Sheet 53°N 06°W, including part of Dublin, Sheet 53°N 08°W, Quaternary Geology.* British Geological Survey and Geological Survey of Ireland, 1:250,000 series.

WOODWORTH, P. L. 1992. Sea level changes. *In:* DAVIES, H. (ed.) *Global Warming and Climatic Change.* Seminar Report 1, Irish Sea Forum, Liverpool University Press, Liverpool, 21–28.

WRIGHT, W. B. & MUFF, H. B. 1904. The pre-glacial raised beach of the south coast of Ireland. *The Scientific Proceedings of the Royal Dublin Society*, New Series, **10**, 250–308.

The floristic record of Ireland's Pleistocene temperate stages

P. COXON[1] & S. WALDREN[2]

[1] Department of Geography, Museum Building, Trinity College, Dublin 2, Ireland
[2] Trinity College Botanic Gardens, Palmerston Park, Dartry, Dublin 6, Ireland

Abstract: The unique flora of Ireland, that includes plants with Atlantic, amphi-Atlantic and Mediterranean affinities, has attracted the attention of many biogeographers. Using a database, this paper compiles the Irish Pleistocene plant fossil record and examines the disappearances of some taxa and the continuity of others. A retreat of some plants to the east since the Tertiary, movement of arctic–alpine species to the north and to the mountains, and the long continuity of Atlantic and amphi-Atlantic taxa, including some not present in Britain, can be identified. The flora of the Irish Middle Pleistocene Gortian temperate stage is examined and its uniqueness, and the long continuity of certain taxa, are discussed. The long record of Atlantic and amphi-Atlantic species, as well as wetland and aquatic taxa, suggests the possibility of proximal refugia for some plants. The implications for *in situ* survival and for migration during and after cold stages are discussed in the light of the compiled fossil record.

Ireland's position on the western fringes of Europe gives it an important biogeographical location and indeed, the Holocene flora of Ireland has long attracted attention because of a number of interesting plant distributions, including those of an amphi-Atlantic nature and the Mediterranean–Atlantic element (Godwin 1975; Webb 1983). This paper examines the Pleistocene vegetational history of Ireland, and in particular that of the Middle Pleistocene Gortian temperate stage (correlated to the continental European Holsteinian and hence probably dating to between 428 and 302 ka BP), in order to compare the Pleistocene floras of Ireland with their Holocene and contemporary counterparts.

Ireland's Pleistocene record

The global Cenozoic sequence, from long marine, lacustrine and terrestrial records exhibits a characteristic pattern with a decline in temperature throughout the Miocene culminating in numerous step-like changes in climatic conditions over the last 2.5 Ma. However, owing to extensive glaciation, the Irish Pleistocene record is fragmentary with only a partial record of glaciations prior to the last one (Midlandian), and likewise a very poor record of the many temperate episodes that Ireland must have experienced.

Figure 1 shows the northwest European record from The Netherlands (Zagwijn 1985, 1992) along with correlations to the British (Gibbard *et al.* 1991; Jones & Keen 1993) and to the Irish sequence (Coxon 1993). The lack of continuous sedimentary records in Ireland, and in particular the absence of biostratigraphically complete organic sequences, has meant that the elucidation of an agreed Pleistocene succession has not been possible.

Ireland's Pleistocene floristic record is therefore incomplete, but what is available can be used to analyse the changing floral composition of the last 2 Ma and to attempt to analyse trends within these changes.

Methodology

All of the published records of Pleistocene plant macro- and microfossils were collated and the details were entered into a database which was then used to identify patterns within the fossil record. The database contains the family, genus, species and authority for recorded fossils, mainly following the nomenclature of Tutin *et al.* (1964–1980) except where more recent names are used by Stace (1991), site-by-site information regarding the locality and the deposit's possible age, and the nature of the fossil record, i.e. the type of data recorded by the relevant author(s). The Holocene distribution of the fossil taxa in Britain and Ireland and the present-day biogeographic ranges of the taxa and their ecology were also included, derived from personal observations and the following sources: Curtis & McGough (1988), Hultén (1958, 1962 and 1970), Jalas & Suominen (1972–1994), Perring & Walters (1962), Jermy *et al.* (1978), Scannell & Synnott (1987), Stace (1991), Tutin *et al.* (1964–1980) and Webb (1983). Obviously, it is often difficult to classify the distribution or ecology of

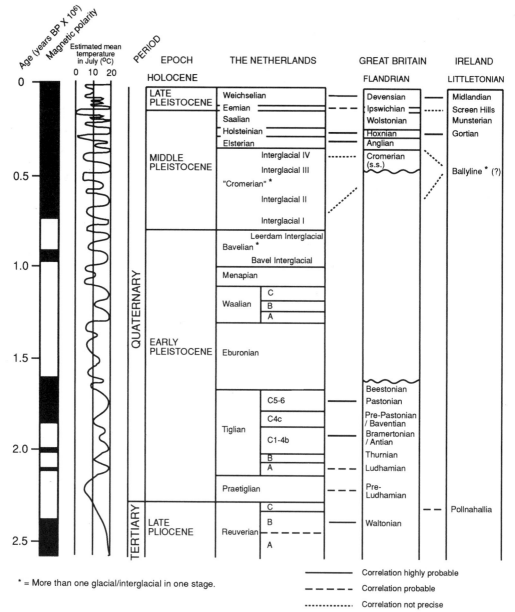

Fig. 1. The stratigraphy of the Quaternary of northwest Europe, showing the probable correlation with the Irish sequence. After Mitchell *et al.* (1973), Zagwijn (1985) and Jones & Keen (1993). From Coxon (1993).

a taxon in a simple enough way to be suitable for database analysis; some may well disagree with our classifications, but to expand the distributional or ecological information for each taxon would severely limit the use of the database. In this context, we use arctic and alpine to describe ecological requirements (or restrictions) rather than geographical ranges: the two are often clearly linked, and there is some overlap of information between categories. Output reports from this database form the basis of the tables presented here.

Ireland's Pleistocene temperate stages

The Irish Pleistocene record has recently been improved with new data on pre-Gortian, Gortian and post-Gortian stages. Allied with new

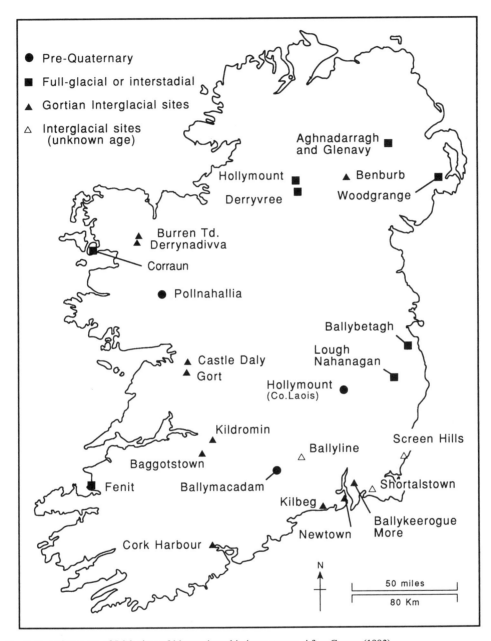

Fig. 2. Location map of Irish sites of biostratigraphic importance. After Coxon (1993).

dating evidence, further sites have been discovered and existing ones have been reinvestigated, giving a more detailed interpretation of a previously poorly documented period (Coxon 1993). However, the record is still very incomplete (see Fig. 1) and detailed floristic records are only known from part of the Middle Pleistocene.

The Late Pliocene and Early Pleistocene

Only one *in situ* deposit of Late Tertiary (or Early Pleistocene) age has been discovered in Ireland and that is at Pollnahallia, between Headford and Tuam, County Galway (Fig. 2). Extensive drilling and field investigations at the site revealed a network of gorges and caves in the limestone, over which are draped windblown sands, and later glacial sediments including till (Coxon & Flegg 1987). Palynological results suggest that the lignite infilling the base of the limestone gorge is probably Pliocene or, less likely, Early Pleistocene in age (Fig. 1).

A group of eight taxa (*Tsuga*, *Sciadopitys*, *Sequoia*, *Taxodium*, Taxodiaceae, *Carya*, *Pterocarya* and *Vitis*) indicate that the Pliocene flora was very different from that in most of the Irish Pleistocene and Holocene. Many of these taxa are only recorded from the Late Tertiary and the earliest temperate stage of the Pleistocene (the Tiglian, see Fig. 1) in NW Europe (Zagwijn 1960), although *Pterocarya* and *Vitis* are also recorded from the Holsteinian and the Hoxnian. They all occupy present-day ranges in America (e.g. *Taxodium*, *Sequoia*), Asia (*Pterocarya*, *Sciadopitys*) or both (*Carya*, *Tsuga*). In some cases, the species involved may be different from those currently extant; in most cases fossil material can only assigned to the generic level. The presence of these genera in the Irish Tertiary (and elsewhere in western Europe) indicates that they were formerly much more widespread than at present, and the species involved may have been replaced by more competitive forms. Except for the vine *Vitis*, all are trees; most are hardy in present Irish climatic conditions, and are well-known in cultivation.

Pre-Gortian (Pleistocene) temperate stages

The extensively karstified surface of Ireland provided sediment traps throughout the Tertiary and the Pleistocene. Such infills have been recorded at numerous sites around Ireland (Watts 1985), including Ballymacadam, County Tipperary (Oligocene) and Hollymount, County Laois (Miocene), see Fig. 2. Large Tertiary and Pleistocene infills have recently been discovered during mineral exploration in the Irish Midlands, but these have yet to be analysed in detail. As shown in Fig. 1, the Early and most of the Middle Pleistocene of Ireland has not been studied and little can be said regarding the floristic composition of this period.

One site of possible Middle Pleistocene age was discovered filling a solution feature in Carboniferous Limestone near Ballyline, County Kilkenny (Fig. 2). The deposit is a laminated, lacustrine clay over 25 m in thickness, and underlies glacial sediments. Preliminary investigations indicate that the Ballyline site may represent part of a Middle Pleistocene temperate stage complex that has been hitherto unrecorded, or that the material is reworked or derived. However, the site has been discussed elsewhere (Coxon & Flegg 1985) and the pollen taxa recorded add nothing to the species known to have occurred in the Gortian stage.

The Gortian temperate stage

The Gortian represents a temperate stage of the Middle Pleistocene recognized at 11 sites, nine of which have substantial records. The localities of these sites are marked on Fig. 2. The duration of the temperate stage recorded at the sites is of variable length and the zone ranges are shown in Fig. 3.

The age of the Gortian interglacial is subject to debate (see Warren 1979, 1985; Watts 1985; Coxon 1993). However, recent attempts have been made to date foraminifera and organic sediments of Gortian age using amino acid racemization and uranium–thorium disequilibrium dating (UTD), respectively. The ratios of the amino acids D-alloisoleucine and L-isoleucine from the foraminifera *Ammonia batavus* (Scourse et al. 1992) suggest that a correlation of the Gortian with the last interglacial is unlikely and that the Gortian was contemporary with the Hoxnian/Holsteinian or an earlier temperate stage, which is consistent with palaeobotanical interpretations (see Watts 1985). UTD has given *minimum* ages of 180 and 191 ka on Gortian material from Burren Townland and > 350 ka (a 'preliminary result') from the Gortian type site at Boleyneendorrish (Heijnis 1992). However, although there are not enough UTD dates on Gortian sites at present to allow us to make firm statements regarding that temperate stage's age, these preliminary dates do strongly suggest that the Gortian is older than the last interglacial (marine oxygen isotope (OI) Stage 5e).

On the basis of the dates cited above and the established biostratigraphy, the Gortian can be placed as a Middle Pleistocene temperate stage with an estimated age lying somewhere between 428 and 198 ka. If the Gortian is the equivalent of the continental Holsteinian temperate stage then it might be placed within a period between OI Stages 9 and 11 (Bowen et al. 1986). This would mean that the Gortian stage is aged between 428 and 302 ka. It is also possible that the Gortian represents OI Stage 7 and hence is aged between 252 and 198 ka, but this is considered less likely as that stage appears poorly developed and the UTD dates from Gort were > 350 ka BP and those from Burren Townland were minimum ages. If the correlation to OI Stage 9 or 11 is correct, then the Irish Pleistocene record of the post-Gortian is very

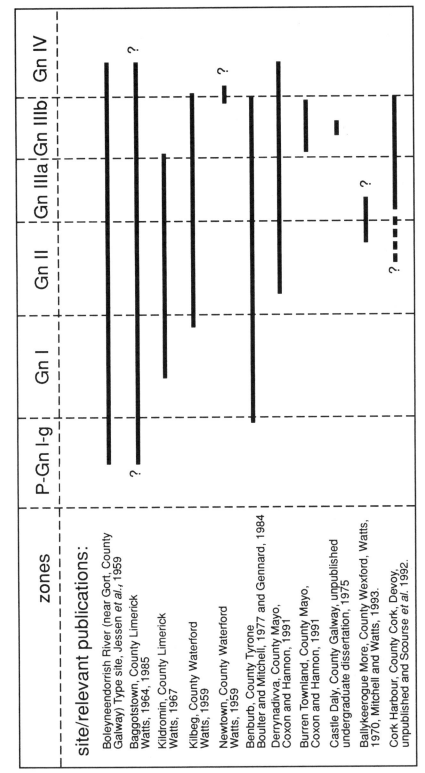

Fig. 3. The pollen zone ranges of Gortian Stage sites. From Coxon (1993).

incomplete and more than one temperate stage may be missing. This is a strong possibility, and some authors are of the opinion that the post-Hoxnian record in Britain is similarly incomplete, with at least one missing temperate episode (Wymer 1985; Jones & Keen 1993). The linking of the Gortian to the Middle Pleistocene allows the floristic record to be compared with that of the Marbellan temperate stage from the southwestern coast of France (see below).

The floristic record of the Gortian stage

The palaeobotanical record of the Gortian provides by far the most complete fossil record of the Irish Pleistocene thanks to the detailed observations of a number of authors (credited in Fig. 3), principally Jessen et al. (1959) and Watts (1959, 1964, 1967). The high quality of the Gortian fossil data has allowed a detailed analysis of this temperate stage, and attempts have been made to compare its record with some continental sites as well as with the Irish Holocene.

Two hundred and fifty one taxa have been recorded from the Gortian, although it must be realized that some identifications are tentative, many are to genus level and a number are to family only. However, the record is particularly useful and allows comparison with the biogeography of the modern Irish flora.

Species recorded from the Gortian that are now absent from Ireland

Of the total of 281 taxa recorded from the Late Pliocene and from the Pleistocene temperate and interstadial stages, 37 are now considered absent from the modern Irish flora (Table 1). Certain elements of the flora have been retreating prior to the Pleistocene, and this has continued right through to the Holocene. Although the general trend of retreat and extirpation is similar to that recorded in the rest of Europe, the Irish record has some distinct anomalies and these will be discussed below.

The Pliocene element As outlined above, eight taxa that can be classified as 'Tertiary relicts' and that were recorded at Pollnahallia are no longer native to Ireland. Seven of these taxa, (*Tsuga, Sciadopitys, Sequoia, Taxodium,* Taxodiaceae, *Carya* and *Vitis*) are not recorded in the Gortian and probably disappeared prior to the Middle Pleistocene (Table 1). A single pollen grain of *Pterocarya* has been recorded at Derrynadivva (Coxon & Hannon 1991) suggesting that this tree may have survived into the Middle Pleistocene in Ireland, as indeed it did elsewhere in Europe (Oldfield & Huckerby 1979).

This leaves 30 additional taxa recorded from the Gortian that are now absent, although three have Holocene records (see Table 1).

Montane/arctic–alpine element A group of five species, now absent from Ireland, occur today in montane/alpine or arctic habitats, presumably because of their ecological requirements, or because they have been eliminated from more favourable habitats by more competitive taxa. All now occur to the north or northeast of their Gortian interglacial distributions. Some, such as *Athyrium distentifolium* and *Lycopodium annotinum*, occur as nearby as the mountains of N Scotland (Jermy et al. 1978). *Abies alba* (Fig. 4) is more strictly a continental montane forest species at present in Europe, occurring in the Alps, Pyrenees, Apennines and Carpathians, but absent from Scandinavia (Jalas & Suominen, 1972–1994). *Picea abies*, by contrast (Fig. 4), has a more northern distribution to the Arctic, but is absent from the Pyrenees and very local in the Apennines (Jalas & Suominen 1972–1994).

Recent extinctions Two species could be included in the previous category, but have good Lateglacial/early Holocene records in Ireland. *Betula nana* is a circumpolar arctic–alpine species (Fig. 5) absent from Ireland since the early Holocene, but still occurring in N Scotland and one northern English station (Perring & Walters 1962). *Pinus sylvestris* was certainly widespread in Ireland as a native species until the 17th century. It is now widely introduced, but may have survived continuously as small isolated populations. These relatively recent retreats are similar to those at the end of the Pliocene, during the Pleistocene and in the Holocene.

Hippophae rhamnoides was also present in the Gortian, and has been recorded from Irish Lateglacial deposits (Watts 1977; Craig 1978). In Britain, as in Scandinavia, *Hippophae* is an important feature of Lateglacial vegetation (Pennington 1974; Godwin 1975). This shrub has been widely introduced in Britain to aid in stabilization of coastal dunes, but is thought to be native only on the east coast of England (Stace 1991); it has been introduced to Ireland for similar reasons but is not considered to be native (Scannell & Synnott 1987). Elsewhere in Europe it is not restricted to coastal sites, but it typically occurs on similarly unstabilized sub-

Table 1. Irish Late Tertiary and Pleistocene taxa no longer native (37 taxa).

Family	Name	Tertiary/Early Pleis.	Gortian	Irish modern	British modern	Biogeography	Ecology
Lycopodiaceae	*Lycopodium annotinum* L.		G	A	L	Circumpolar	Montane
Lycopodiaceae	*Diphasiastrum complanatum* (L.) Holub (?)		G	A	VL	N Europe	Montane
Woodsiaceae	*Athyrium distentifolium* Tausch ex Opiz		G	A	L	Amphi-Atlantic	Arctic–alpine
Azollaceae	*Azolla filiculoides* Lam.		G	A (int)	A (int)	American	Aquatic
Pinaceae	*Abies* cf. *alba* Miller		G	A	A	C and S Europe	Montane forest
Pinaceae	*Picea abies* (L.) Karsten		G	A	A	N Continental	Boreal forest
Pinaceae	*Picea* sp.		G	A	A		
Pinaceae	*Tsuga* sp.	T			A	America, C and E Asia	Montane forest
Pinaceae	*Pinus sylvestris* L.		G	A/H (+ int)	L	N Continental	Woodland
Sciadopityaceae	*Sciadopitys* sp. (type)	T		A	A	E Asian	
Taxodiaceae	*Sequoia* sp. (type)	T		A	A	American	
Taxodiaceae	*Taxodium* sp.	T		A	A	American	
Taxodiaceae	Taxodiaceae undiff.	T		A	A	American	
Nymphaeaceae	*Nuphar* cf. *pumila* (Timm) DC.		G	A	L	Amphi-Atlantic?	Aquatic
Nymphaeaceae	*Brasenia* cf. *schreberi* J.F. Gmel.		G	A	A	Widespread disjunct	Aquatic
Juglandaceae	*Pterocarya* cf. *fraxinifolia* (Poiret) Spach	T	G	A	A	W Asia	Moist woodland
Juglandaceae	*Carya* sp.	T		A	A	America, 1 in E Asia	
Fagaceae	*Fagus sylvatica* L.		G	A (int)	L (int)	Europe	Woodland
Betulaceae	*Betula nana* L.		G	A/H	L	Circumpolar	Arctic–alpine
Betulaceae	*Carpinus betulus* L.		G	A (int)	L (int)	Europe, W. Asia	Woodland
Caryophyllaceae	*Dianthus* sp.		G	A	A		
Polygonaceae	*Persicaria bistorta* (L.) Samp.		G	A (int)	L	N and C Europe, W and C Asia	Grassland
Tiliaceae	*Tilia* sp.		G	A (int)	L		Woodland
Brassicaceae	*Thlaspi caerulescens* J.S. & C. Presl.		G	A	VL	Circumpolar	Arctic–alpine
Ericaceae	*Rhododendron ponticum* L.		G	A (int)	A (int)	C Europe, Portugal	Woodland
Ericaceae	*Bruckenthalia spiculifolia* (Salisb.) Reichenb.		G	A	A	E Europe, W Asia	Montane
Primulaceae	*Lysimachia punctata* L.		G	A (int)	A	C Europe, W Asia	Wetland
Eleagnaceae	*Hippophae rhamnoides* L.		G	A/H (int)	L (int)	Europe, temperate Asia	Open scrub
Viscaceae	*Viscum album* L.		G	A	L	Europe, temperate Asia	Parasite
Buxaceae	*Buxus sempervirens* L.		G	A	VL	Europe, N Africa, W Asia	Scrub/Woodland
Vitaceae	*Vitis* sp.	T		A	A		
Apiaceae	*Astrantia* cf. *minor* L.		G	A	A	Europe, W Asia	Montane
Menyanthaceae	*Nymphoides* cf. *cordata* (Ell.) Fern.		G	A	A	American	Aquatic
Menyanthaceae	*Nymphoides* cf. *peltata* Kuntze		G	A (int)	L(int?)	Europe, Asia	Aquatic
Plantaginaceae	*Plantago media* L.		G	A (int?)	W	Europe ?	Grassland
Cyperaceae	*Eleocharis ovata* (Roth) Roemer & Schultes		G	A	A	C Europe (W Asia?)	Wetland
Cyperaceae	*Eleocharis* cf. *carniolica* Koch		G	A	A	E Europe, W Asia	Wetland

T, Late Tertiary/Early Pleistocene; G, Gortian; A, absent; H, Holocene record; R, rare; (V)L, (very)local; W, widespread; Int, introduced

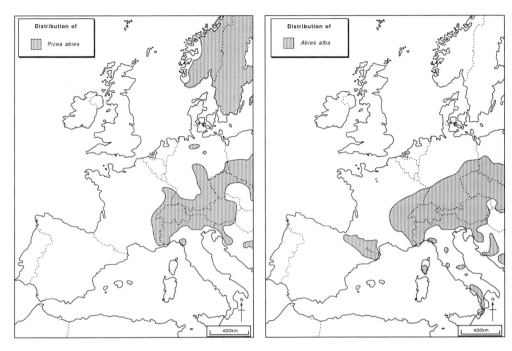

Fig. 4. The Western European distribution of *Picea abies* and *Abies alba* (based on Jalas & Suominen 1972–1994).

strates such as river gravels. Development of complete vegetation cover may have eliminated this species, and, like many other continental species it has subsequently failed to recolonize Ireland.

American species Two species (*Azolla filiculoides* and *Nymphoides cordata*) are now absent from Europe (although introduced), occurring only in America. To these may be added *Brasenia schreberi*, which now occurs mainly in the southern United States and northern Mexico. All three are aquatic species. They may perhaps be thought of as former amphi-Atlantic species which have now been greatly restricted in range (see discussion below). *Brasenia* is interesting because it is still very widely distributed but disjunct over large areas (Fig. 6), occurring chiefly in north and central America, central and southern Africa, Japan, Sumatra and Australia (Hultén 1970). This may be a Tertiary relict which has undergone a great fragmentation of range similar to the Pliocene element described above. Hultén (1970) suggests that the interglacial record indicates that this was formerly a circumpolar species, but presumably with some southern outliers.

Continental species Twenty species were present in the Gortian flora, but their present distribution is to the east or southeast. Some occur well to the east, such as *Rhododendron ponticum*, which has its main station in Bulgaria, northern Turkey, western Caucasia and Lebanon, with outliers in Portugal and Spain (Fig. 7; Chamberlain 1982). The anomaly of the Portuguese population is further emphasized by Irish interglacial pollen more closely resembling that of the Caucasian populations (Coxon & Hannon 1991; Coxon *et al.* 1994). Some of the species that have shown this eastern retreat occur as near as Wales, e.g. *Persicaria* (=*Polygonum*) *bistorta* (Perring & Walters 1962), now considered to be introduced to all of its Irish stations (Scannell & Synnott 1987). However, it is obvious that marine barriers alone cannot explain why species have failed to return to Ireland: *Carpinus betulus* reaches SE England but is absent from Wales and W Britain (Perring & Walters 1962); *Eleocharis ovata* occurs near Paris but has not penetrated as far as the English Channel. Eco-logical requirements in the Holocene may have been at least as important as marine barriers in preventing these species from returning. Insularity alone cannot explain the low numbers of species in the present Irish flora

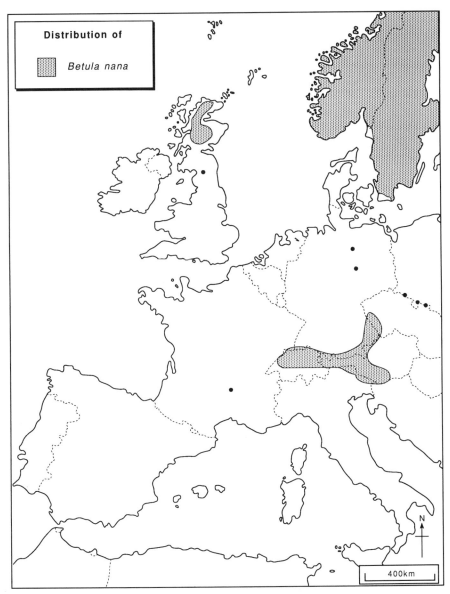

Fig. 5. The Western European distribution of *Betula nana* (based on Perring & Walters 1962; Jalas & Suominen 1972–1994).

(Webb 1983; Bennett 1995). However, other species that are absent from Britain but occur widely on the continent do reach as far as the English Channel.

Of the Irish interglacial species now absent, at least 29 are known to be hardy or have been successfully cultivated in Ireland; climatic considerations alone may not therefore determine the composition of the flora.

Species recorded from the Gortian that are now absent from Britain

Sixteen taxa are known to have occurred in Ireland during the Middle Pleistocene but are now absent from Britain (see Table 2). Most of these are also now absent from Ireland, only three taxa being present in Ireland but absent from Britain (*Daboecia cantabrica, Erica mack-*

Fig. 6. The distribution of *Brasenia schreberi* (based on Hultén 1970).

aiana, *Saxifraga* sect *Gymnopera*, represented by *S. spathularis* and *S. hirsuta*). These are all Atlantic species and will be discussed below.

The arctic–alpine/montane element of the Gortian flora

Nineteen of the interglacial taxa can be thought of as montane or arctic–alpine taxa (Table 3). Seven of these are absent from the modern Irish flora, although one has been shown to occur in the early Holocene (*Betula nana*, Fig. 5, discussed above). Three of these (*Astrantia minor*, *Bruckenthalia spiculifolia* and *Abies alba*) are now found well to the east of the British Isles and may be thought of as continental montane species. Many of the arctic and arctic–alpine species occur locally in Ireland. Some such as *Saxifraga nivalis*, occur as isolated populations in Ireland (Curtis & McGough 1988), but are more widespread in Britain and to the north and east of Ireland. Two species (*Armeria maritima* and *Plantago maritima*) today occur in both montane and coastal habitats. Their ecology suggests a requirement for open habitats, and Pennington (1974) has suggested that their modern distribution is the result of fragmentation of more widespread Lateglacial ranges. These and other truly arctic–alpine taxa are

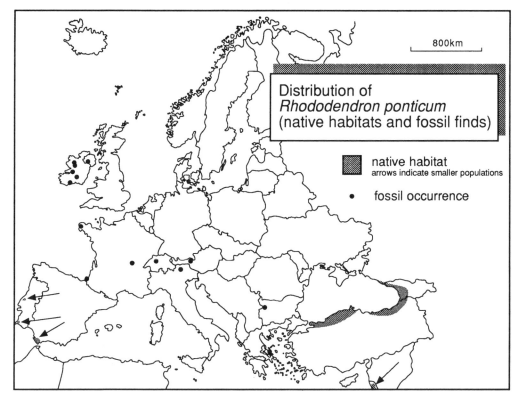

Fig. 7. The Holocene and Pleistocene distribution of *Rhododendron ponticum*. After Cross (1975) and Chamberlain (1982).

probably incapable of tolerating competition for resources by taller-growing vegetation, and therefore the establishment of scrub and woodlands during climatic amelioration may be responsible for displacing some of the arctic–alpine element of the interglacial Irish flora.

Widespread occurrences of arctic–alpine species in Ireland may indicate the presence of tundra-like conditions during cold stages. There has clearly been a loss of the Pleistocene arctic–alpine flora, which possibly reflects climatic amelioration. Tundra-like conditions have also been reported from the Pays Basque by Oldfield & Huckerby (1979), but a true arctic–alpine flora is absent: in the Weichselian cold stage only the currently alpine *Rhododendron ferrugineum* is truly indicative of cold conditions. *Armeria maritima*, *Calluna* and *Vaccinium myrtillus* also occur, but these may all be found in both mountains and lowlands of Ireland today (see discussion of *Armeria* above). The absence of taxa such as *Betula nana* and arctic–alpine species of Lycopodiaceae and Saxifragaceae suggests that full arctic–alpine-type flora did not develop in the Pays Basque cold stages and that conditions there were milder than those found in Ireland and Britain.

Leaving absences from the Irish flora and concentrating on the fossil record of the Gortian, a fascinating continuity within the flora becomes apparent, especially in the following group.

Atlantic and amphi-Atlantic species recorded from the Gortian

This group of 34 taxa is shown in Table 4 and only two of them are now absent from the Irish flora. *Athyrium distentifolium* has already been dealt with as an arctic–alpine species which is now absent from Ireland and occurs to the north (Scotland, Iceland and Scandinavia) or in alpine areas of Europe, e.g. Pyrenees, Cantabrian mountains, Alps (Jalas & Suominen 1972–1994). The only Gortian record of this taxon is as a spore from Benburb (Gennard 1984). The other species, *Nuphar pumila*, is much less easy to explain, as it occurs in Welsh, Northern

Table 2. Gortian fossil records of species no longer native in Britain (16 taxa).

Family	Name	Irish modern	British modern	Biogeography	Ecology
Azollaceae	*Azolla filiculoides* Lam.	A (int)	A (int)	American	Aquatic
Pinaceae	*Abies* cf. *alba* Miller	A	A	Continental Europe	Montane
Pinaceae	*Picea abies* (L.) Karsten	A	A	N Continental	Boreal forest
Pinaceae	*Picea* sp.	A	A		
Nymphaeaceae	*Brasenia* cf. *schreberi* J.F. Gmel.	A	A	Widespread disjunct	Aquatic
Juglandaceae	*Pterocarya* cf. *fraxinifolia* (Poiret) Spach	A	A	W. Asia	Moist woodland
Ericaceae	*Rhododendron ponticum* L.	A (int)	A (int)	C Europe, Portugal	Woodlands
Ericaceae	*Daboecia cantabrica* (Hudson) K. Koch	L	A	Atlantic	Acid open
Ericaceae	*Erica mackaiana* Bab.	L	A	Atlantic	Bogs
Ericaceae	*Bruckenthalia spiculifolia* (Salisb.) Reichenb.	A	A	E Europe, W Asia	Montane
Primulaceae	*Lysimachia punctata* L.	A (int)	A	C Europe, W Asia	Wetland
Saxifragaceae	*Saxifraga* cf. sect *Gymnopera*	L	A	Atlantic	Montane/rocks
Apiaceae	*Astrantia* cf. *minor* L.	A	A	Europe, W Asia	Montane
Menyanthaceae	*Nymphoides* cf. *cordata* (Ell.) Fern.	A	A (int)	American	Aquatic
Cyperaceae	*Eleocharis ovata* (Roth) Roemer & Schultes	A	A	C Europe (W Asia?)	Wetland
Cyperaceae	*Eleocharis* cf. *carniolica* Koch	A	A	E Europe, W Asia	Wetland

A, absent; L, local; int, introduced

Table 3. Irish interglacial arctic, alpine and montane taxa (19 taxa).

Family	Name	Irish modern	British modern	Biogeography	Ecology
Lycopodiaceae	*Huperzia selago* (L.) Bernh. ex Schrank & C. Martius	W	L	Circumpolar	Montane
Lycopodiaceae	*Lycopodium clavatum* L.	L	L	Circumpolar	Montane
Lycopodiaceae	*Lycopodium annotinum* L.	A	L	Circumpolar	N Montane
Lycopodiaceae	*Diphasiastrum alpinum* (L.) Holub	L	L	Amphi-Atlantic	Arctic–alpine
Selaginellaceae	*Selaginella selaginoides* (L.) P. Beauv.	W	L	Amphi-Atlantic	Arctic–alpine
Woodsiaceae	*Athyrium distentifolium* Tausch ex Opiz	A	L	Amphi-Atlantic	Arctic–alpine
Pinaceae	*Abies* cf. *alba* Miller	A	A	C and S Europe	Montane forest
Ranunculaceae	*Thalictrum alpinum* L.	L	L	Circumpolar	Arctic–alpine
Betulaceae	*Betula nana* L.	A/H	L	Circumpolar	Arctic–alpine
Plumbaginaceae	*Armeria maritima* (Miller) Willd.	L	L	Amphi-Atlantic	Halophyte/montane
Salicaceae	*Salix herbacea* L.	L	L	Amphi-Atlantic	Arctic–alpine
Brassicaceae	*Thlaspi caerulescens* J.S. & C. Presl.	A	VL	Circumpolar	Arctic–alpine
Ericaceae	*Vaccinium vitis-idaea* L.	W	W	Circumpolar	Montane
Ericaceae	*Bruckenthalia spiculifolia* (Salisb.) Reichenb.	A	A	E Europe, W Asia	Montane
Saxifragaceae	*Saxifraga nivalis* L.	R	VL	Circumpolar	Arctic
Saxifragaceae	*Saxifraga stellaris* L.	L	W	Amphi-Atlantic	Arctic–alpine
Rosaceae	*Sorbus aucuparia* L.	W	W	Circumpolar	Montane
Apiaceae	*Astrantia* cf. *minor* L.	A	A	Europe, W Asia	Montane
Plantaginaceae	*Plantago maritima* L.	L	L	Europe (W Asia?)	Halophyte/montane

A, Absent; (V)L, (very) local; R, rare; W, widespread; H, Holocene record

English and Scottish upland lakes (Perring & Walters 1962). Ireland has plenty of lakes in hill country, but many are in areas of acidic soils, and perhaps water quality may prevent this species from recolonizing Ireland. It may also be that the range of this species is gradually contracting in a manner similar to that already described for the aquatics *Azolla filiculoides*, *Nymphoides cordata*, and, most dramatically, *Brasenia schreberi*. It is also possible that *N. pumila* has been confused with the widespread and variable *N. lutea* (W. A. Watts, pers. comm.).

The presence of certain Atlantic species in Ireland but not in Britain has attracted much attention, and it is interesting that at least some of these species are known to have been present in the Gortian. *Daboecia cantabrica*, *Erica mackaiana* (Mitchell & Watts 1970; Godwin 1975) and *Saxifraga* sect *Gymnopera* (represented by *S. spathularis* and *S. hirsuta*) are examples and their Holocene distributions are shown in Figs 8 and 9.

Of similar interest are interglacial records of species with a modern European distribution centred on Britain and Ireland, including *Erio-*

Table 4. Irish interglacial Atlantic and amphi-Atlantic taxa (34 taxa), including interglacial presence in the Pays Basques.

Family	Name	Irish modern	British modern	Marbellan	Biogeography	Ecology
Lycopodiaceae	*Diphasiastrum alpinum* (L.) Holub	L	L		Amphi-Atlantic	Arctic–alpine
Selaginellaceae	*Selaginella selaginoides* (L.) P. Beauv.	W	L		Amphi-Atlantic	Arctic–alpine
Isoetaceae	*Isoetes lacustris* L.	W	L	>M<	Amphi-Atlantic	Aquatic
Isoetaceae	*Isoetes echinospora* Durieu	VL	L		Amphi-Atlantic	Aquatic
Osmundaceae	*Osmunda regalis* L.	W	L		Amphi-Atlantic	Wetland
Marsileaceae	*Pilularia globulifera* L.	L	L	M	Amphi-Atlantic/Widespread	Lake margin
Hymenophyllaceae	*Trichomanes speciosum* Willd.	VL	VL	M	Sub-Atlantic	Humid microclimate
Hymenophyllaceae	*Hymenophyllum tunbridgense* (L.) Smith	L	L		Atlantic	Humid microclimate
Hymenophyllaceae	*Hymenophyllum wilsonii* Hook.	L	L		Atlantic (amphi-Atlantic?)	Humid microclimate
Woodsiaceae	*Athyrium distentifolium* Tausch ex Opiz	A	L		Atlantic	Arctic–alpine
Nymphaeaceae	*Nuphar* cf. *pumila* (Timm) DC.	A	L		Amphi-Atlantic?	Aquatic
Ranunculaceae	*Ranunculus flammula* L.	W	W		Amphi-Atlantic?	Wetland
Ranunculaceae	*Ranunculus hederaceus* L.	W	W		European (Amphi-Atlantic?)	Aquatic
Myricaceae	*Myrica gale* L.	W	W		Amphi-Atlantic	Bogs
Caryophyllaceae	*Silene vulgaris* (Moench) Garcke	W	W	M	Amphi-Atlantic?	Grassland/woodland
Ericaceae	*Daboecia cantabrica* (Hudson) K. Koch	L	A		Atlantic	Acid open
Ericaceae	*Erica ciliaris* L.	R	VL	M	Atlantic	Bogs
Ericaceae	*Erica mackaiana* Bab.	L	A	M	Atlantic	Bogs
Ericaceae	*Erica cinerea* L.	W	W	M	Sub-Atlantic	Bogs
Saxifragaceae	*Saxifraga stellaris* L.	L	W		Amphi-Atlantic	Arctic–alpine
Saxifragaceae	*Saxifraga granulata* L.	R (int)	L		Sub-Atlantic	Meadow
Saxifragaceae	*Saxifraga* cf. sect *Gymnopera*	L	A		Atlantic	Montane/rocks
Fabaceae	*Ulex* sp.				Sub-Atlantic	Scrub
Haloragaceae	*Myriophyllum alterniflorum* DC.	W	W	M	Amphi-Atlantic	Aquatic
Lamiaceae	*Clinopodium vulgare* L.	Int?	W		Amphi-Atlantic	Scrub/grassland
Caprifoliaceae	*Viburnum opulus* L.	W	W	M	Amphi-Atlantic?	Woodland/scrub
Potamogetonaceae	*Potamogeton filiformis* Pers.	L	L		Amphi-Atlantic	Aquatic
Najadaceae	*Najas flexilis* (Willd.) Rostkov & W. Schmidt	VL	VL	M	Amphi-Atlantic	Aquatic
Eriocaulaceae	*Eriocaulon* cf. *aquaticum* L.	L	VR		Amphi-Atlantic	Aquatic
Juncaceae	*Juncus bulbosus* sens. lat. L.	W	W		Amphi-Atlantic	Wetland
Juncaceae	*Juncus effusus* L.	W	W	M	Amphi-Atlantic	Wetland
Juncaceae	*Juncus conglomeratus* L.	W	W		Amphi-Atlantic	Wetland
Juncaceae	*Luzula campestris*/*multiflora*	W	W	>M	Amphi-Atlantic?	Grassland
Cyperaceae	*Cladium mariscus* (L.) Pohl	W	L	M	Amphi-Atlantic?	Wetland

A, absent; (V)L, (very) local; R, rare; W, widespread; Int, introduced; M, Marbellan record; >M, pre Marbellan; M<, post Marbellan

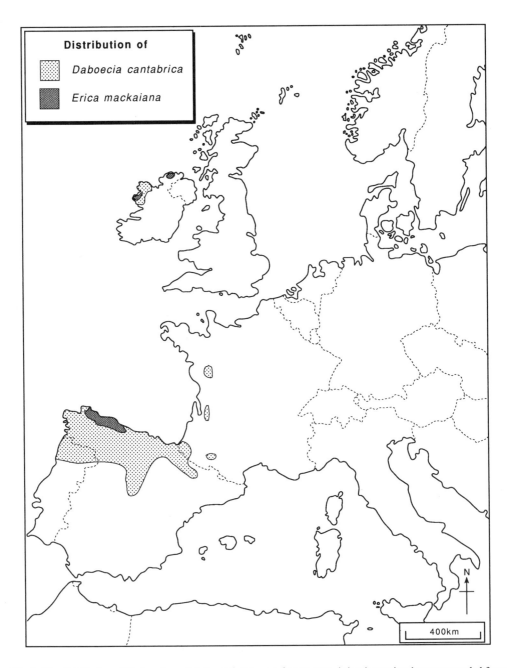

Fig. 8. The distribution of *Daboecia cantabrica* and *Erica mackaiana*, two Atlantic species that are recorded from the Gortian temperate stage (based on Webb 1983).

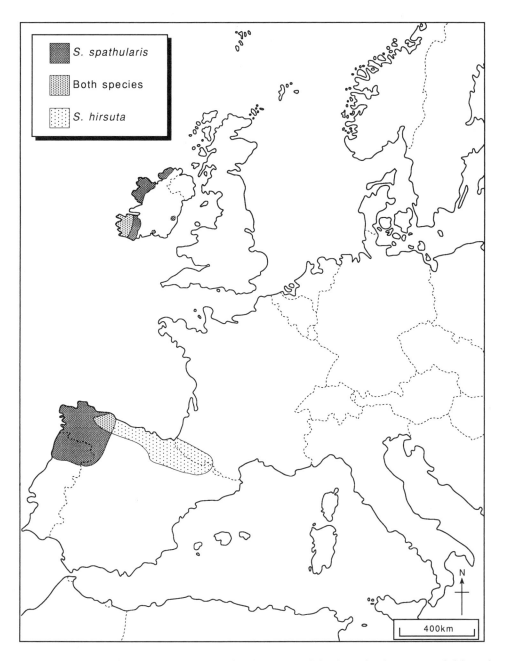

Fig. 9. The distribution of *Saxifraga spathularis* and *S. hirsuta*, two Atlantic species that are recorded from the Gortian temperate stage (as *Saxifraga* sect. *Gymnopera*) (based on Waldren & Scally 1993).

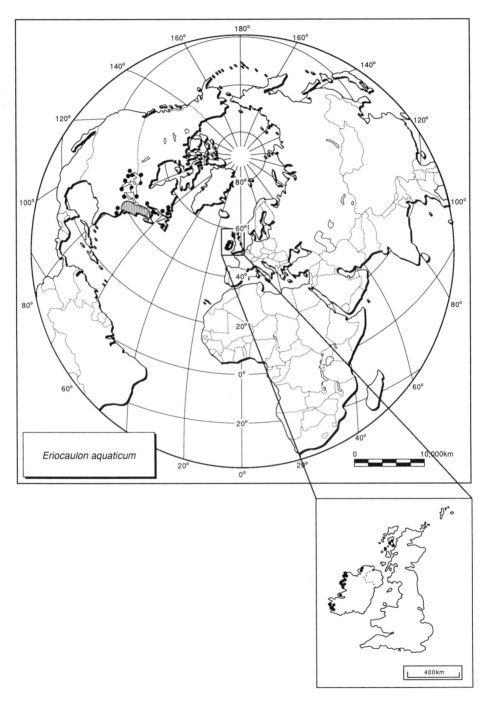

Fig. 10. The global distribution of *Eriocaulon aquaticum* (based on Hultén 1958; Perring & Walters 1962).

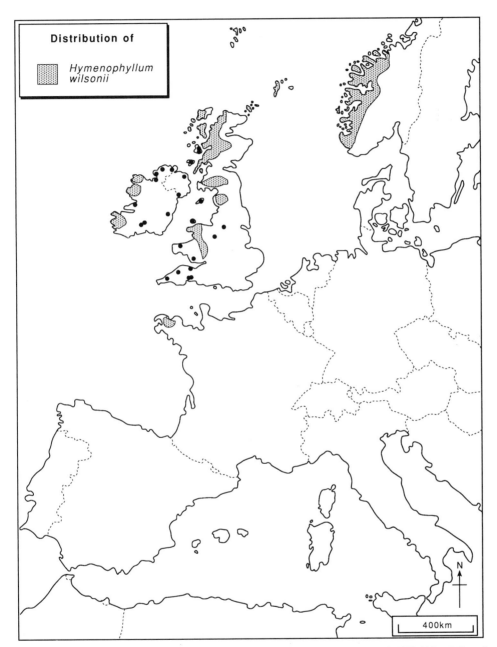

Fig. 11. The European distribution of *Hymenophyllum wilsonii* (based on Jermy *et al.* 1978; Jalas & Suominen 1972–1994).

caulon aquaticum (Fig. 10), *Najas flexilis*, *Hymenophyllum wilsonii* (Fig. 11) and *Trichomanes speciosum*. The first two are amphi-Atlantic plants with uneven distributions, both being far more widespread in America than in Europe. For *Najas* at least, there is clear evidence that the species has retreated westwards, as fossil records extend to central Europe (Hultén 1958; Godwin 1975). *Eriocaulon* and *Najas* are both aquatics with a modern amphi-Atlantic distribution and therefore show some similarities with *Brasenia schreberi* and *Nymphoides cordata*, aquatics now absent from Europe but still present widely in America, as discussed above (see Table 1). *Hymenophyllum* and *Trichomanes* both have a strong requirement for high atmospheric humidity; the latter is a widespread species but *Hymenophyllum* is more frequent in Britain and Ireland than elsewhere, and its distribution is centred on the British Isles (N Atlantic Islands, Brittany, Norway; Jalas & Suominen (1972–1994); Fig. 11). *Hymenophyllum tunbridgense* may also fall into this group, but uncertainties surround the identification of this taxon; Richards & Evans (1972) consider it an Atlantic taxon, but similar or identical taxa occur in the Caribbean and Australasia.

None of the Atlantic and few of the amphi-Atlantic species present in the Irish interglacial are therefore absent from the modern Irish flora. The long record of many of these species along the Atlantic seaboard of southwestern France (Huckerby & Oldfield 1976; Oldfield & Huckerby 1979) is also important, showing that here too there has been great continuity of the flora during the Pleistocene. The Marbellan Interglacial has been correlated to the Holsteinian within the Middle Pleistocene (Oldfield & Huckerby 1979) and, as such, the presence of so many of the Gortian Atlantic and amphi-Atlantic species is notable (Table 4).

Of the 34 Atlantic and amphi-Atlantic taxa, over half are wetland or aquatic species. Three are epilithic/epiphytic with a requirement for high atmospheric humidity, and the remainder are characteristic of open habitats, being arctic–alpine, montane or grassland/open scrub species.

The post-Gortian (Late Pleistocene) floristic record in Ireland

Although this paper has concentrated on the vegetational record of the Middle Pleistocene Gortian temperate stage, it is worth noting that limited information of varying detail is also available from the post-Gortian (the taxa recorded from these sites were included in the database).

The lack of an Irish stratotype for the last interglacial (OI Stage 5e) has caused considerable confusion and disagreement amongst those working on the Irish Pleistocene (Coxon 1993). However, OI Stage 5e deposits may have been identified from a redeposited ball of organic sediment within the sands and gravels of the Screen Hills moraine (McCabe & Coxon 1993). The organic material contains a *Carpinus*-rich pollen assemblage with the following taxa (figures are percentages of the sum of pollen excluding lower plants and aquatics (ΣP) = 517): *Betula* 8%, *Pinus* 16%, *Quercus* 2%, *Carpinus* 14%, *Picea* 9.5%, *Corylus* 14%, *Salix* 1.4%, *Ilex* 3.5%, Ericaceae undiff. 4%, Umbelliferae 9% and Rosaceae 6%. Also present in low percentages are *Alnus*, *Fraxinus*, *Acer*, *Myrica*, *Juniperus*, *Hedera* and Gramineae.

The Screen Hills pollen assemblage strongly resembles the latter parts of continental Eemian sequences. The similarity of the Screen Hills assemblage with the Eemian and Ipswichian biostratigraphical record is discussed further by McCabe & Coxon (1993). The redeposited peat does not provide a useful stratotype nor does it add significantly to our knowledge of Late Pleistocene vegetational history, as the record does not include specific identifications of taxa. However, the implication of the biostratigraphical information is that deposits of equivalent age to the Eemian (OI Sub-stage 5e) dated to 122–132 ka BP might be discovered in the future, and that *Carpinus* may have been an important component of the Irish Late Pleistocene flora, as it was elsewhere in Europe.

Late Pleistocene deposits post-dating OI Stage 5e have been recorded from a number of Irish localities, including Fenit (probably OI Stages 5c and 5d (Heijnis *et al.* 1993)) and Aghnadarragh (OI Stages 3 and 4 (McCabe *et al.* 1987)). The floristic record of these sites has been discussed in the relevant papers and by Coxon (1993).

Discussion

The long Pleistocene and Holocene records of Atlantic species, and especially uneven amphi-Atlantic species such as *Najas* and *Eriocaulon*, suggests that a glacial refuge may have existed somewhere on land to the south and/or southeast of Ireland, as has often been suggested. *Erica ciliaris*, *E. mackaiana* and *Daboecia cantabrica* have all been recorded from Pleistocene deposits in both Ireland and the Pays Basques (both during the Marbellan Interglacial and before it), indicating a long and possibly continuous existence of these taxa in, or proximal to, this area. This may be additionally supported by the apparent westward restriction in the European range of *Najas*, and the difference in chromosome number of European and the majority of American *Eriocaulon aquaticum* (Webb 1983). The apparently continual western European restriction of the Atlantic species is in contrast to the many continental species, which have often retreated a long way to the east since the Middle Pleistocene, and the

Table 5. *Holocene Irish plants not native in Britain (15 taxa), based on Webb (1983).*

Family	Name	Gortian record	Nearest station	Minimum disjunction (km)	Biogeography	Ecology
Orchidaceae	*Neotinea maculata* (Desf.) Stearn		Isle of Man	350	Med/Atlantic	Grassland
Liliaceae	*Simethis planifolia* (L.) Gren.		Brittany	550	Med/Atlantic	Heathland
Ericaceae	*Arbutus unedo* L.		Brittany	550	Med/Atlantic	Woodland/scrub
Asteraceae	*Inula salicina* L.		Normandy	700	Atlantic	Wetland
Ericaceae	*Daboecia cantabrica* (Hudson) K.Koch	G	W France	900	Atlantic	Acid open
Saxifragaceae	*Saxifraga spathularis* Brot.	?	NW Spain	900	Atlantic	Rock fissures
Saxifragaceae	*Saxifraga hirsuta* L.	?	N Spain	950	Atlantic	Shaded rocks
Lentibulariaceae	*Pinguicula grandiflora* Lam.		N Spain	950	Amphi-Atlantic	Open flushes
Ericaceae	*Erica erigena* R. Ross		near Bordeaux	1100	Med/Atlantic	Lake margins/heathland
Ericaceae	*Erica mackaiana* Bab.	G	NW Spain	1100	Atlantic	Bogs
Caryophyllaceae	*Minuartia recurva* (All.) Schinz & Thell.		NE Portugal	1100	Amphi-Atlantic	Alpine
Scrophulariaceae	*Euphrasia salisburgensis* Funck		Vosges	1250	Amphi-Atlantic	Montane??
Caryophyllaceae	*Arenaria ciliata* L.		Jura	1350	Amphi-Atlantic	Arctic–alpine
Clusiaceae	*Hypericum canadense* L.		Newfoundland	3250	Amphi-Atlantic	Wetland
Iridaceae	*Sisyrynchium bermudiana* L.		Newfoundland	3250	Amphi-Atlantic?	Wetland

G, Gortian record; ?, Gortian pollen for *Saxifraga* sect. *Gymnopera* could refer to either species.

arctic–alpines which have retreated to the north and/or to more mountainous areas further south and east. None of the Atlantic species present in the Irish Middle Pleistocene is absent from the present Irish flora, and neither are any of the currently amphi–Atlantic species, save for the arctic–alpine *Athyrium distentifolium* and the anomalous *Nuphar pumila*. The taxa have survived repeated cold stages to produce local populations not only in the Holocene but also during the Middle Pleistocene. This continuity provides strong evidence for survival throughout the Pleistocene in areas close to the current range of these taxa.

However, the absence of three previously amphi-Atlantic species (*Azolla filiculoides*, *Brasenia schreberi* and *Nymphoides cordata*) suggests that any glacial refuge may have been relatively limited in extent. In addition, one of the Atlantic/Mediterranean species now present in Ireland but absent from Britain, *Erica erigena*, may have been introduced in historical times (Foss *et al.* 1987; Foss & Doyle 1988), while another, *E. mackaiana*, at present rarely sets seed in Ireland. However, the statements made by these authors are based on a lack of pollen evidence from early Holocene sediments which does not necessarily prove the absence of the species. The existence of numerous marine trading routes between Ireland and SW Europe in historical times may be responsible for the introduction of several other taxa which are now very local components of the modern Irish flora (T. F. G. Curtis, pers. comm.). The apparent absence of true arctic–alpine species from the Pays Basque during full-glacial conditions (e.g. the Weichselian), and the long continuity of western species in the area (Oldfield 1962; Oldfield & Huckerby 1979) may, together with the Irish data, suggest that somewhat milder conditions existed somewhere on the western seaboard of Europe continuously throughout the Pleistocene. The Atlantic species, of which various Ericaceae are such an obvious feature, may therefore have existed continuously in Western Europe since at least the Middle Pleistocene. If these taxa were able to survive in some refuge proximal to their present distributions, their ecological requirements may help to suggest the habitats that occurred in any such refuge. As discussed above, many are wetland or aquatic species, and many of the others are plants of open habitats. This suggests a mixture of pools and marshes, together with some well drained open habitats. It might be argued that fossil preservation would be favoured by wetland conditions, and our interpretation of the Irish Pleistocene flora may be biased by preservation of predominantly wetland and aquatic species. However, many of the continental species which have also left fossil records have been extirpated from Ireland since the Middle Pleistocene. It is perhaps significant that none of the Gortian Atlantic/amphi-Atlantic taxa (Table 4) are woodland species or trees; many of the Gortian woodland species are now eastern or continental in distribution (e.g. *Fagus sylvatica*, *Abies alba*, *Buxus sempervirens* and *Carpinus betulus*). It seems clear that ecological conditions in any glacial refuge were not suitable for these taxa. The relative paucity of the present Irish arctic–alpine flora may be explained by the gradual extirpation of many of this ecological group since the last glaciation, as witnessed by the Holocene extinction of *Betula nana*, and the very restricted modern distributions of other arctic–alpines, including *Saxifraga nivalis*.

An extreme western glacial refuge may help to explain why many of the Atlantic species present in the Irish Holocene are absent from Britain (see Table 5). However, there is no Pleistocene record for some of these taxa, for example *Erica erigena* and *Arbutus unedo*. Both would have had difficulty surviving cold stage conditions as they are front sensitive (see Pennington 1974), and *E. erigena* may be a deliberate introduction, as described above. Other taxa listed in Table 5 have no Pleistocene record, though this does not mean that they were absent. Some would leave little evidence and others may yet be discovered. The Pleistocene fossil record is obviously incomplete, and there is no fossil record for much of the present Irish flora. Many may yet be discovered in the Irish Pleistocene flora with the location of suitable sediments.

Webb (1983) listed 67 modern Irish species whose distribution can be considered Atlantic (including sub-Atlantic, Atlantic–Mediterranean and amphi-Atlantic). They are not ecologically homogeneous, but of the 67 only six can be thought of as woodland species, and at least four of these grow equally well amongst well shaded rocks in the absence of tree canopies; *Saxifraga hirsuta* provides a typical example (see Waldren & Scally 1993). Of the rest, many are either wetland species or plants of relatively open habitats (grassland, heathland, coastal dunes and shingle, rock outcrops). Some of these taxa may also have been present in Irish interglacial stages and survived glaciation in refugia as described above, but left no record of their temperate stage occurrence. On this point, it is important to note that about two-thirds of the modern Irish Atlantic species are either local or rare (Webb 1983). It is perhaps also interesting to note that relatively few are strictly coastal species, and their present distributions are therefore not determined by the presence of saline habitats.

Although many of the Irish Atlantic species are characteristic of wetlands and other relatively open habitats, it cannot be assumed that they have *all* survived the last (and possibly repeated) glaciations *in situ* or in proximal refugia. Many of the taxa listed by Webb (1983) also occur in Britain, and those which do so are not separated from continental populations by anything like the disjunctions shown in Table 5. It is likely that different mechanisms of survival, dispersal and recolonization apply to the Irish Atlantic taxa. However, the great disjunctions of Irish Atlantic taxa not recorded from Britain (Webb 1983; Table 5) and the long record of other Irish Atlantic taxa through successive glacial stages strongly suggest that at least some of the Atlantic taxa have survived in, or close to, Ireland through at least the last (Midlandian) glaciation. The Gortian (=Hoxnian) flora of Ireland was different from that of Britain, with high percentages of *Ilex, Taxus, Rhododendron* and the occurrence of *Daboecia* and *Erica mackaiana*; none of the latter three are recorded from the numerous British Hoxnian deposits studied to date (Godwin 1975; Watts 1985). This provides strong evidence that interglacial, as well as modern Irish floras were distinct from their British counterparts.

One explanation for the modern Irish flora being distinct from that of Britain is the suggestion that some of the Irish Atlantic element migrated along the Atlantic seaboard at times of lower sea-level, and found their furthest outpost in SW Ireland (see Pennington 1974). However, many of this group of species do not occur in SW Ireland, the presence of the Ericaceae from this group in Connemara being a good example. Webb (1983) suggested that *in situ* survival, proximal survival in refugia, and long-distance transport may all have been involved in providing Ireland with species that are absent from Britain. He considers that at least some of this group could have survived the last glaciation *in situ*, but this would seem impossible for the Munsterian Glaciation, when the country was apparently overrun with ice (McCabe 1987). It is possible that many of the Irish Atlantic (*sensu lato*) species survived the most severe cold stages (e.g. the Munsterian) in proximal refugia, and then recolonized the country in the subsequent temperate stages when their distributions diverged for ecological reasons. The regions that were possibly free of Midlandian ice in Ireland (McCabe 1985) very closely match the modern distributions of most taxa listed in Table 5, with ice-free areas in the south and southwest, the west and northwest. In the Holocene, these taxa may have spread out somewhat from any *in situ* refuge, but in some cases certainly not to the extent of their former Gortian distributions. This might help to explain the very localized modern Irish distributions of the Atlantic species, and the presence of both Gortian and modern Irish species which are unknown in Britain. However, it should be emphasized that the palaeogeography of Midlandian ice cover is still the subject of considerable debate (e.g. O'Cofaigh 1993), the extent of the Munsterian ice is far from clear (McCabe 1987; O'Cofaigh 1993), and we have little idea of the palaeoclimatic conditions surrounding the last ice-sheets (see Conclusions).

Submarine contours suggest no possibility for the colonization of Ireland by Atlantic species from the Iberian peninsula, where many of the specialized elements of the modern Irish flora may be found (Table 5). The only possible Late Midlandian link would be a northern land-bridge with Britain (Devoy 1985), although changes in the sea-bed through ice scouring and sediment deposition during the Late Midlandian remain unknown. Our uncertainty regarding the timing of Late Midlandian sea-level rise and of Late Midlandian palaeoenvironment makes it impossible to reconstruct possible migration routes with any certainty.

It may be that Ireland's insularity has been at least as important in preventing certain elements of the Irish flora from colonizing Britain as it has been in limiting the spread of British species into Ireland. It may be of great significance that two of the modern Irish amphi-Atlantic species (*Eriocaulon aquaticum* and *Spiranthes romanzoffianum*) also occur in NW Scotland and the Isle of Man; they may have (re-?) colonized Britain from Ireland via this northern channel in early Holocene times. However, other Irish Atlantic taxa do indeed occur locally in SW Britain, examples being *Euphorbia hyberna* and *Erica ciliaris*. It is obvious that generalizations are often inaccurate and difficult to sustain, and in many cases plants with roughly similar modern distributions and ecologies may have had very different interglacial and Holocene histories.

Conclusions

The Holocene colonization of Ireland by plants and animals, and in particular the occurrence of the Lusitanian element, has been the subject of a great amount of speculation since the 19th century (Godwin 1975; Sleeman *et al.* 1983; Preece *et al.* 1986). In addition, Mitchell &

Watts (1970) have made use of the temperate-stage floral record in order to analyse the Pleistocene history of the Ericaceae in Ireland. The general conclusions outlined here are similar to those of other authors who have looked at the subject of the colonization of Ireland by plants (see Webb 1983; Preece et al. 1986) and the problems regarding the existence and location of glacial refuges, the location of migration routes, the timing of such migration and the mechanism of colonization (and recolonization) remain. There is a general lack of agreement regarding both per-glacial survival with intensely cold climatic conditions and the possibility of colonization from distant refuges at times of rapidly rising sea-level (Webb 1983; Preece et al. 1986), which has arisen for very good reasons. It is clear that certain floristic elements are present in the Irish flora and that we have yet to determine how and why they have come to be where they are.

However, the plant fossil record of the Irish Middle Pleistocene contains important elements that suggest long continuity with rare and very disjunct (in the Holocene) populations having been present since the Gortian, possibly between 428 and 302 ka BP.

The method of survival and/or recolonization of these plant populations can be summarized as follows:

1. The plants survived Pleistocene cold stages (including proximal glaciation) *in situ* in disjunct populations. For many of the species, such survival is not probable.
2. They survived in refugia (in disjunct populations) possibly close to their present distribution (e.g. as argued by Mitchell & Watts (1970) for the Ericaceae).
3. They survived in refugia as widespread distributions in Europe (to the south and southeast) then recolonized at the end of cold stages and subsequently became disjunct.
4. They survived in isolated localities, in disjunct populations, and subsequently managed to re-occupy ecologically suitable niches.
5. They managed continual recolonization from distant glacial refugia. The long presence of Atlantic and amphi-Atlantic taxa makes this seem unlikely to be a major factor.

In situ survival, and possibly the ability to withstand glaciation, is unlikely for some of the species, but others may have been able to survive, perhaps along the Atlantic seaboard and in valleys close to the sea. Recolonization is evidenced by the former widespread fossil distributions of some species and evidence of their early Holocene presence, e.g. that of *Daboecia cantabrica* in SW France. Other species show former widespread distributions that have shrunk throughout the Pleistocene as the plants failed to recolonize (e.g. *Rhododendron ponticum*), eventually disappearing from Europe or Ireland. The disappearance of some taxa after long Pleistocene records suggests that the refugia were limited in extent, or that it was not possible to maintain some of the taxa in some refuges. As commented on above, many of the taxa suggest that any refuge may have been open and damp with wetlands. Such sites may have existed well to the southeast of Ireland and acted as glacial refuges.

Our understanding of the *in situ* survival or survival in areas proximal to present (Holocene) distributions is hindered by three very important factors. The first is our poor (but improving) understanding of the climate of cold stages, especially of their seasonality and microclimates. Steep climatic gradients, with marked seasonal changes proximal to the large ice-sheets are a distinct possibility (e.g. Harrison et al. 1992), which may have allowed refuges to exist further north than might be expected. Evidence is already available indicating palaeoenvironmental conditions in important offshore locations from critical time periods. For example, temperatures at least as warm as today have been recorded for the Lateglacial interstadial (between 13.3 and 11 ka BP) in the Bay of Biscay (Duplessy et al. 1981). Further research should allow better climatic reconstructions to be made for regions that were acting as important corridors of plant migration.

The second factor is our lack of information on undisturbed modern plant distributions and the co-existence of species: modern species distributions which are subjected to different biotic inputs (including large anthropogenic effects and different climatic controls) are unlikely to be good analogues for Pleistocene fossil assemblages. The third problem is one that bedevils attempts to understand how recolonization might have occurred, and that is the problem of elucidating sea-level change during the cold and temperate stages and hence being able to reconstruct the palaeogeography of western Europe (e.g. Devoy 1985). The latter problem is compounded rather than solved as our understanding of sea level history becomes more sophisticated. A rapidly rising sea-level some 13 ka BP would leave very little time for recolonization, and in Ireland's case such a rise may account for about 50% of the 'absences' in the present flora (Webb 1983). It may be that forebulge effects around previously isostatically depressed coastlines (e.g. McCabe 1987; Eyles & McCabe 1989a, b; Devoy 1995) may hold the key to some potential 'land-bridges' and that such migration routes may yet make a valid reappearance in the geological literature.

Before any further useful conclusions can be reached it will be necessary to locate and date the postulated areas of southern refuges or the migration routes which may be recorded in organic deposits on the continental shelf or in coastal locations in southern France, Portugal and Spain.

The floral history of Ireland probably still holds many surprises. The large disjunction between many Irish species and their continental distributions (Table 5), and the presence of some of these plants in the Middle Pleistocene, argues for continuity and proximal survival (if not *in situ* survival). However, the ability of plants to move into areas that have been deglaciated, possibly given rapid climatic amelioration or strong seasonal variation, may be far greater than has been realized to date. Either way, certain groups of taxa have been able to survive along the western edge of Europe throughout the Pleistocene and have recolonized Ireland to produce disjunct populations, possibly on numerous occasions.

The authors would like to thank Professor Frank Oldfield for providing information on the Pays Basque fossil record. We would also like to thank sincerely Professors G. F. Mitchell, W. A. Watts and D. A. Webb for comments on the manuscript of this paper. One of the authors (P.C.) would like to take this opportunity of thanking Dr Rendel Williams for giving him an interest in Quaternary geology, and Professor R. G. West FRS for his considerable help and friendship during the time that he was P.C.'s PhD supervisor.

References

BENNETT, K. D. 1995. Insularity and the Quaternary tree and shrub flora of the British Isles. *This volume*.

BOULTER, M. C. & MITCHELL, I. 1977. Middle Pleistocene (Gortian) deposits from Benburb, Northern Ireland. *Irish Naturalists' Journal*, **19**, 2–3.

BOWEN, D. Q., RICHMOND, G. M., FULLERTON, D. S., SIBRAVA, V., FULTON, R. J. & VELICHKO, A. A. 1986. Correlation of Quaternary Glaciations in the Northern Hemisphere. *Quaternary Science Reviews*, **5**, 509–510.

CHAMBERLAIN, D. F. 1982. A revision of *Rhododendron*. II Subgenus Hymenanthes. *Notes from the Royal Botanic Garden Edinburgh*, **32**, 209–486.

COXON, P. 1993. Irish Pleistocene Biostratigraphy. *Irish Journal of Earth Sciences*, **12**, 83–105.

—— & FLEGG, A. 1985. A Middle Pleistocene interglacial deposit from Ballyline, Co. Kilkenny. *Proceedings of the Royal Irish Academy*, **85B**, 107–120.

—— & —— 1987. A Late Pliocene/Early Pleistocene deposit at Pollnahallia, near Headford, Co. Galway. *Proceedings of the Royal Irish Academy*, **87B**, 15–42.

—— & HANNON, G. 1991. The interglacial deposits at Derrynadivva and Burren Townland. *In:* COXON, P. (ed.) *Field Guide to the Quaternary of North Mayo*. Irish Association for Quaternary Studies (IQUA), 24–36.

——, —— & FOSS, P. 1994. Climatic deterioration and the abrupt end of the Gortian Interglacial recorded in interglacial sediments from Derrynadivva and Burren Townland, near Castlebar, County Mayo, Ireland. *Journal of Quaternary Science*, **9**, 33–46.

CRAIG, A. J. 1978. Pollen percentage and influx analyses in south-east Ireland: A contribution to the ecological history of the late Glacial period. *Journal of Ecology*, **66**, 297–324.

CROSS, J. R. 1975. Biological flora of the British Isles. No. 137 *Rhododendron ponticum* L. *Journal of Ecology*, **63**, 345–364.

CURTIS, T. F. G. & McGOUGH, N. 1988. *The Irish Red Data Book*, 1. *Vascular Plants*. Stationery Office, Dublin.

DEVOY, R. J. 1985. The problems of a Late Quaternary landbridge between Britain and Ireland. *Quaternary Science Reviews*, **4**, 43–58.

—— 1995. Deglaciation, Earth crustal behaviour and sea-level changes in the determination of insularity: a perspective from Ireland. *This volume*.

DUPLESSY, J. C., DELIBRIAS, J. L., TURON, J. L., PUJOL, C. & DUPRAT, J. 1981. Deglacial warming of the northeastern Atlantic Ocean: Correlation with the paleoclimatic evolution of the European continent. *Palaeogeography Palaeoclimatology Palaeoecology*, **35**, 121–144.

EYLES, N. & McCABE, A. M. 1989a. The Late Devensian (<22,000 BP) Irish Sea Basin: The sedimentary record of a collapsed ice sheet margin. *Quaternary Science Reviews*, **8**, 307–351.

—— & —— 1989b. Glaciomarine facies within subglacial tunnel valleys: the sedimentary record of glacio-isostatic downwarping in the Irish Sea Basin. *Sedimentology*, **36**, 431–448.

FOSS, P. J. & DOYLE, G. J. 1988. A palynological study of the Irish Ericaceae and *Empetrum*. *Pollen et Spores*, **30**, 151–178.

——, —— & NELSON, E. C. 1987. The distribution of *Erica erigena* R.Ross in Ireland. *Watsonia*, **16**, 311–327.

GENNARD, D. E. 1984. A palaeoecological study of the interglacial deposit at Benburb, Co. Tyrone. *Proceedings of the Royal Irish Academy*, **84B**, 43–56.

GIBBARD, P. L., WEST, R. G., ZAGWIJN, W. H. *ET AL.*

1991. Early and Early Middle Pleistocene correlations in the southern North Sea basin. *Quaternary Science Reviews*, **10**, 23–52.

GODWIN, H. 1975. *The History of the British Flora*. 2nd edn, Cambridge University Press.

HARRISON, S. P., PRENTICE, I. C. & BARTLEIN, P. J. 1992. Influence of insolation and glaciation on atmospheric circulation in the North Atlantic sector: Implications of general circulation model experiments for the Late Quaternary climatology of Europe. *Quaternary Science Reviews*, **11**, 283–299.

HEIJNIS, H. 1992. *Uranium/Thorium Dating of Late Pleistocene Peat Deposits in N.W. Europe*. PhD thesis. Wiskunde en Natuurwetenschappen aan de Rijksuniversiteit Groningen, The Netherlands.

——, RUDDOCK, J. & COXON, P. 1993. A uranium-thorium dated Late Eemian or Early Midlandian organic deposit from near Kilfenora between Spa and Fenit, Co. Kerry, Ireland. *Journal of Quaternary Science*, **8** (1) 31–43.

HUCKERBY, E. & OLDFIELD, F. 1976. The Quaternary vegetational history of the French Pays Basque. II Plant macrofossils and additional pollen-analytical data. *New Phytologist*, **77**, 499–526.

HULTÉN, E. 1958. The amphi-Atlantic plants and their phytogeographical connections. *Kungl Svenska Vetenskapsakademiensis Handlingar Series*, **4** (7), 1–340.

—— 1962. The circumpolar plants, I. Vascular cryptograms, conifers, monocotyledons. *Kungl Svenska Vetenskapsakademiensis Handlingar Series*, **4** (8), 1–275.

—— 1970. The circumpolar plants, II. Dicotyledons. *Kungl Svenska Vetenskapsakademiensis Handlingar Series*, **4** (13), 1–463.

JALAS, J. & SUOMINEN, J. 1972–1994. *Atlas Florae Europeae I-X. Pteridophyta to Cruciferae*. The Committee for Mapping the Flora of Europe and Societas Biologica Fennica Vanano, Helsinki.

JERMY, A. C., ARNOLD, H. R., FARRELL, L. & PERRING, F. H. 1978. *Atlas of Ferns of the British Isles*. Botanical Society of the British Isles & British Pteridological Society, London.

JESSEN, K., ANDERSEN, S. T. & FARRINGTON, A. 1959. The interglacial deposit near Gort, Co. Galway, Ireland. *Proceedings of the Royal Irish Academy*, **60B**, 1–77.

JONES, R. L. & KEEN, D. H. 1993. *Pleistocene Environments of the British Isles*. Routledge, Chapman & Hall, London.

MCCABE, A. M. 1985. Glacial Geomorphology. *In:* EDWARDS, K. J. & WARREN, W. P. (eds), *The Quaternary History of Ireland*, Academic Press, London, 67–93.

—— 1987. Quaternary deposits and glacial stratigraphy in Ireland. *Quaternary Science Reviews*, **6**, 259–299.

—— & COXON, P. 1993. A *Carpinus*-dominated interglacial peat ball within glaciomarine delta sediments, Blackwater, Co. Wexford: evidence for part of the last interglacial cycle? *Proceedings of the Geologists' Association*, **114**, 201–207.

——, COOPE R. J., GENNARD, D. E., & DOUGHTY, P. 1987. Freshwater organic deposits and stratified sediments between Early and Late Midlandian (Devensian) till sheets, at Aghnadarragh, County Antrim, Northern Ireland. *Journal of Quaternary Science*, **2**, 11–33.

MITCHELL, G. F. & WATTS, W. A. 1970. The history of the Ericaceae in Ireland during the Quaternary Epoch. *In:* WALKER, D. & WEST, R. G. (eds) *Studies in the Vegetational History of the British Isles*. Cambridge University Press.

—— & —— 1993. Notes on an interglacial deposit in Ballykeerogemore Townland and an interstadial deposit in Battlestown Townland, both in County Wexford. *Irish Journal of Earth Sciences*, **12**, 107–117.

——, PENNY, L. F., SHOTTON, F. W. & WEST, R. G. 1973. *A Correlation of Quaternary Deposits in the British Isles*. Geological Society of London, Special Report, **4**.

O'COFAIGH, C. 1993. *Sedimentology of Late Pleistocene Glacimarine and Shallow Marine Deposits from the South Coast of Ireland*. MSc thesis, University of Dublin, Trinity College.

OLDFIELD, F. 1962. Quaternary plant records from the Pays Basque. I. Le Moura, Mouligna, Marbella. *Bulletin du Centre d'Etudes et Recherches Scientifiques, Biarritz*, **2**, 211–217.

—— & HUCKERBY, E. 1979. The Quaternary palaeobotany of the French Pays Basque: A summary. *Pollen et Spores*, **21** (3), 337–360.

PENNINGTON, W. 1974. *The History of British Vegetation*. 2nd edn, English Universities Press, London.

PERRING, F. H. & WALTERS, S. M. 1962. *Atlas of the British Flora*. Thomas Nelson, London.

PREECE, R. C., COXON, P. & ROBINSON, J. E. 1986. New biostratigraphic evidence of the Post-glacial colonization of Ireland and for Mesolithic forest disturbance. *Journal of Biogeography*, **13**, 487–509.

RICHARDS, P. W. & EVANS, G. B. 1972. Biological flora of the British Isles. *Hymenophyllum. Journal of Ecology*, **60**, 245–268.

SCANNELL, M. J. P. & SYNNOTT, D. M. 1987. *Census Catalogue of the Flora of Ireland*, 2nd edn, Stationery Office, Dublin.

SCOURSE, J. D., ALLEN, J. R. M., AUSTIN, W. E. N., COXON, P., DEVOY, R. J. N. & SEJRUP, H. P. 1992. New evidence on the age and significance of the Gortian Temperate Stage: A preliminary report on the Cork Harbour site. *Proceedings of the Royal Irish Academy*, **92B**, 21–43.

SLEEMAN, D. P., DEVOY, R. J. & WOODMAN, P. C. 1983. *Proceedings of the Postglacial Colonization Conference*. University College Cork, 15–16 October 1983. Occasional Publication of the Irish Biogeographical Society **1**.

STACE, C. A. 1991. *New Flora of the British Isles*. Cambridge University Press, Cambridge.

TUTIN, T. G., HEYWOOD, V. H., BURGES, N. A., MOORE, D. M., VALENTINE, D. H., WALTERS, S. M. & WEBB, D. A. 1964–1980. *Flora Europaea*. Volumes 1–5. Cambridge University Press.

WALDREN, S. & SCALLY, L. 1993. Ecological factors controlling the distribution of *Saxifraga spathu-*

laris Brot. and *S. hirsuta* L. in Ireland. *In:* COSTELLO, M. J. & KELLY, K. S. (eds) *Biogeography of Ireland: Past, Present and Future.* Occasional Publication of the Irish Biogeographical Society, **2**, 45–55.

WARREN, W. P. 1979. The stratigraphic position and age of the Gortian Interglacial deposits. *Bulletin of the Geological Survey of Ireland*, **2**, 315–332.

—— 1985. Stratigraphy. *In:* EDWARDS, K. J. & WARREN, W. P. (eds) *The Quaternary History of Ireland,* Academic Press, London, 39–65.

WATTS, W. A. 1959. Interglacial deposits at Kilbeg and Newtown, Co. Waterford. *Proceedings of the Royal Irish Academy*, **60B**, 79–134.

—— 1964. Interglacial deposits at Baggotstown, near Bruff, Co. Limerick. *Proceedings of the Royal Irish Academy*, **63B**, 167–189.

—— 1967. Interglacial deposits in Kildromin Townland, near Herbertstown, Co. Limerick. *Proceedings of the Royal Irish Academy*, **65B**, 339–348.

—— 1970. Tertiary and interglacial floras in Ireland. *In:* STEPHENS, N. & GLASSCOCK, R. E. (eds) *Irish Geographical Studies.* Queen's University, Belfast, 17–33.

—— 1977. The Late Devensian vegetation of Ireland. *Philosophical Transactions of the Royal Society of London*, **B280**, 273–293.

—— 1985. Quaternary vegetation cycles. *In:* EDWARDS, K. J. & WARREN, W. P. (eds) *The Quaternary History of Ireland.* Academic Press, London, 155–185.

WEBB, D. A. 1983. The flora of Ireland in its European context. *Journal of Life Sciences, Royal Dublin Society*, **4**, 143–160.

WYMER, J. J. 1985. *The Palaeolithic sites of East Anglia.* GeoBooks, Norwich.

ZAGWIJN, W. H. 1960. Aspects of the Pliocene and Early Pleistocene vegetation in the Netherlands. *Mededelingen van de Geologische Stichting CIII*, **5**, 1–78.

—— 1985. An outline of the Quaternary stratigraphy of The Netherlands. *Geologie en Mijnbouw*, **64**, 17–24.

—— 1992. The beginning of the Ice Age in Europe and its major subdivisions. *Quaternary Science Reviews*, **11**, 583–591.

Index

Abies alba, 176, 248, 249, 250, 252, 254, 262
Acer campestre, 176
Acer platanoides, 176
Acinonyx pardinensis, 142
Aghnadarragh, County Antrim, 118, 121, 260
Alces alces, 117, 118, 119, 121
Aldeburgh, Suffolk, 32
Aldingbourne Raised Beach Gravels, 37, 40, 41
Allocricetus bursae, 134
Alnus glutinosa, 176, 177
American species, 250
amino acid racemization, 66, 97, 104, 128, 246
Ammonia batavus, 246
amphi-Atlantic elements, 243, 262–3
Anancus arvernensis, 144
Anglesey, 176, 225
Anglian Stage, 31, 32, 65
Anglian/Elsterian Stage, 10, 11, 20
Angulus distortus, 23, 98
Antian Stage, 9, 24
Apodemus sylvaticus, 112, 116, 120, 122, 133, 135, 151, 181
Arbutus unedo, 173, 261, 262
arctic–alpine/montane element, 252
Ardleigh, Essex, 10
Armeria maritima, 252, 253, 254
Arromanches, Normandy, 163
Arun, River, 16, 55, 56, 60
Arvicola terrestris, 115, 116, 117, 118, 120
Astralium rugosum, 70, 155
Astrantia minor, 249, 252, 254
Atapuerca, Spain, 142
Athyrium distentifolium, 248, 253, 262
Atlantic elements, 253, 262
aurochs, 117, 118, 121, 122
Aveley, Essex, 68, 132
Axial Channel, 30, 35, 38, 42
Azolla filiculoides, 249, 250, 254, 260, 262

Bacon Hole Cave, Gower, 117, 118, 134, 135, 136, 157
Bacton, Norfolk, 143, 144
badger, 113, 116, 119, 120, 135
Bakkersdam, The Netherlands, 104
Balderton Sand and Gravel, 133, 147
Ballycotton Bay, County Cork, 189
Ballyline, County Kilkenny, 246
Ballymacadam, County Tipperary, 246
Baltic Ice Lake, 175
bank vole, 112, 116, 120, 133, 135
Bantry Bay, SW Ireland, 223
Barling/Dammer Wick Gravel, 34, 35, 36, 40, 41
Barnham, Suffolk, 114, 115
Barrington, Cambridgeshire, 135, 137, 145, 157
bathymetric deeps, 226–7, 232
Bavelian Stage, 10, 101
Baventian Stage, 9
beach height, factors affecting, 164–5
beaver, 112, 115, 116, 120, 122, 132
Beeches Pit, Suffolk, 114
Belcroute, 70
Belgrandia marginata, 101, 102
Belle Hougue Caves, 70, 163, 165–6
Belvédère Interglacial, 94
Bembridge Raised Beach deposits, 17, 37, 40, 64, 66, 67, 69, 81, 95, 163–4
Benburb, Ireland, 253
Berberis sempervirens, 176
Berry Head, 69, 70
Betula nana, 248, 249, 251, 252, 253, 254, 262
Betula pubescens spp. *tortuosa*, 175
Bielsbeck, Yorkshire, 131
Bilzingsleben, Germany, 114
bison, 113, 117, 118, 119, 122, 134, 135
Bison bonasus, 119

Bison priscus, 115, 117, 118, 121, 135
Bittium reticulatum, 96, 98, 104
Black Park Gravel, 34
Black Rock Raised Beach, 68
Blackwater, 34
Bobbitshole, Ipswich, Suffolk, 128, 130, 135
Bolboforma costata, 7
Boleyneendorrish, Ireland, 246
Boreal Province, 98
Bos primigenius, 117, 118, 119, 121, 132, 135
Boxgrove, Sussex, 17, 64, 70–1
 deposits, 67, 81, 95, 113, 114, 144, 146, 147
 human site, 130, 131, 137
 Raised Beach gravels, 37, 40, 41
Boyn Hill Gravel, 34, 43
Bramertonian Stage, 9, 94, 101
Bramford Road, Suffolk, 137
Brasenia schreberi, 249, 250, 254, 260, 262
Brentford, London, 135
Brighton Raised Beach, 64, 68, 131, 132, 133
brown bear, 116, 118, 120, 122, 132, 133, 134, 144
brown hare, 120, 151
Bruckenthalia spiculifolia, 249, 252, 254
Bruhnes/Matuyama boundary, 10, 142
Brundon, Suffolk, 131, 132
Bubalus murrensis, 115
Bufo bufo, 114, 120
Bufo calamita, 114, 120, 121
Bure Valley Beds, 101, 105
Burren Townland, 246
Burtle Beds, Somerset, 133
Buxus sempervirens, 249, 262

Canis etruscus, 143
Canis lupus, 112, 115, 116, 118, 120, 122, 132, 133, 134, 135, 142, 143, 144
Canis mosbachensis, 142, 143
Cap Gris Nez, 21, 38
Capreolus capreolus, 113, 117, 119, 121, 131, 132, 157, 161
'*Cardium* sands of Lo', 98
Carpinus betulus, 176, 249, 250, 262
Carya, 246, 248, 249
Castor fiber, 112, 116, 120, 132
cave bear, 133
Cefn Cave, North Wales, 135
Celtic Deep, 225–6, 236, 237
Celtic Province, 92, 96, 98, 99
Celtic Sea, 76, 82, 189, 190, 193, 197, 199
 sea-levels, model, Irish and Celtic seas, *see under* sea-level changes
Cerastoderma edule, 11, 96, 97, 106
Cervus elaphus, 113, 115, 117, 118, 120, 122, 132, 134, 135
 Jersey, 152, 157, 166
Cervus elaphus jerseyensis, 157
Chalk ridge, 32, 63, 65–70, 105, 130, 136
 see also land-bridges
Channel, the, *see* English Channel
Channel River system, 17, 21, 22, 23
cheetah, 142
Chesil Beach, 65
Chillesford Crag Member, 9
Clacton-on-Sea, 34, 39, 42, 43, 95, 96, 101–2, 104, 105, 114, 133, 146
clawless otter, 131, 133
Clethrionomys glareolus, 112, 116, 120, 133, 135
coastal changes, 6, 27, 222–3, 232, 235, 236, 237–8
 see also shorelines
coccolith assemblages, 81
Coelodonta antiquitatis, 116, 118, 119, 122, 133, 134
Colchester earthquake, 42
Colchester Formation, 10
common hamster, 112, 115, 116, 133, 134
common vole, 112, 116, 119, 151

continental shelf, inner
  evolution, 57
  subaerial exposure, 48, 55, 61
Coralline Crag Formation, 5, 7, 30
Corbicula, 90, 133, 134, 136, 137
Corbicula fluminalis, 90, 101, 102, 103, 115, 132, 135
Corbridge Cave, Devon, 70
Cornus sanguinea, 176
Corton Cliff, 142
Corylus avellana, 176, 177
Côte des Abers, 71
Crayford, Kent, 115, 133–4, 135, 137, 143
Cricetus cricetus, 112, 115, 116, 133, 134
Crocidura, 117, 131, 132
Crocidura russula, 116, 119
Crocuta crocuta, 112, 115, 116, 118, 135, 142, 144, 145, 147
Cromer, 30, 142
  see also West Runton
Cromer Forest Bed Formation, 9, 113, 141, 143
'Cromerian Complex', 10, 17, 19, 31, 81, 95, 101
Cromerian Stage, 10, 19, 64, 66, 95, 101, 111, 136, 141
Crouch, River, gravels, 37, 40, 41
crustal rebound, 187, 190–9, 200, 201, 202
Cuckmere river, 55
Cudmore Grove, 114, 115, 131, 133, 134, 136, 137
Cyrnaonyx sp, 131, 132, 133

Daboecia cantabrica, 252, 254, 255, 256, 260, 261, 262, 264
Dama dama, 113, 115, 117, 119, 133, 134, 135, 136, 157
Dammer Wick gravels, 40, 41
deglaciation, effects of, 11, 187–90, 201, 202, 210–11
Denmark, fauna, 98, 99, 119, 121, 176
Derrynadivva, Ireland, 248
Devensian, 20, 37, 39, 55, 119, 122, 130, 136–7, 159, 161
Devensian/Weichselian glacial stage, 11, 27, 38,42, 55
Devil's Hole, Nevada, 161
Dicrostonyx torquatus, 118, 123, 134
Dnepr river, 92
Dogger Bank, 191
dogs, 143, 144
Dömnitz, 94
Dorst glacial, 10
Doveholes, Derbyshire, 144
Dover Strait, *see* Strait of Dover
Drenthe Sub-stage, 23
Durdham Down, Somerset, 135
dwarf red deer of Jersey, 151–66
dwarfed mammals, 151–2, 168
dwarfism, 166–8

Easington Raised Beach, Durham, 98
East Anglia, 7, 19, 22, 30, 31–2
Eastbourne, Sussex, 135
Eastern Torrs Quarry Cave, Devon, 135, 157
East Hyde, Essex, 102, 104
Eburonian Stage, 10, 141
Eem Formation, 38
Eemian/Ipswichian, 15, 23, 94, 96, 98, 153–9
Elaphe longissima, 121
electron spin resonance (ESR), 66–7
Eleocharis ovata, 249, 250, 254
elephant, 137, 151
elk, 113, 117, 118, 119, 121
Elphidiella hannai, 7
Elphidiella oregonense, 8
Elsterian Stage, 32, 33, 39
Elsterian/Anglian Stage, 17, 18, 20–1, 33, 39
Emys orbicularis, 114, 115, 120, 121, 132, 136
English Channel, 10, 15–18, 20–1, 24, 65, 71, 92
  bathymetry, 83–7
  North Sea connection, 11, 42, 43, 82, 86, 87, 130
  North Sea separation from, 63–4
  as route for Rhine–Thames drainage, 38, 39, 40
  submerged valleys, 47, 48
    see also submerged valley, English Channel
  tidal dynamics modelling, Pleistocene, 80–7
Ensis americanus, 90, 92

Equus ferus, 116, 117, 118, 119, 132, 134
Equus hydruntinus, 116, 117
Erica ciliaris, 255, 260, 263
Erica erigena, 261, 262
Erica mackaiana, 252, 254, 255, 256, 260, 261, 262, 263
Erinaceus europaeus, 112, 116, 119, 120
Eriocaulon aquaticum, 255, 258, 259–60, 263
Euonymus europaeus, 176
Euphorbia hyberna, 263
Eurasian jaguar, 143
extinctions, 102, 119, 121–3, 177, 248

Faeroes, flora, 178
Fagus sylvatica, 176, 249, 262
'Faille de Landrethun', 42
fallow deer, 113, 115, 117, 119, 133, 135, 136, 157
Felis pardoides, 142
Felis silvestris, 112, 116, 119, 120, 135
Felixstowe, Suffolk, 42
Fenit, Ireland, 260
field vole, 135, 151
Flanders, land connection with, 65
floristic record of Pleistocene Ireland, 243–6, 260–5
  species now absent from Britain, 251–2
  species now absent from Ireland, 248–51
fluvial drainage, central southern England, 56
fluvial species as biogeographic indicators, 93
flying squirrel, 113
forebulge, 21, 23, 24, 65, 196, 198, 210, 212, 214
  advance of and sea-levels, 219, 220, 221, 222, 227, 236
  Irish Sea region, 195, 199–200, 201, 202
  limit and sea-levels, 232, 233, 235
  peak, 217, 227–8, 230, 232, 233, 237–8
  retreat of and sea-levels, 214, 220, 234
Fosse Dangeard, 15, 16, 20, 33, 39, 42
Foulness Sands, 35, 37
fox, 135
Frangula alnus, 176
Fransche Kamp, The Netherlands, 94
Fraxinus excelsior, 176
frogs, absence, 122
Fulbeck, Lincolnshire, 135

giant deer, 119, 123, 132, 134, 135, 236
giant ox, 132, 134. 135
glacial refuge for flora, 262, 264
glacio-eustasy and sea-level, 3, 11, 57, 60–1
  modelling, 213, 214, 217, 229, 237
glacio-isostasy and sea-level, 24, 65, 199, 237, 238
  Irish Sea region, 187–90, 193, 198–200, 202
  modelling, 210, 213, 214–16, 217, 227, 235, 237
glacio-lacustrine sediments, 17–18, 21, 23
glacio-marine sediments, 18, 23, 187, 189–90
global ice volumes, 5, 61
global sea-levels, 3–5, 8–9, 10, 11, 213, 229
Gombaszoeg, 143
Gortian Stage, 243, 246–65
Gotland, 174
Goyet, Belgium, 145
Gray's Thurrock, Essex, 144
Great European (Ur-Frisia) Delta, 9–10, 11
ground squirrel, 113, 115, 116, 118, 134
Gulf of Corryvreckan, 227
Gulo gulo, 118, 122

Haisboro-Terschelling Rise, 41, 42
hare, 112, 116, 120, 135
Harkstead, Suffolk, 131, 132
Hebridean islands, 174, 177, 197
hedgehog, 112, 116, 119, 120
Herzeele, 19, 22, 64, 65, 66, 67, 98, 103, 104
Hinia pygmaea, 98
Hinia reticulata, 96, 98, 104
Hippophae rhamnoides, 176, 248, 249
hippopotamus, 98, 103, 115, 117, 122, 132, 133, 135, 136, 137, 151, 197
Hippopotamus major, 21

INDEX

Histon Road, Cambridge, 128, 133, 157
Hollesley, 32
Holocene, 27, 35, 37, 38, 39, 42, 55, 60, 76–80, 87, 98–9, 103, 104, 119, 122–3, 173–9, 195
Holsteinian Stage, 21, 33, 39, 64, 94, 95, 96, 97, 103
Holsteinian/Hoxnian Stage, 22, 39, 40, 42
Hollymount, County Laois, 246
Homo sapiens, 116, 118, 120, 136–7
Homotherium latidens, 112, 115, 142, 143
Honiton, Devon, 135
Hoogeveen Interstadial, 94
Hopes' Nose, Torbay, 69, 71
horse, 103, 113, 116, 117, 118, 119, 132, 133, 134, 135, 136, 137, 147
house mouse, 151
Hoxne, Suffolk, 97, 113, 137, 146
Hoxnian Stage, 33, 34, 66, 94
Hoxnian/Holsteinian Stage, 11
humans, 113, 117, 118, 119, 120, 122, 130–3, 134, 136, 137, 182
Hummelsbüttel, 97
Huon Peninsula, Papua New Guinea, 3, 5
Hurd Deep, 16, 47
Hutton cave, Somerset, 134
hyaenas, 143–6
Hyaena prisca, 145
hydro-isostasy, 42, 65, 190, 213, 216–17, 227, 229–32, 233, 237
Hyla arborea, 114, 121
Hymenophyllum tunbridgense, 260
Hymenophyllum wilsonii, 255, 259
Hypnum cupressiforme, 54–5

ice loading and unloading, 196, 199, 210, 212
ice marginal sediments, 22
ice sheet, continental, extension, 18–19
icebergs as glacigenic agent, 222
Ijmuiden Formation, 31
Ilex aquifolium, 176
Ilford, Essex, 115, 131, 133, 157
infilled valley, evolution model, 57–60, 61
Inner Silver Pit, 97
'insular British Mousterian of Acheulian tradition', 137
Ipswichian, 11, 66, 68, 94
 see also Eemian
Ireland, 174, 189, 190, 193, 200
 fauna, 111–12, 118, 120, 121–3, 135, 181–3
 flora, 176, 177, 203
 land-bridge, see land-bridges
 see also floristic record of Pleistocene Ireland
Irish Sea
 bathymetry, 184–7
 deposits, 189, 190, 223, 225, 226, 236
 sea-levels, model, Irish and Celtic seas, see under sea-level changes
 shorelines, 195, 196, 201, 203
Irish Sea region, land uplift, 191–2
Irish 'shelly' till, 189
island extinctions of flora, 177–8, 179
Isle of Man, 174, 177, 210, 223, 225, 238
Isle of Wight, 174, 176, 177
Isles of Scilly, 189, 209
isostasy, 21, 187, 188, 191–2, 210, 237
Itteringham, Norfolk, 117

Jersey, 64, 70, 71, 152
 connection to, and isolation from mainland, 71, 153, 159, 161, 163–6
 dwarf red deer *see* dwarf red deer of Jersey
Joint Mitnor Cave, Devon, 135, 137, 143, 144, 145, 146, 157
Juniperus communis, 175

Kempton Park/East Tilbury Marshes Formation, 37, 69
Kentish Knock sandbank, 34
Kent's Cavern, Devon, 137, 143, 145, 146, 161
Kesgrave Group, 5

kettle-holes, 226–7, 232, 236
Kieschoölite Formation, 7
Kirkdale Cave, Yorkshire, 135, 144, 145, 146, 157
Kreftenheye Formation, 23, 34, 35, 38, 39
Kruisschans Formation, 7, 94

La Cotte, red deer, 152–3, 159–61, 166
Lacerta vivipara, 114, 120, 123
Lagurus lagurus, 134
Lake Missoula, 20
La Londe, Normandy, 17
land-bridges, 119, 136, 181, 187
 Ireland to Britain, 122, 234, 236, 237–8
  biogeographical evidence, 181–3, 203
  forebulge operation, 199–200, 201, 202
  geomorphological and geological record, 183–7
  land-sea changes and crustal rebound, 190–9
  role of deglaciation, 187–90, 201, 202
 Ireland to Scotland, 227
 Strait of Dover, 75, 76, 80, 82, 89
Last Interglacial
 see Eemian or Ipswichian
Late-glacial, 119, 212
Late Pliocene, 9, 94, 99, 101, 104, 144, 245, 248
Lauenburg Clay Formation, 21
Lawford Pit, 144
'*Leda myalis* Bed', 95
Leerdam interglacial, 10, 101
Lemmus lemmus, 118, 123, 132, 134
leopard, 113, 143, 144
Lepus timidus, 118, 120, 122, 151
L'Escale, Southern France, 145
Levallois technique, 132, 134, 137
Lexden, Essex, 131
Linge glacial, 10
lion, 116, 118, 119, 132, 134, 135, 142, 143, 144, 147
Lithoglyphus jahni, 101
Lithoglyphus naticoides, 90–1
Littorina saxatilis, 67, 95, 97
Llandudno, north Wales, 223
Lo, Belgium, 103
Lobourg, River, 16, 21, 30, 31, 35, 43, 47
Lower Thames, 67–8
Luchtbal Formation, 7
Ludhamian Stage, 8, 94
Lunel Viel, southern France, 145
Lusitanian elements, 11, 92, 98, 99, 100, 181, 263
Lusitanian Province, 92, 96, 98
Lycopodium annotinum, 248, 249, 254
Lynch Hill Gravel, 35
Lynx lynx, 119
Lyrodiscus, 103, 105

Maas, River, 10, 17, 21, 24
Maassluis Formation, 63, 101
Maastricht Belvédère, 115
Macaca sylvanus, 112, 115
Macoma balthica, 104
Maidenhall, Suffolk, 131
malacological evidence for insularity, 89, 92–4, 104–6
 dispersal and migration, 89–92
Malin Sea, 212, 224
mammalian evidence for insularity, 127–38, 147–8
mammoth, 116, 131–2, 133, 134, 136, 137
 miniature, 151–2
Mammuthus meridionalis, 21, 43
Mammuthus primigenius, 116, 118, 119, 121, 134
Mammuthus trogontherii, 10, 112, 132
Maplin Sands, 36, 40, 41
Marbellan temperate stage, 248, 260
March Gravels, 98
Margaritifera auricularia, 103
Marsworth Lower Channel, 130, 131, 133, 137
Mauer, Germany, 145
Mediterranean Province, 98, 99
Mediterranean–Atlantic elements, 243
Medway, River, 30, 31, 37, 40, 42

Megaloceros giganteus, 115, 117, 118, 119, 122, 123, 132, 134, 135, 236
Meles meles, 113, 116, 119, 120, 135
Menapian Stage, 10
Mercuria confusa, 95, 96, 98
Merksem Member, 94
Meuse, River, 31, 35, 38, 93, 103
Microtus agrestis, 115, 116, 118, 120, 133, 135, 151
Microtus arvalis, 112, 116, 119, 151
Microtus oeconomus, 112, 115, 116, 118, 132, 133, 134
Midlandian, 121–2, 209, 210, 263
Miesenheim I, Germany, 113
Milton Hill Caves, Somerset, 135, 157
Minchin Hole Cave, Gower, 135, 164
molluscs, 23, 39, 134, 135, 136, 155
  dispersal and migration, 64, 89–92
  fluvial provinciality, 92–4
  fossil record, 94–106
Montane/arctic-alpine element, 248
Mosbach, Germany, 2, 143, 145
Mundesley, Norfolk, 143
Munsterian, 263
Mus musculus, 151
musk-ox, 118, 121, 133, 134, 137
Myrica gale, 176, 178, 255
Mytilaster minima, 22, 97, 98

Najas flexilis, 255, 259–60
Nar Valley, 95, 96, 97, 104, 105
narrow-nosed rhinoceros, 115, 116, 132, 134, 135
Natrix maura/tessellata, 121
natterjack, 114, 120, 121
Neanderthals, 132, 137–8
Neede, The Netherlands, 21, 103, 114–5
Neogloboquadrina atlantica, 8
Neptunea contraria, 95
Netherlands, 7, 10, 11, 22, 23, 24, 39, 160, 161
  fauna, 93, 94, 95, 98, 99, 100, 101, 104, 105, 115
Newcastle, 226
Newhaven, Sussex, 55
Newlands Cross, County Dublin, 122, 182, 183
Noordbergum, The Netherlands, 10, 95
North Sea, 212, 216, 226
  marine biogeography, 92
  see also southern North Sea/North Sea Basin
North Sea borehole E8/4, 101
North Sea Drift Formation, 17
North Sea–Atlantic Ocean marine linkage, 7, 8–9, 11, 89, 104
Northern Palaeovalley, 47, 55, 56, 60
northern vole, 132, 133, 134, 136
Northfleet, Kent, 131, 132
Norway lemming, 118, 132, 134
Norwich Crag Formation, 8–9, 94, 101, 104
Norwich Member, 9
Nucella lapillus, 67, 69, 95
Nuphar pumila, 249, 253–4, 255, 262
Nymphoides cordata, 249, 250, 254, 260, 262

Ochotona pusilla, 118, 122
Olivola, 143
Oostende Formation, 104
Oosterhout Formation, 7, 63
Orford Ness, 31, 32, 33, 42
Orkney, 174
Ostend, Norfolk, 146, 147
Ostrea edulis, 96, 98
Ouse, River (Sussex), 55
Ovibos moschatus, 118, 121, 133, 134
Owers Bank, 48, 55, 57

Pachycrocuta brevirostris, 142, 143
Pachycrocuta perrieri, 142, 144–5
Pakefield, Suffolk, 142
palaeocoasts, changes in, 232, 235, 236, 237–8
palaeogeography, Holocene, 173–5
Palaeoloxodon antiquus, 115, 116, 119, 132, 135, 136
palaeotidal modelling, 75–87

Palling, Norfolk, 142
Panthera gombaszoegensis, 112, 142, 143
Panthera leo, 115, 116, 118, 119, 122, 132, 134, 135, 142, 143
Panthera pardus, 113, 142, 143
Pas-de-Calais, 15, 16, 19
Pastonian Stage, 9, 19
Paviland Cave, South Wales, 137
peat, English Channel, 54–5
Peckham, London 135
Peelo Formation, 18, 21, 32
Pelobates fuscus, 114, 121
Pennington, Hampshire, 64, 68, 69, 71
periglacial environments and mass movement, 57
Persicaria bistorta, 249, 250
Petauria voigtstedtensis, 113
Petralona, Greece, 145
Pholas dactylus, 51, 54, 98
Picea abies, 176, 248, 249, 250, 254
pika, 118, 132
pine vole, 112, 115, 116, 117, 119
Pinus sylvestris, 175–6, 177, 248, 249
Pisidium clessini, 101, 102
Pitymys subterraneus, 115, 116, 117, 119
Plantago maritima, 252
Pliocene, 7–8, 30, 31, 94, 101, 248
Plio-Pleistocene boundary, 9
Pollnahallia, Ireland, 245–6
pond tortoise, 114, 115, 118, 120, 121, 132, 136
Pontnewydd Cave, 131, 132, 133, 137, 143, 144, 146
Populus tremula, 175
Portballintrae, 192
Portelet, Jersey, 70
Portland, Dorset, 64, 68, 69, 70, 71, 163
Post-glacial 181–3, 186, 202, *see also* Holocene
Potamopyrgus jenkinsi, 89–90
Praetiglian Stage, 8, 11, 144
Pre-Gortian, 246
Pre-Ludhamian Stage, 8, 144
Pre-Pastonian a, 95
Priory Bay gravel, 37, 40, 41
proboscidians, dwarfed, 151
proglacial bulge
  sea-levels and, 210, 212
  see also forebulge
proglacial lake, North Sea, 19–21, 32–3, 64, 130
  overflow, 20, 21–2, 23, 24, 32, 39, 42, 43, 82, 105
Pseudodama spp, 157
Pterocarya, 175, 246, 248, 249, 254
Purfleet, Essex, 68
pygmy shrew, 112, 116, 120, 122, 123

Quercus, 54, 176, 177, 178, 192
Quercus pubescens, 176
Quinton, Birmingham, 133

raised beaches
  dating, 66–7
  sea-level records, 65–70, 209–10
Rana esculenta, 121
Rana lessonae, 114, 121
Rangifer tarandus, 118, 119, 122, 133, 134, 157
  platyrhyncus, 168
Rannoch Moor, ice loading focus, 214, 216, 217, 231
Ravenscliff Cave, South Wales, 135
Raygill Fissure, Yorkshire
Red Crag Formation, 7, 8, 9, 30, 94, 144
red deer, 113, 117, 120, 122, 132, 134, 135, 152, 161, 166
  see also dwarf red deer of Jersey
red squirrel, 120, 123, 151
reindeer, 118, 119, 122, 123, 133, 134, 136, 137, 157
Reuverian Stage, 7, 8
Rhamnus catharticus, 176, 177
'Rhenish' molluscan fauna, 64, 102, 104
Rhine, River, 23, 24, 39
  drainage systems, 28–35, 38, 39, 40–2, 43
  fauna, 90, 91, 93, 103
  provinciality, 101

INDEX

sediments, 17, 18, 35, 38
Rhine–IJssel course, 35
Rhine–Meuse system, 35, 43
Rhine–Thames river system, 38, 39, 40–2, 43, 64, 101
Rhododendron ferrugineum, 253
Rhododendron ponticum, 181, 249, 250, 254, 262, 264
Ridge Acre Boulder Clay, 133
rivers, evolution, 9, 17, 27, 57
Rocourt Soil, 104
roe deer, 113, 119, 121, 122, 132, 157
Rosmalen, 10
Rushmere-Kesgrave Ridge, 31

'Saalian Complex', 94
Saalian/Wolstonian Stage, 22, 23, 34, 40, 152, 157, 161
sabretooth cat, 112, 115, 143
Saiga tatarica, 118, 122
St George's Channel, 192, 202, 226, 227, 234, 236
Sandford Hill, 146
Sand Hole Formation, 97
Sangatte Raised Beach, 15, 64, 66, 67, 68, 69
Saxifraga hirsuta, 252, 254, 257, 261, 262
Saxifraga nivalis, 253, 254, 262
Saxifraga spathularis, 252, 254, 257, 261
Scheldt, River, 21, 22, 23, 90
  deposits, 17, 90, 93, 103, 104
Schöningen, 94, 97, 114
Sciadopitys, 246, 248, 249
Scilly Isles, 218, 220
  see also Isles of Scilly
Sciurus vulgaris, 120, 123, 151
Scotland, 192, 193, 196, 209, 212, 235
  connection with Europe, 173–4
  connection with Ireland, 186
  flora, 176
Screen Hills pollen assemblage, 260
sea-level changes, 8, 23, 24, 39, 161–2, 163, 165, 173, 174
  effect on insularity, 7, 40, 42, 55, 57, 63, 70–1
  English Channel, 16, 60–1, 63
  Irish Sea region, 187–90, 193, 198–200, 202, 237
  model, Irish and Celtic seas, 209–14, 227–38
    glacio-eustasy modelling, 213, 214, 217, 229, 237
    glacio-isostasy modelling, 210, 213, 214–16, 217, 227, 235, 237
    hydro-isostasy modelling, 213, 216–17, 227, 229–32, 233, 237
    indications from present sea-bed, 217–27
  northern North Sea, 212
  southern North Sea, 5–6, 9, 63
  Strait of Dover, 71
  see also glacio-eustasy and sea-level; glacio-isostasy and sea-level; global sea-level
sea-levels
  current, 61
  height of raised beaches related to, 65–70, 209–10
Seine, River, 16, 17, 38
Selsey, 64, 68, 69, 114, 115, 131, 132, 134, 136, 163
Selva Vecchia, Verona, 142
Semisalsa stagnorum, 95, 96, 97, 98
Sequoia, 246, 248, 249
Sewerby Raised Beach, 98, 135
Shepperton Gravel, 39
Sheringham, Norfolk, 9
Shetland, 174, 175, 176, 177, 178
Shetland–Orkney channel, as dispersion barrier, 92
shorelines
  creation and preservation, 189–90, 191, 192–9, 200, 201
  see also coastal changes
short-tailed field vole, 133
Shottisham, 32
Sidestrand, Norfolk, 143
Slades Green, *see* Crayford
Slindon Sands, 81, 82, 87
small-toothed mammoth, 132
smooth newt, 123
snakes, 122
Snape, Suffolk, 32

Solent River, 16, 17, 47, 55, 56, 57, 67, 68, 69
Sorbus aucuparia, 175, 254
Sorbus torminalis, 176, 178
Sorex araneus, 116, 120, 122
Sorex minutus, 112, 116, 120, 122
Southchurch/Asheldham/Mersea Island/Wigborough Gravel, 34
Southern Bight, 7, 9, 11, 64, 80, 106
southern North Sea/North Sea Basin, 3, 6–7, 17–20, 23, 40, 80
  connection with Atlantic Ocean, 7, 8, 11, 89, 104
    closure, 8–9, 10, 11, 63–4
  connection with English Channel, 11, 42, 43, 82, 86, 87, 130
  drainage into, 27, 30, 31, 35, 39, 42
  fauna, 11, 23, 92, 94, 98, 101, 106
  glacio-isostatic depression, 23, 24
  ice-dammed (proglacial) lake, 19, 20, 22, 31, 64, 130
    overflow, 20, 21–2, 23, 24, 32, 39, 42, 43, 82, 105
  sea-level changes, 5–6, 9, 63
  sediments, 19–20, 23, 35, 41
  tectonic downwarping/uplifting, 17, 43, 65
spadefoot toad, 114, 121
Speeton Shell Bed, 98
Spermophilus citellus, 115, 116
Spermophilus dietrichi, 113
Spermophilus primigenius, 134
Sphaerium icenicum, 101
Sphaerium rosmalense, 101
Spiranthes romanzoffianum, 263
spotted hyaena, 112, 115, 116, 118, 135, 142, 144, 147
St Erth, Cornwall, 16
Stanton Harcourt, Oxfordshire, 128, 130, 131, 132, 133, 137
Stanton Harcourt Gravel, 133
Star Carr, Yorkshire, 119
Steinheim, Germany, 114, 115, 131, 136
Stephanorhinus hemitoechus, 115, 116, 119, 132, 134, 135, 136
Stephanorhinus kirchbergensis, 115, 116, 117, 133
steppe lemming, 134
Steyne Wood Clay, 67, 70–1, 81, 82, 87
Stoke Tunnel, Suffolk, 114, 115, 131, 132
Stone, Hampshire, 64, 68–9, 163
Stradbroke borehole, 8
straight-tusked elephant, 115, 116, 119, 132, 135
Strait of Dover, 11, 15, 42, 47, 65, 71, 86
  breaching, 87
  closure, 63–4, 84, 86, 87, 94
  as drainage route, 38, 39, 41, 43
  as fluvial route, 27, 106
  formation, 15, 19–23, 30, 32
  land-bridge, 75, 76, 80, 82, 89
  open or closed status
    marine fauna as evidence for, 92, 94
    palaeotidal implications, 76–87
  as submerged Rhine–Thames valley, 38, 39, 41, 42, 43
Stránska Skála, 143
striped hyaena, 145
Stump Cross Cave, Yorkshire, 136
Stutton, Suffolk, 114, 115, 131, 132
submerged valley, English Channel, 34, 47, 48–61
Sudbury Formation, 10
Suffolk cliffs, retreat, 11
Sugworth, near Oxford, 101, 111, 114
Summertown–Radley terrace, 130
Sus scrofa, 113, 117, 119, 120, 122, 135
Süssenborn, Germany, 143
Sussex coastal plain, 65
Svarte Bank Formation, 5, 18
Swalecliff, Kent, 135
Swanscombe, Kent, 34, 97
  fauna, 65, 101, 102, 104, 105, 113, 114, 143, 144, 146
  human site, 130, 131, 136, 137
Swanton Morley, Norfolk, 114, 135, 157

Tancarville, France, 70
Tanousia runtoniana, 101
Taplow Gravel, 133
Taxodiaceae, 175, 246, 248, 249
Taxodium, 246, 248, 249

Taxus baccata, 176
Tegelen Formation, 101
Texel, The Netherlands, 23
Thames, River, 10, 21, 23, 24, 67–8
  drainage systems, 28–35, 38, 39, 40–2, 43
  fauna, 93, 101, 105
  Pleistocene evolution, 27–35
  provinciality, 101
  sediments, 18, 35–7, 38, 69
  structure and sedimentation, 42–3
Thames–Lobourg confluence, 38
Thames–Medway gravels, 31, 34, 36, 37, 40, 41
Thames–Medway river system, 31–3, 35, 37, 38, 39, 43
Thames–Rhine river system, 33, 38, 39, 40–2, 43, 64, 101
Thatcham, Berkshire, 119
Thatcher Rock, Torquay, 69
Theodoxus danubialis, 65, 97, 101, 102, 103, 104
Theodoxus fluviatilis, 103, 104
Theodoxus serratiliniformis see *Theodoxus danubialis*
thermoluminescence dates, 66–7, 152, 163
thermophilous flora, 192
Thurian Stage, 9
tidal modelling, 75–87
tidal sand ridges, Celtic Sea, 218–22, 235, 237
tidal scour cauldrons, 224, 226–7, 234, 236
Tiglian Stage, 8–11, 19, 24, 63, 94, 101, 104–5
Tilia cordata, 176, 249
Tillingham *see* East Hyde
Torbay, 64, 68, 69, 70, 71, 163
Torcourt Cave, 146
Tornewton, Devon, 131, 132, 133, 134, 135, 137, 143, 144, 145, 146, 157
Tottenhill, Norfolk, 97
Tournouerina belnensis, 101
Tourville, France, 70
Trafalgar Square site, 135, 137, 157
transgression
  Eemian, 11, 71, 98, 103, 106, 132, 136, 160–6
  Holocene, 11, 57, 98, 103, 106, 122, 173–4, 181–203, 209–38
  Holsteinian/Hoxnian, 19, 22, 39, 42, 43, 95, 97, 102–3, 105
tree frog, 114, 121
tree and shrub flora
  anthropogenic influences, 176–8
  dispersal characteristics, 175–6, 178
  extinctions, 177–8, 179
  insularity and, 173–5, 178–9
Trichomanes speciosum, 255, 259, 260
Triturus vulgaris, 114, 120, 123
Trogontherium cuvieri, 121, 115
Tsuga, 175, 246, 248, 249

U-series dates, 66, 136, 155, 246
Ulmus laevis, 176
Untermassfeld, Germany, 143
Upper Holland Gravel, 31
Ursus arctos, 116, 118, 120, 122, 132, 133, 134, 135, 142, 144
Ursus deningeri, 112, 142, 144
Ursus spelaeus, 116, 142, 144, 146
Ushant, 226

Vaccinium myrtillus, 253
Vallonnet, Southern France, 142, 144
Valvata goldfussiana, 101
Valvata naticina, 101
Venerupis aurea, 98, 104

vertebrate faunas, insularity and, 111–12, 141–8
  Britain versus Europe, 111–16, 117, 118–19, 121
  Britain versus Ireland, 111–12, 118, 120, 121–3
Viburnum lantana, 176, 178
Viburnum opulus, 177, 255
Victoria Cave, Yorkshire, 135, 137, 157
Villafranchian, 142, 143
Vitis, 246, 248, 249
viviparous lizard, 123
Viviparus diluvianus, 101, 102
Viviparus glacialis, 101, 105
Viviparus medius, 101
Viviparus viviparus gibbus, 101
Voigtstedt, Germany, 112–13, 131
Vulpes vulpes, 116, 120, 135

Waalian Stage, 9, 10, 95
Waardenberg, The Netherlands, 10
Wacken, 94, 97, 104
Wales, northwest coast sea-levels, 212
Waltonian, 94
Walton-on-the Naze, Essex, 7, 34, 135
Warwickshire Avon Terrace, 4, 133
Wash, the 67, 68, 98
water frog, 121
Waterhall Farm, Hertfordshire, 135
water vole, 112, 115, 116, 117, 118, 120, 133, 135
Weald–Artois anticlinorium, 15, 17, 42, 43
Weald–Artois Chalk ridge, 32, 43, 63, 65–70, 105, 130, 136
Weichselian Stage, 23, 39, 40, 41, 42
Weimar-Ehringsdorf, Germany, 115
Weser river, 31
West Runton, Norfolk, 10, 95, 101, 131
West Runton Freshwater Bed, 111, 112–13, 114, 142, 143, 144, 146
Westbury-sup-Mendip, 113, 114, 142, 143, 144, 145, 146–7
Westerhoven, The Netherlands, 10
Western Isles, 168, 174, 176, 177
Westkapelle Ground Formation, 8
West Thurrock, Essex, 68, 131, 132
Weybourne Crag, 9, 95, 101, 105
white-toothed shrew, 116, 117, 119, 132, 133
wild boar, 113, 117, 119, 120, 122
wild cat, 116, 120, 135
Wimereux, 21
Winterton Shoal Formation, 5, 17, 31
Wissant, 15, 20, 21, 22, 34, 42, 43, 64
wolf, 112, 116, 118, 120, 122, 132, 133, 134, 135
Wolstonian, 128, 144, 147
wolverine, 136
wood mouse, 112, 116, 120, 122, 123, 133, 135, 151, 181, 182
woodland decline, 177–8
Woodston, Peterborough, 94, 95, 96, 97, 105
woolly mammoth, 118–19, 121, 136, 137
woolly rhinoceros, 116, 117, 118, 119, 122, 133, 134, 136, 137
Wrangel Island, NE Siberia, 169

Xenocyon 144
Xenocyon falconeri, 143
Xenocyon lycaonoides, 142, 143

Yarmouth Roads Formation, 5, 17, 31

Zelzate, Belgium, 104
Zuurland borehole, 94